Bose–Einstein Condensates and Atom Lasers

Bose–Einstein Condensates and Atom Lasers

Edited by

Sergio Martellucci

University of Rome "Tor Vergata"
Rome, Italy

Arthur N. Chester

Hughes Research Laboratories, Inc.
Malibu, California

Alain Aspect

Institut d'Optique
Orsay, France

and

Massimo Inguscio

University of Florence
Florence, Italy

Springer Science+Business Media, LLC

Library of Congress Cataloging-in-Publication Data

Bose–Einstein condensates and atom lasers/edited by S. Martellucci ... [et al.].
p. cm.
"Proceedings of the 27th Course of the International School of Quantum Electronics on Bose–Einstein Condensates and Atom Lasers, held October 19–24, 1999, in Erice, Sicily, Italy"—T.p. verso.
Includes bibliographical references and index.

1. Bose–Einstein condensation—Congresses. 2. Quantum optics—Congresses. 3. Lasers—Congresses. I. Martellucci, S. II. Course of the International School of Quantum Electronics on Bose–Einstein Condensates and Atom Lasers (27th: 1999: Erice, Italy)

QC175.45.B65 B66 2000
530.4′2—dc21 00-061084

Proceedings of the 27th Course of the International School of Quantum Electronics on Bose–Einstein Condensates and Atom Lasers, held October 19–24, 1999, in Erice, Sicily, Italy

ISBN 978-1-4757-8658-3 ISBN 978-0-306-47103-2 (eBook)
DOI 10.1007/978-0-306-47103-2

Originally published by Kluwer Academic/Plenum Publishers, New York in 2000.
Softcover reprint of the hardcover 1st edition 2000

http://www.wkap.nl

10 9 8 7 6 5 4 3 2 1

A C.I.P. record for this book is available from the Library of Congress

Preface

Since the experimental demonstration in 1995 of Bose-Einstein condensation (BEC) in dilute atomic gases there has been an explosion of studies of the properties of this novel macroscopic quantum system. A matter wave analogous to the optical laser is becoming a reality. The purpose of this volume is to cover the methods used to produce these new samples of coherent atoms, to manipulate them and to study their properties. Emphasis has been given to anticipated developments of a new type of sources, more and more similar to the various lasers (pulsed, CW, mode locked...) and likely to produce a revolution similar to the laser revolution of 40 years ago!

Recently there have been several new proposed applications of atom optics, and it is possible to foresee an increasing demand for atom lasers, sensors and associated instrumentation. Consequently, the chapters cover current developments in the basic techniques, materials and applications in the field of the generation of coherent beams of atoms. In October 1999, an international group of scientists convened in Erice, Sicily, for a meeting on the subject of "Bose-Einstein Condensates and Atom Lasers". This Conference was the 27th Course of the International School of Quantum Electronics, under the auspices of the "Ettore Majorana Centre and Foundation for Scientific Culture". This book presents the Proceedings of this Conference, providing a fundamental introduction to the topic as well as reports on recent research results. The aim of the Conference was to bring together some of the world's acknowledged scientists who have as a common link the use of instrumentation, techniques and procedures related to the fields of Bose-Einstein Condensates and Atom Lasers. Most of the lecturers attended all the lectures and devoted their spare hours to stimulating discussions. We would like to thank them all for their admirable contributions.

The Conference also took advantage of a very active audience; most of the participants were active researchers in the field and contributed with discussions and seminars. Some of these seminars are also included in these Proceedings.

The Conference was an important opportunity to discuss the latest developments and emerging perspectives on the use of experimental techniques for Bose-Einstein Condensates and Atom Lasers. The Chapters in these Proceedings are not ordered exactly according to the chronology of the Conference, but they give a fairly complete accounting of the Conference lectures with the exception of the informal panel discussions. The contributions presented at the Conference are written as extended, review-like papers to provide a broad and representative coverage of the fields of Bose-Einstein Condensates and Atom Lasers. We did not modify the original manuscripts in editing this book, except to assist in uniformity of style.

We are grateful to Prof. Carlo Bellecci for contributing to the secretarial organisation of the Conference, to our editor at Kluwer Academic/Plenum Publishers, Joanna Lawrence, for outstanding professional support. We also greatly appreciate the expert help from our assistants, Rosalia Caruso, Laura Cemoli, Maila Lanzini and the support of Eugenio Chiarati for much of the computer processing work. This International School was held under the auspices of Prof. Antonino Zichichi, Director of the "Ettore Majorana" Centre and Foundation for Scientific Culture, Erice, Italy. Finally, we acknowledge with gratitude the generous financial support of the organizations who sponsored the Conference: the Italian Ministry of Education, the Italian Ministry of University and Scientific Research, the Sicilian Regional Parliament, the Italian Research Group on Quantum Electronics and Plasma Physics (G. N. E. Q. P.) of the National Research Council (C.N.R.), the University of Rome "Tor Vergata", the I.N.F.M. (Florence), the French Embassy in Italy and the European Laboratory for Non Linear Spectroscopy (LENS).

Arthur N. Chester
President and General Manager
HRL Laboratories, LLC.
Malibu, California (USA)

Sergio Martellucci
Professor of Quantum Electronics
University of Rome "Tor Vergata"
Rome (Italy)

Alain Aspect
Professeur à l'Ecole Politechnique
Directeur de Recherche CNRS
Orsay (France)

Massimo Inguscio
Professor of Atomic Physics
University of Florence
Florence (Italy)

May, 2000

Contents

EXPERIMENTAL STUDIES OF BOSE-EINSTEIN CONDENSATES IN SODIUM

W. Ketterle 1

MEASUREMENT OF THE RELATIVE PHASE BETWEEN TWO BOSE-EINSTEIN CONDENSATES

D. S. Hall 31

INTERTWINED BOSE-EINSTEIN CONDENSATES

D. S. Hall 43

COHERENT ATOM OPTICS WITH BOSE-EINSTEIN CONDENSATES

K. Helmerson 55

NON-LINEAR ATOM OPTICS WITH BOSE-EINSTEIN CONDENSATES

K. Helmerson 65

MOMENTUM DISTRIBUTION OF A BOSE CONDENSED TRAPPED GAS

S. Stringari, L. Pitaevskii, D. M. Stamper-Kurn, and F. Zambelli 77

ATOM OPTICS WITH BOSE-EINSTEIN CONDENSATES

S. Burger, K. Bongs, K. Sengstock, and W. Ertmer 97

GENERATING AND MANIPULATING ATOM LASER BEAMS

T. Esslinger, I. Bloch, M. Greiner, and T. W. Hänsch 117

MULTIPLE ^{87}RB CONDENSATES AND ATOM LASERS BY RF COUPLING

F. Minardi, C. Fort, P. Maddaloni, and M. Inguscio 129

THEORY OF A PULSED RF ATOM LASER

J. Schneider, and A. Schenzle 141

THE ATOMIC FABRY-PEROT INTERFEROMETER

I. Carusotto, and G. C. La Rocca 153

RF-INDUCED EVAPORATIVE COOLING AND BEC IN A HIGH MAGNETIC FIELD

P. Bouyer, V. Boyer, S. G. Murdoch, G. Delannoy, Y. Le Coq, A. Aspect, and M. Lécrivain 165

DISSIPATIVE DYNAMICS OF AN OPEN BOSE-EINSTEIN CONDENSATE

F. T. Arecchi, J. Bragard, and L. M. Castellano 187

NON-GROUND-STATE BOSE-EINSTEIN CONDENSATION

V. S. Bagnato, E. P. Yukolova, and V. I. Yukalov 201

TOWARDS A TWO SPECIES BOSE-EINSTEIN CONDENSATE

E. Arimondo 213

ATOM INTERFEROMETRY WITH ULTRA-COLD ATOMS

M. Kasevich 231

CLASSICAL AND QUANTUM JOSEPHSON EFFECTS WITH BOSE-EINSTEIN CONDENSATES

A. Smerzi 249

JOSEPHSON QUBITS FOR QUANTUM COMPUTATION

G. Falci, R. Fazio, E. Paladino, and U. Weiss 265

ADDRESSING SINGLE SITES OF A CO_2-LASER OPTICAL LATTICE

F. S. Cataliotti, R. Scheunemann, T. W. Hänsch, and M. Weitz 275

SCISSORS MODE AND SUPERFLUIDITY OF A TRAPPED BOSE-EINSTEIN CONDENSED GAS

O. M. Maragò, S. A. Hopkins, J. Arlt, E. Hodby,
G. Hechenblaikner, and C. J. Foot 285

EXPERIMENTS WITH POTASSIUM ISOTOPES

C. Fort 291

EQUILIBRIUM STATE AND EXCITATIONS IN TRAPPED FERMI VAPOURS

A. Minguzzi, and M. P. Tosi 301

PHOTOASSOCIATIVE SPECTROSCOPY OF Cs_2

C. Drag, B. Laburthe Tolra, D. Comparat, A. Fioretti, A. Crubellier,
O. Dulieu, F. Masnou-Seeuws, S. Guibal, and P. Pillet 313

INDEX 323

Experimental Studies Of Bose-Einstein Condensates In Sodium

W. KETTERLE
Department of Physics and Research Laboratory of Electronics, Massachusetts Institute of Technology, Cambridge, Massachusetts 02139, USA

1. INTRODUCTION

1.1 Bose-Einstein Condensation in a Gas

When a gas of bosonic atoms is cooled below a critical temperature T_c, a large fraction of the atoms condenses into the lowest quantum state. This phenomenon was first predicted by Albert Einstein in 1925[1] and is a consequence of quantum statistics. Atoms at temperature T and mass m can be regarded as quantum-mechanical wavepackets which have an extension on the order of a thermal de Broglie wavelength $\lambda_{dB}=(2\pi\hbar^2/mk_BT)^{1/2}$. When atoms are cooled to the point where λ_{dB} is comparable to the interatomic separation, the atomic wavepackets "overlap" and the indistinguishability of particles becomes important. At this temperature, bosons undergo a quantum-mechanical phase transition and form a Bose-Einstein condensate, a coherent cloud of atoms all occupying the same quantum state. The relation between the transition temperature and the peak atomic density n is given by $n\lambda_{dB}^3 = 2.612$.

With the realization of Bose-Einstein condensation (BEC) in dilute atomic gases[2-6], several long-standing goals were achieved. First, by cooling atoms into the lowest energy state, one exerts ultimate control over the motion and position of atoms, limited only by Heisenberg's uncertainty

Bose-Einstein Condensates and Atom Lasers
Edited by Martellucci *et al.*, Kluwer Academic/Plenum Publishers, 2000

relation. Second, a coherent macroscopic sample of atoms all occupying the same quantum state was generated, leading to the realization of atom lasers, devices which generate coherent matter waves. Third, degenerate quantum gases were produced with properties quite different from the quantum liquids ^3He and ^4He. This provides a testing ground for many-body theories of the dilute Bose gas which were developed many decades ago, but never tested experimentally[7]. BEC of dilute atomic gases is a macroscopic quantum phenomenon with similarities to superfluidity, superconductivity and the laser[8]. More generally, atomic Bose-Einstein condensates are a new "nanokelvin" laboratory where interactions and collisions at ultralow energy can be studied. The quest for Bose-Einstein condensation has a long history and is nicely summarized in various contributions to the 1998 Varenna summer school[9-11].

1.2 Basic Techniques

The realization of Bose-Einstein condensation requires techniques to cool gases to sub-microkelvin temperatures and atom traps to confine them at high densities and to keep them away from the hot walls of the vacuum chamber. Over the last 15 years, such techniques were developed in the atomic and low temperature physics communities[9-11]. Our experiment uses a multi-stage cooling process to cool hot sodium vapor down to temperatures where the atoms form a condensate. A beam of sodium atoms is created in an oven at a density of about 10^{14} atoms per cm^3, similar to the eventual density of the condensate. While the density is almost unchanged, the gas is cooled by nine orders of magnitude from 600 K to 1 μK in several stages: first by slowing the atomic beam, followed by optical trapping and laser cooling, then by magnetic trapping and evaporative cooling.

Table 1. Multi-stage cooling to BEC in the MIT sodium experiment. Through a combination of optical and evaporative cooling the temperature of a gas is reduced by a factor of 10^9, while the density at the BEC transition is similar to the initial density in the atomic oven (all numbers are approximate). In each step shown, the phase-space density or ground state population increases by about 10^6.

	Temperature	Density (cm^{-3})	Phase-space density
Oven	500 K	10^{14}	10^{-13}
Laser cooling	50 μK	10^{11}	10^{-6}
Evaporative cooling	500 nK	10^{14}	1
BEC			10^7

Tab. 1 shows how these cooling techniques together reduce the temperature of the atoms by a factor of a billion. The phase space density enhancement is almost equally distributed between laser cooling and evaporative cooling, providing about six orders of magnitude each. Bose-Einstein condensation can be regarded as "free cooling," as it increases the quantum occupancy by another factor of about a million without any extra effort. This reflects one important aspect of BEC: the fractional population of the ground state is no longer inversely proportional to the number of states with energies smaller than $k_B T$, but quickly approaches unity when the sample is cooled below the transition temperature.

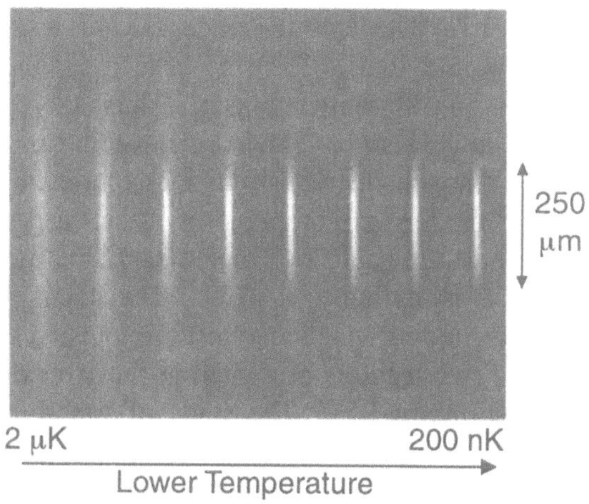

Figure 1. Phase contrast images of a trapped sodium gas across the BEC phase transition. As the final radio frequency used in evaporative cooling is lowered, the temperature is reduced (left to right). Images show the onset of Bose condensation, the growth of the condensate fraction and contraction of the thermal wings, and finally a pure condensate with no discernible thermal fraction. The axial and radial frequencies are about 17 and 230 Hz, respectively.

Atom clouds can either be observed by absorptive or dispersive techniques. In the first case, the shadow cast by the atom cloud is imaged onto a CCD camera. In the latter case, dispersively scattered photons are collected creating an image of the spatially varying index of refraction. This non-destructive technique was used to observe the BEC phase transition directly in the spatial domain[12]. Fig. 1 shows a series of such spatial images above and below the phase transition. They show the sudden appearance of a high-density core of atoms in the center of the distribution – the Bose-Einstein condensate. Lowering the temperature further, the condensate number grows and the thermal wings of the distribution become shorter.

Finally, the temperature drops to the point where only the central peak remains.

Similarly, the BEC phase transition can be observed by imaging the shadow cast by an atom cloud which expands ballistically after suddenly switching off the magnetic trap. The signature of BEC is the sudden appearance of a slow component with anisotropic expansion[2,3]. This can be regarded as observing BEC in momentum space.

1.3 Atom Lasers and Quantum Fluids

Research on gaseous BEC can be divided into two areas: In the first (which could be labeled "The atomic condensate as a coherent gas", or "Atom lasers"), one would like to have as little interaction as possible - almost like photons in an optical laser. Thus the experiments are preferentially done at low densities. The Bose-Einstein condensate serves as an intense source of ultracold coherent atoms for experiments in atom optics, in precision studies or for explorations of basic aspects of quantum mechanics. The second area could be labeled as "BEC as a new quantum fluid" or "BEC as a many-body system". The focus here is on the interactions between the atoms which are most pronounced at high densities.

In the spirit of the two lectures presented at the workshop in Erice, we will illustrate both aspects of BEC. The study of sound, multi-component condensates and the evidence for a critical velocity are part of our study of BEC as a new quantum fluid. Coherent matter wave amplification is at the heart of atom lasers. Our studies of light scattering from a Bose condensate link both aspects together: Light scattering was used to imprint phonons into the condensate, but also to study its coherence properties which are relevant for atom lasers.

2. STUDIES OF SOUND IN A BOSE-EINSTEIN CONDENSATE

2.1 The Nature of Collective Excitations

While Bose-Einstein condensates are produced and probed using the tools of atomic physics, their connection to decades-old condensed-matter physics is most evident in the study of sound. The nature of sound in a homogeneous condensate depends on the hierarchy of three length scales (shown in Tab. 2). In an inhomogeneous, trapped Bose gas there is an additional length scale: the length of the condensate l_c. This divides the description of condensate excitations into three regimes:

Table 2. The nature of collective excitations of the condensate and the thermal cloud in the homogeneous case. Various regimes are distinguished according to a hierarchy of length scales: the reduced wavelength of the excitation $\lambda\!\!\!^{-}_{ex}$; the healing length ξ which is given by the scattering length a and the condensate density n_0 as $\xi=(8\pi a n_0)^{-1/2}$ and characterizes the strength of the interaction; and the mean-free path for collisions between quasi-particles l_{mfp}.

Regime	Length scales	Condensate	Thermal cloud
collisionless	$\lambda\!\!\!^{-}_{ex} < \xi, l_{mfp}$	Ballistic	ballistic
collisionless	$\xi < \lambda\!\!\!^{-}_{ex} < l_{mfp}$	zero sound	ballistic
hydrodynamic	$\xi < l_{mfp} < \lambda\!\!\!^{-}_{ex}$	second sound	first sound

- For long wavelengths of the excitation, which approach the size of the condensate ($\lambda\!\!\!^{-}_{ex} \approx l_c$), the excitation spectrum becomes discretized, i.e. the collective modes of the system are standing sound waves at specific frequencies. It is interesting to note that at high densities (Thomas-Fermi limit), both the speed of Bogoliubov sound and the length of the condensate scale as $(an_0)^{1/2}$. Thus, the frequency of the collective excitation $\omega \propto c_0/l_c$, is independent of the speed of sound c_0. Such discrete collective excitations have been studied both at Boulder and at MIT for a few modes and over a wide range of temperatures[13-16].
- For excitations of wavelengths smaller than all dimensions of the condensate, $\lambda\!\!\!^{-}_{ex} << l_c$, the condensate can be treated as locally homogeneous. The excitation of sound in this regime was accomplished by light scattering, i.e. by optically imprinting phonons into the condensate. This will be discussed in Sect. 6.
- For Bose condensates in anisotropic potentials there is an intermediate regime in which the wavelength of the excitation is larger than the size of the condensate in one or two dimensions, but smaller than the size of the condensate in the other directions. In this case, the axial discretization of the collective modes is not apparent, and thus the pulses propagate as sound waves. The connection between this phonon picture and the aforementioned discrete spectrum was laid out by Stringari[17]. Our study of the propagation of sound along the axial direction of a cigar-shaped condensate falls into this regime[18].

One of our lectures in Erice focused on various aspects of sound. In these notes, we refer to our comprehensive discussion in the Varenna book[9] and present only very recent results. This includes the optical excitation of phonons which will be discussed in Sect. 6, and the study of shape oscillations of higher multipolarity. They belong to the regime of low-lying discrete excitations and will be discussed in the following section.

2.2 Surface Excitations in a Bose-Einstein Condensate

Collective modes which have no radial nodes and are localized close to the surface of the condensate are called surface modes. In a semiclassical picture these excitations can be considered the mesoscopic counterpart of tidal waves at the macroscopic level. Those excitations are of special interest since they show a crossover between collective and single-particle behavior, which is crucial for the existence of a non-zero critical rotational velocity. Furthermore, they probe the surface region of the condensate where the density of the thermal cloud is peaked, and should be sensitive to the interactions between condensed and non-condensed atoms[19,20]

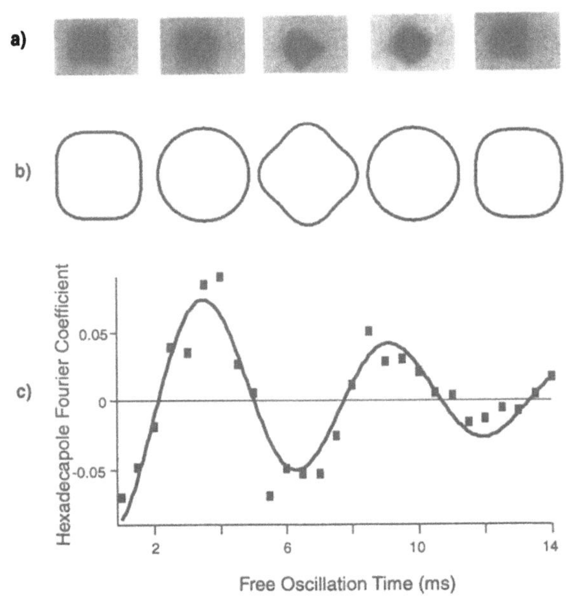

Figure 2. Observation of a standing hexadecapole excitation of a Bose-Einstein condensate. Absorption images of a condensate driven to excite the $m = 4$ mode for various hold times (1, 2, 3.5, 4.5 and 6.5 ms from left to right) in the magnetic trap (a). The shape oscillations of a pure $m = 4$ mode are schematically depicted in (b) for one cycle. Fig. c shows the oscillation of the $m = 4$ Fourier coefficient, obtained by Fourier analyzing the contours of clouds like those in Fig. a. Figure taken from Ref. [21].

Since these modes don't have cylindrical symmetry, they cannot be excited by modulating the currents in the coils of our *dc* magnetic trap. We have therefore developed a method to create perturbations with high spatial and temporal control by optical means. The magnetic trapping potential was perturbed with light from a Nd:YAG laser (emitting at 1064 nm) traveling parallel to the axis of the trap and focused near the center of the magnetic trap. Because of the low intensity of the laser beam and the large detuning from the sodium resonance, heating from spontaneous scattering was negligible. The laser beam was red-detuned from the sodium resonance and therefore gave rise to an attractive dipole potential. The 1 mm Rayleigh range of the beam waist was considerably longer than the 220 μm axial extent of the condensate. Therefore, the laser only created radial inhomogeneities in the trapping potential, leaving the axial motion almost undisturbed.

The spatial and temporal control of the beam was achieved with two crossed acousto-optic deflectors. Using the two-axis deflector arbitrary laser patterns could be scanned in a plane transverse to the propagation of the laser beam.

The scan rate was chosen to be 10 kHz, which is much larger than the trapping frequencies. Thus, the atoms experienced a time-averaged potential that is superimposed upon the magnetic trap potential. The rapid scan created a pattern of two or four points which had the correct symmetry to excite quadrupolar or hexadecapolar surface oscillations[21]. A temporal modulation of the intensity or a rotation of the whole pattern resulted in standing and rotating waves, respectively (Fig. 2). Within the accuracy of measurement, the observed excitation frequencies ν_{exc} were in agreement with the theoretical prediction of Stringari[22]: $\nu_{ext} = \sqrt{m}\nu_r$ where ν_r is the radial trapping frequency and m = 2 for the quadrupole mode and m= 4 for the hexadecapole mode. This novel method should be useful for exciting even higher-lying excitations. Very recently, it has been used to excite vortices in a condensate[23].

3. SPINOR BOSE-EINSTEIN CONDENSATES

3.1 Ground-State Spin Domains

In a magnetic trap, the atomic spin adiabatically follows the direction of the magnetic field. Thus, although alkali atoms have internal spin, alkali Bose-Einstein condensates are described by a scalar order parameter similar to the spinless superfluid ^{4}He. One exception is the two-component condensate which was discovered by the Boulder group, when they trapped atoms in both the upper and lower hyperfine states[24] of ^{87}Rb. This observation was surprising because a large rate of inelastic collisions had been expected for this system. The suppression of these spinflip collisions turned out to result from a fortuitous equality in the scattering lengths in the two hyperfine states.

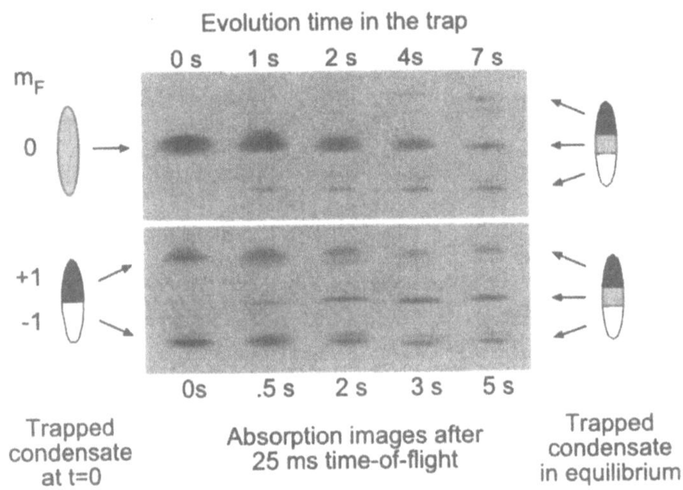

Figure 3. Observation of coupled spinor condensates. The components of the condensate are states with different orientations (m_F =-1,0,1) of the total spin F = 1 confined in an optical trap. These components are coupled by an anti-ferromagnetic spinflip interaction which drives them into an equilibrium domain structure. After a variable holding time, the condensate was analyzed by time-of-flight absorption imaging. During the ballistic expansion, a magnetic field gradient acted as a Stern-Gerlach filter and separated the components with different spin orientation as indicated by the arrows. The upper image shows how a condensate, initially in a pure m_F = 0 state, developed a spin domain structure. The same equilibrium state was reached when the condensate started in an equal mixture of m_F = +1 and -1 states[26,31].

A more general method for creating multi-component condensates is to employ an optical trap that can confine condensates with arbitrary orientations of the spin, thus liberating the spin orientation as a new degree of freedom[25]. Our group used an optical trap to study condensates with arbitrary population in the three orientations m_F = 1, 0, -1 of the ground state of sodium which has a total spin[26] F = 1. These condensates have a three-component vectorial order parameter. A variety of new phenomena have been predicted for such spinor condensates such as spin textures, spin waves, and the coupling between atomic spin and superfluid flow[27-29]. Such phenomena cannot occur in condensates with a single component order parameter such as in ^{4}He and more closely resemble the complex features of the superfluid phases of ^{3}He.

If the components are not coupled (i.e. their populations don't change), they can be regarded as multi-species condensates ("condensate alloys"). Both the group in Boulder and our group have studied the dynamics of the phase separation of these components[30,31]. By selecting two of the three states of the F = 1 spinor condensates, we could realize two-component condensates which were either miscible or immiscible[26]. Multi-component condensates are promising systems for the study of interpenetrating superfluids, a longstanding goal since the early attempts in 1953 using ^{4}He-^{6}He mixtures[32]. New phenomena arise when the three components are coupled by spinflip collisions, as displayed in Fig. 3.

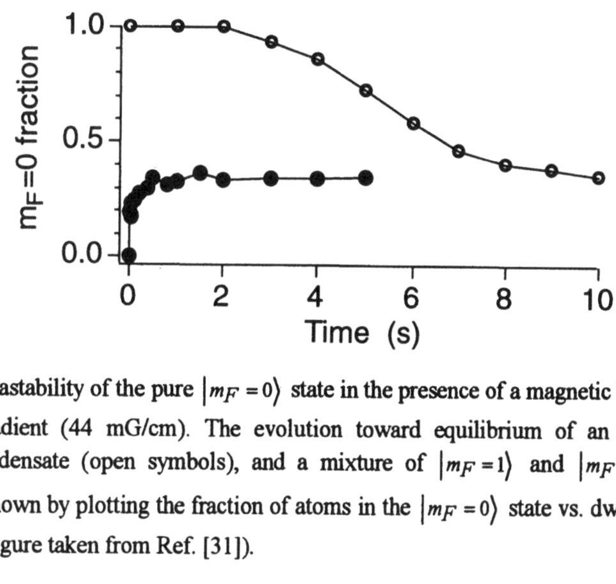

Figure 4. Metastability of the pure $|m_F = 0\rangle$ state in the presence of a magnetic bias field (250 mG) and gradient (44 mG/cm). The evolution toward equilibrium of an initially pure $|m_F = 0\rangle$ condensate (open symbols), and a mixture of $|m_F = 1\rangle$ and $|m_F = -1\rangle$ (closed symbols) is shown by plotting the fraction of atoms in the $|m_F = 0\rangle$ state vs. dwell time in the optical trap (figure taken from Ref. [31]).

3.2 Metastable Bose-Einstein Condensates

During our studies of the spinor ground states we encountered two different types of metastability which we investigated in more detail[31]. In one case, a two-component condensate in the m_F = 0 and m_F = +1 hyperfine states was stable in spin composition, but spontaneously formed a metastable spatial arrangement of spin domains. In the other, a single component m_F = 0 condensate was metastable in spin composition with respect to the development of m_F = + 1, -1 ground-state spin domains (see Figs. 3 and 4). In both cases, the energy barriers which caused the metastability were much smaller than the temperature of the gas (as low as 0.1 nK compared to 100 nK) which would suggest a rapid thermal relaxation. However, since the thermal energy is only available to non-condensed atoms, this thermal relaxation was slowed considerably due to the high condensate fraction and the extreme diluteness of the non-condensed cloud. As a result, one can study (sub-)nanokelvin physics in a condensate although the true temperature of the total system is much higher!

The upper part of Fig. 4 shows the time evolution towards equilibrium for a condensate initially prepared in the $m_F = 0$ state. For the coldest and most dilute condensates metastability of up to 5 s was observed. In contrast, when the system was prepared in an equal mixture of $m_F = 1$ and $m_F = -1$ atoms the fraction of atoms in the $m_F = 0$ state grew without delay, arriving at equilibrium within just 200 ms. This difference can be understood by considering a spin-relaxation collision, in which two $|m_F = 0\rangle$ atoms collide to produce an $|m_F = 1\rangle$ and an $|m_F = -1\rangle$ atom. In the presence of a magnetic field B_0, quadratic Zeeman shifts cause the energy of the two $|m_F = 0\rangle$ atoms to be lower than that of the $|m_F = 1\rangle$ and $|m_F = -1\rangle$ atoms by 2×390 Hz $\times$ $(B_0/G)^2$.

Due to this activation energy, condensate atoms in the $|m_F = 0\rangle$ state could not undergo spin-relaxation collisions. Thus, even though the creation of $|m_F = 1\rangle$ and $|m_F = -1\rangle$ spin domains at the ends of the condensate is energetically favored *globally* in the presence of a magnetic field gradient, the $|m_F = 0\rangle$ condensate cannot overcome the *local* energy barrier for spin-relaxation. In contrast, condensate atoms in the $|m_F = 1\rangle$ and $|m_F = -1\rangle$ states can directly lower their energy through such collisions, and equilibrate quickly.

Metastability of the spatial distribution was observed in a system of $|m_F = 1\rangle$ and $|m_F = 0\rangle$ atoms. Condensates in these hyperfine states are immiscible due to an anti-ferromagnetic interaction[26]. When an equal mixture of these states was prepared by appropriate rf pulses, the system underwent rapid phase separation and developed alternating layers of $|m_F = 0\rangle$ and $|m_F = 1\rangle$ spin domains of about 40 μm in thickness which were metastable for 20 seconds in the absence of magnetic field gradients. This many-domain spin distribution is a macroscopically occupied excited state; the ground state of the two-component system would contain only one domain each of $|m_F = 0\rangle$ and $|m_F = 1\rangle$ atoms, minimizing the surface energy of the domain walls[33]. For the many-domain state to decay directly to the ground state, the two condensate components would have to either overlap, or else pass by each other without overlapping. Such motion is energetically forbidden: the former due to the anti-ferromagnetic repulsive mean-field energy (typically 50 Hz or 2.5 nK[26]), and the latter due to the kinetic energy required to vary the condensate wavefunction radially (the condensate is about 5 μm wide, yielding a kinetic energy barrier of 50 - 100 Hz). For this reason, the observed excited states are metastable.

3.3 Quantum Tunneling Across Spin Domains in a Bose-Einstein Condensate

The observation of metastable spin domains in optically trapped $F = 1$ spinor Bose-Einstein condensates of sodium (previous section) raised the question of how thermal equilibrium would ultimately be achieved. Besides thermally activated processes we observed quantum tunneling as equilibration process. For the study of this process, spinor condensates were prepared which consisted of only two spin domains in the $|m_F = 0\rangle$ and $|m_F = +1\rangle$ states. Those domains are immiscible due to their anti-ferromagnetic interaction. When a field gradient was added which made it energetically favorable for the two domains to change sides, quantum tunneling was observed (Fig. 5).

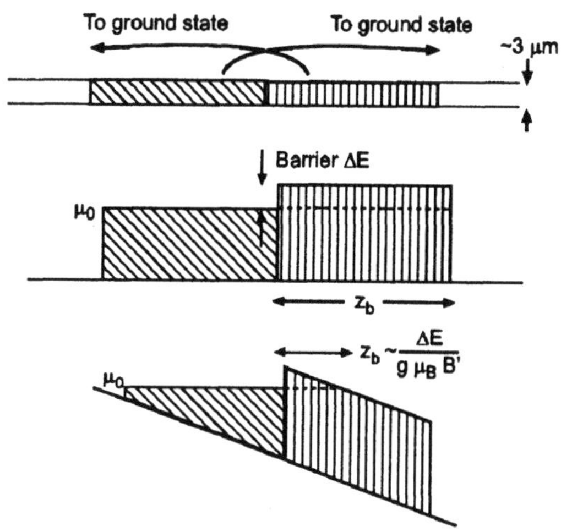

Figure 5. Study of quantum tunneling in a two-component Bose-Einstein condensate. The $m_F = 1$ (grey) and $m_F = 0$ atoms form two immiscible spin domains (upper figure) because the $m_F = 1$ atoms experience an energy barrier due to the repulsive mean field interaction (middle figure). When a longitudinal magnetic field gradient B' is applied, the effective potential for the $m_F = 1$ atoms changes to the lower figure, and the spin domains can rearrange themselves by quantum tunneling through each other.

The tunneling barriers are created intrinsically by the mean-field repulsion between two immiscible components of a quantum fluid. These energy barriers are naturally of nanokelvin-scale height, and their width is typically of micron-scale and can be simply varied by the application of a weak magnetic field gradient. The triangular tunneling burner is similar to the situation for field emission of an electron from a metal surface. The latter situation is described by the Fowler-Nordheim equation. A similar description was developed for the tunneling process in the spinor condensates and agreed well with the measurements[34].

The tunneling rates are a sensitive probe of the boundary between spin domains. As the field was lowered below about 1 G, the time for relaxation to the ground state was dramatically shortened. The analysis indicated an unpredicted spin structure in the boundary between spin domains. In the domain walls, the wavefunctions of the various components have large derivatives which gives rise to a surface energy. This situation makes it energetically favorable to admix the third $|m_F = -1\rangle$ component which is prohibited in the bulk of either domain. Below 1 G, the fraction of m_F = -1 atoms in the barrier becomes large enough to weaken the effective repulsion between the spin domains and to speed up the tunneling process.

4. EVIDENCE FOR A CRITICAL VELOCITY IN A BEC

The existence of a macroscopic order parameter implies superfluidity of gaseous condensates. Observing frictionless flow is a challenge given the small size of the system and its metastability. We have taken a step towards this goal by studying dissipation when an object was moved through the condensate[35]. This is in direct analogy with the well-known argument by Landau[36] and the vibrating wire experiments in superfluid helium[37]. Instead of dragging a massive macroscopic object through the condensate we used a blue detuned laser beam which repelled atoms from its focus to create a moving boundary condition.

The beam created a "hole" with a diameter of 13 μm which was scanned back and forth along the long axis of the cigar-shaped condensate (Thomas-Fermi diameters of 45 and 150 μm in the radial and axial directions, respectively). After exposing the condensate to the scanning laser beam for about one second, the final temperature was determined. As a function of the velocity of the scanning beam, we could distinguish two regimes of heating separated by a critical velocity.

For low velocities, no or little dissipation was observed, and the condensate appeared immune to the presence of the scanning laser beam. For higher velocities, the heating increased, until at a velocity of about 6 mm/s the condensate was almost completely depleted after the stirring. The crossover between these two regimes was quite pronounced and occurred at a critical velocity of about 1.6 mm/s which was a factor of roughly four smaller than the speed of sound at the peak density of the condensate (Fig. 6).

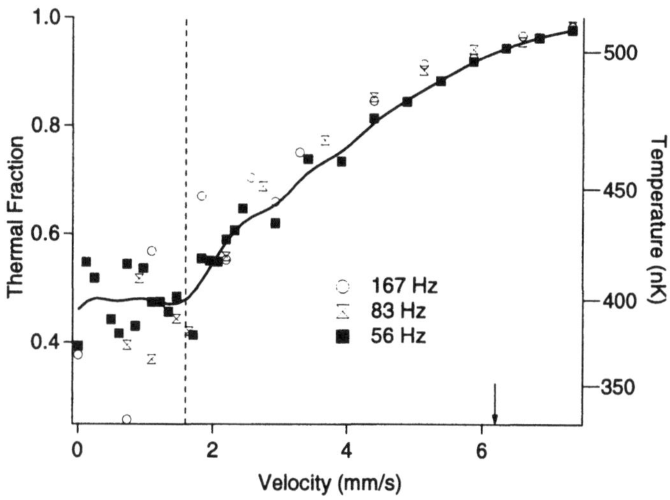

Figure 6. Evidence for a critical velocity. Shown is the final temperature after a laser beam was scanned through the condensate at variable velocity for 900 ms using different scan frequencies. The dashed line separates the regimes of low and high dissipation. The peak sound velocity is marked by an arrow. The data series for 83 and 167 Hz showed large shot-to-shot fluctuations at velocities below 2 mm/sec. The solid line is a smoothing spline fit to the 56 Hz data set to guide the eye. Figure taken from Ref. [35].

These observations are in qualitative agreement with numerical calculations based on the non-linear Schroedinger equation which predict that heating at subsonic velocities is due to the onset of vortex nucleation[38-40]. Because of surface effects and the non-zero temperature, we expect additional corrections leading to dissipation at even lower velocities and a smooth crossover between low and high dissipation. More precise measurements of the heating should allow us to study these finite-size and finite-temperature effects.

5. ATOM LASERS

In an ideal gas, Bose condensed atoms would all occupy the same single-particle ground-state wavefunction. This picture is largely valid even when weak interactions are included. They lead to admixtures of other configurations of typically 1 % or less for the alkali condensates. This is in contrast to liquid helium where this correction (called quantum depletion) is about[36] 90 %. This means that even for an interacting dilute-gas condensate, the atoms can all be regarded as occupying the same single-particle wavefunction (with 99 % weight). Consequently, gaseous Bose-Einstein condensates can serve as sources of coherent atomic beams called atom lasers. Essential aspects of atom laser are *coherence*, the *output coupling* and the *gain process*.

The *coherence* of the condensate was demonstrated by our group in 1997 when two condensates in a double-well potential were released from the trap and allowed to expand, resulting in a high-contrast interference pattern in the overlap region[41].

Interference between many condensates was observed by the group at Yale, when they generated condensates in a multiple-well optical potential and saw interference between them. The observed temporal oscillations were related to Josephson oscillations[42]. Coherence in multi-component condensates was demonstrated at Boulder[43]. A recent spectroscopic measurement[44] of the coherence length of a condensate is discussed in Sect. 6.

An *output coupler* for a Bose-Einstein condensate was realized by our group in 1997 by using pulsed radio-frequency radiation to flip the spin of a fraction of the condensed atoms into an untrapped state which fell downward by gravity[45]. The atoms were shown to be coherent, and the system constituted a pulsed atom laser. Recently, there has been a lot of excitement about more advanced output couplers. The group in Munich was able to expose a magnetically shielded condensate to continuous radio-frequency radiation and realized a cw output coupler[46]. The Gaithersburg group replaced the radio-frequency transition by an optical Raman transition. The photon recoil pushed the atoms out of the trap horizontally, realizing a directional output coupler[47]. The above-mentioned self pulsing atom emission from an array of condensates at Yale can be regarded as a "mode-locked" atom laser[42].

The *gain mechanism* of an atom laser is analogous to that in the optical laser: bosonic stimulation by the coherent matter wave. Bosonic stimulation was observed in the formation of the condensate at MIT[48], during the four-wave mixing experiments at Gaithersburg[49] and most dramatically in the build-up of "superradiant" pulses of matter-waves[50]. More recently, the superradiant atom amplification was used to realize phase-coherent amplification of matter waves (see Sect. 7).

What will atom lasers be used for? The Gaithersburg group has used the condensate as a superior atom source with its high brightness, small *rms* momentum and excellent initial spatial localization. Their experiments include several studies of diffraction of atoms by light[51], an important element in atom interferometers. In the optical domain, the laser is crucial for nonlinear optics. Similarly, atom lasers are crucial for non-linear atom optics. In contrast to photons, however, atoms don't need a non-linear medium - their interactions provide the non-linearity. A beautiful example is the recent experiment in Gaithersburg, where three condensates collided and formed a fourth condensate by four-wave mixing[49].

Condensates can be a highly nonlinear medium not only for matter waves, but also for light. This was dramatically demonstrated recently by a group at the Rowland institute in Cambridge, when they slowed the speed of light to 17 m/s using the condensate as a dense cold medium[52]. Ultimately, atom lasers may replace conventional atomic beams in applications like precision measurements of fundamental constants, tests of fundamental symmetries, atom optics (in particular, atom interferometry and atom holography) and precise deposition of atoms.

6. LIGHT SCATTERING FROM BOSE-EINSTEIN CONDENSATES

In the early '90s, before Bose-Einstein condensation was realized in atomic gases, there were lively debates about what a condensate would look like. Some researchers thought it would absorb all the light and would therefore be "pitch black", some predicted it would be "transparent" (due to superradiant line broadening[53]), others predicted that it would reflect the light due to polaritons[54] and be "shiny" like a mirror.

All the observations of Bose condensates have employed scattering or absorption of laser light. Until recently, the observations were done in regimes where the Bose condensate could be regarded as a cold dilute cloud of atoms that scatters light as ordinary atoms do. On resonance, the condensate strongly absorbed the light, giving rise to the well-known "shadow pictures" of expanding condensates where the condensate appeared black. For off-resonant light, the absorption could be made negligibly small, and the condensate acted as a dispersive medium bending the light like a glass sphere. This regime has been used for non-destructive in-situ imaging of Bose-Einstein condensates (see Fig. 1).

Our group has recently looked more closely at how coherent atoms interact with coherent light, or to be more precise, how the coherence of a condensate plays a role in the interaction with coherent light. Light scattering imparts momentum to the condensate and creates an excitation (Fig. 7). Consequently, the coherence and collective nature of excitations in the condensate can strongly affect the optical properties. As we discuss here, the use of light scattering to characterize atomic Bose condensates is analogous to the use of neutron scattering in the case of superfluid helium[55,56].

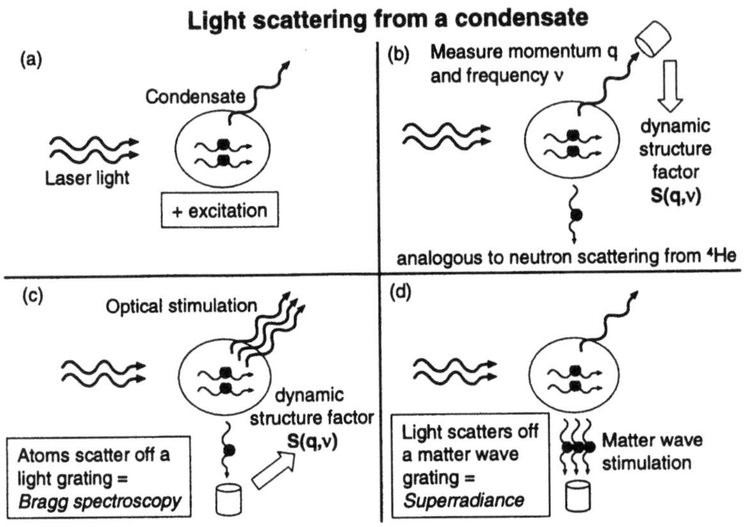

Figure 7. Light scattering from a Bose-Einstein condensate. When a photon is scattered, it transfers momentum to the condensate and creates an excitation (upper left). Therefore, an analysis of the scattered light allows the determination of the dynamic structure factor, in close analogy to neutron scattering experiments with superfluid helium (upper right). For sufficiently large momentum transfer, an atom is scattered out of the condensate. The signal is greatly increased by stimulating the light scattering by a second laser beam and detecting the scattered atoms (lower left) - this is the scheme for Bragg spectroscopy. Light scattering can also be stimulated by adding a coherent atomic field (lower right). This led to superradiant scattering of light and atoms and was used to amplify matter waves.

The simplest light scattering experiment would involve only a single laser beam illuminating the condensate and analysis of the scattered photons. However, the light scattered from a sample containing only 10^7 atoms is hard to detect when it is distributed over the full solid angle. Therefore, we used a second laser beam to stimulate scattering of light with a frequency and direction pre-determined by the laser beam rather than post-determined by analyzing scattered light (Fig. 7). This scheme, which we call Bragg spectroscopy, establishes a high-resolution spectroscopic tool for Bose-Einstein condensates which is sensitive to the momentum distribution of the trapped condensate as well as the effects of interactions[44]. We studied Bragg scattering in two regimes differing by the amount of momentum transfer.

Bogoliubov theory predicts that for a momentum transfer which is smaller than the speed of sound (times the atomic mass) phonons are excited, whereas for larger momentum transfer, the excitations are free-particle like (Fig. 8). In the regime of large momentum transfer, the impulse approximation is valid, and the resonance shows a Doppler broadening due to the zero-point motion of the condensate, i.e. it can be used to measure the momentum distribution of condensates as pursued[56,57] for superfluid ^{4}He. More generally, Bragg spectroscopy can be used to determine the dynamic structure factor $S(\mathbf{q}, \nu)$ over a wide range of frequencies ν and momentum transfers[58] $\mathbf{q}$.

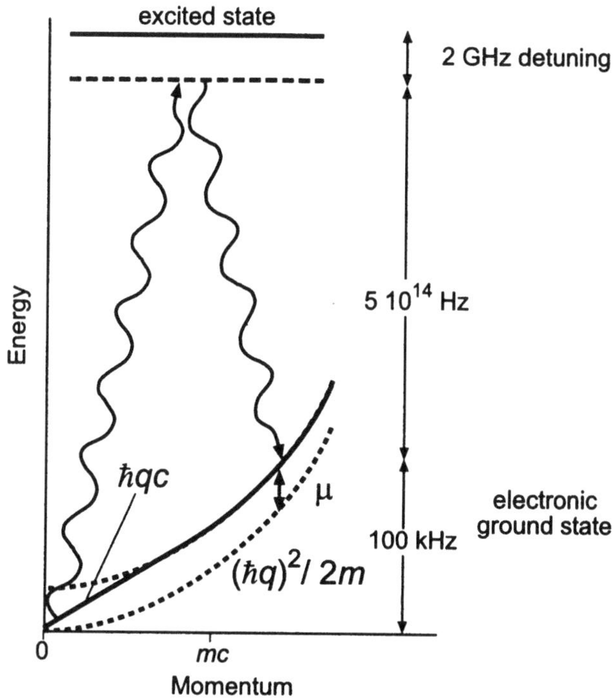

Figure 8. Probing the dispersion relation of a Bose-Einstein condensate by off-resonant light scattering. The microscopic excitation spectrum in a homogeneous weakly-interacting Bose-Einstein condensate is given by the Bogoliubov dispersion relation (solid line). For small momenta $\hbar q$, such that $\hbar q \ll mc$ the dispersion relation is phonon-like (linear). Here m is the mass, and the speed of sound c is related to the interaction energy μ by $\mu = mc^2$. For large momenta $\hbar q \gg mc$ it is particle-like (quadratic) offset from the energy of a free-particle excitation by a mean-field shift of μ ($\approx$ 5 kHz for our experiments). Excitations can be created optically by stimulated light scattering using two laser beams which are both far detuned (about 2 GHz) from the atomic resonance. Momentum and energy are provided by absorption of one photon to a virtual excited level, followed by stimulated emission of a second, lower energy photon.

Bragg spectroscopy was realized by exposing the condensate to two off-resonant laser beams with a frequency difference ν. The intersecting beams formed a moving interference pattern from which atoms could scatter when the Bragg condition was fulfilled (i.e. energy and momentum were conserved). The momentum transfer q is given by q =

$2\hbar k \sin(\vartheta/2)$, where ϑ is the angle between the two laser beams with wavevector k. Figs. 9 and 10 summarize the results for large and small scattering angles, probing both the phonon and free-particle regime. In the regime of low momentum transfer, light scattering was observed to be dramatically reduced (Fig. 10).

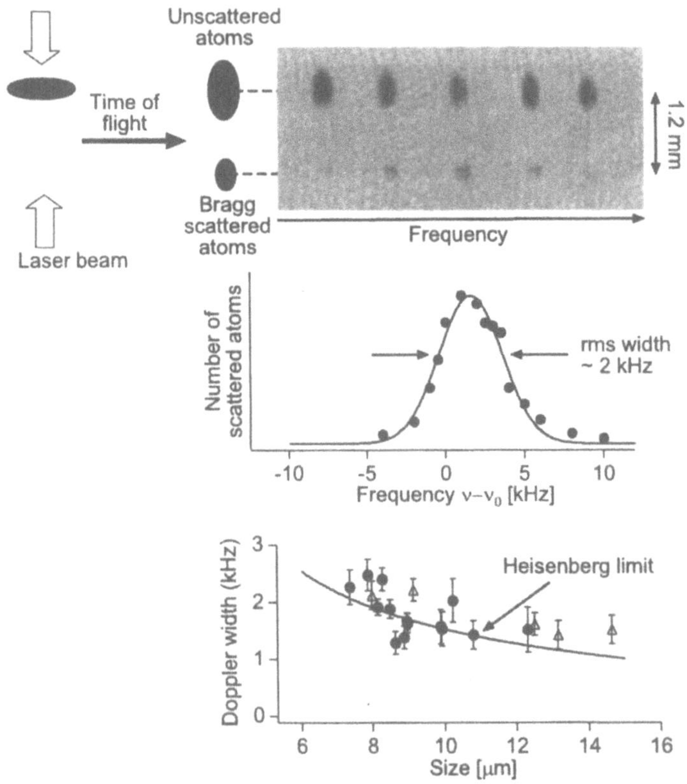

Figure 9. Measuring the momentum distribution of a condensate[44]. A condensate was exposed to two counterpropagating laser beams. Atoms absorbed a photon from one beam and were stimulated to re-emit it by the other beam, resulting in the transfer of recoil momentum to the atoms, as observed in ballistic expansion using absorption imaging after 20 ms time-of-flight (upper part). The number of Bragg scattered atoms showed a narrow resonance when the difference frequency between the two laser beams was varied (upper and middle part). The width of the resonance is caused by Doppler broadening and is therefore proportional to the condensate's momentum uncertainty Δp. It was determined for various sizes Δx of the condensate. The agreement with the Heisenberg limit $\Delta p \approx \hbar/\Delta x$ proves that the Doppler width of the resonance is only due to the zero-point motion of the condensate, or equivalently, that the coherence length of the condensate is equal to its physical size. This demonstrates that a condensate is one "coherent matter wave"! An analogous measurement in the time domain has been done by the group in Gaithersburg[59].

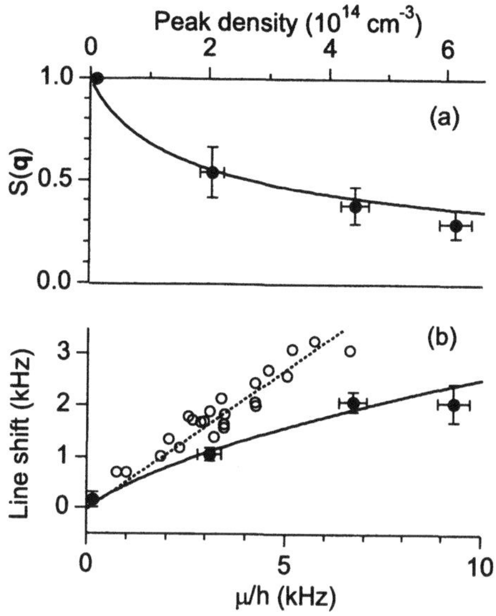

Figure 10. (a) Static structure factor $S(\mathbf{q})$ and (b) shift of the line center from the free particle resonance. $S(\mathbf{q})$ is the ratio of the line strength at a given chemical potential μ to that observed for free particles. As the density and μ increase, the structure factor is reduced, and the Bragg resonance is shifted upward in frequency. Solid lines are predictions of a local-density approximation for light scattering by 14 degrees. The dotted line indicates a mean-field shift of $4\mu/7h$ as measured in the free-particle regime using a scattering angle of 180 degrees. Figure taken from Ref. [60].

In this regime, where atoms cannot absorb momentum "individually" but only collectively, the suppression arises from destructive interference of two excitation paths. The suppression provides dramatic evidence for the presence of correlated momentum excitations in the many-body condensate wavefunction. A similar suppression would occur for spontaneous scattering from a sufficiently dense condensate - turning a "pitch-black" condensate transparent!

A condensate which reflected the incident light was encountered when it was illuminated with a single intense laser beam[50]. When a condensate has scattered a photon, an imprint is left in the form of long-lived excitations. These excitations form a periodic density modulation which diffracts light into the same direction as the first scattered photon. The more photons have been scattered into a certain direction, the larger is the density modulation left behind and the larger the increase of the scattering rate into this direction. This self-acceleration of scattering can be described as bosonic stimulation of the scattering by the population of the final (quasi-particle) state (see Fig. 7).

The gain for this process is highest when the light is scattered along the long axis of the cylindrically shaped condensate and leads to the generation of directed beams of atoms (Fig. 11). This is accompanied by directed emission of light – a new form of superradiance where a density modulation spontaneously develops which makes the condensate "reflect" light like a mirror.

Figure 11. Superradiance and matter-wave amplification in a Bose-Einstein condensate[50]. The figure shows the velocity distribution of the atoms after a condensate (highest peak at the back of each image) was illuminated by laser light. Normal Rayleigh scattering (rear image) is random, and the velocity distribution was smeared out in the direction of the incident light. In contrast, collective "superradiant" scattering created several highly directional "atom laser" beams (front image). The images are absorption images of a cloud which expanded ballistically after the light scattering. The two horizontal axes represent two velocity components and the vertical axis is the column density of the observed atoms. The closest spacing between the peaks is $\sqrt{2}$ times the single-photon recoil velocity.

7. PHASE-COHERENT AMPLIFICATION OF MATTER WAVES

Atom amplification differs from light amplification in one important aspect. Since the total number of atoms is conserved (in contrast to photons), the active medium of a matter wave amplifier has to include a reservoir of atoms. One also needs a coupling mechanism which transfers atoms from the reservoir to an input mode while conserving energy and momentum.

The superradiance discussed in the previous section can act as a matter wave amplifier. The momentum required to transfer atoms from the condensate at rest to the input mode is provided by light scattering. Refs. [61, 62] discussed that a condensate pumped by an off-resonant laser beam acts as a matter wave amplifier which can amplify input matter waves within the momentum range which can be reached by scattering a single pump photon.

The inversion in this matter wave amplifier is most apparent in the dressed atom picture where the condensate at rest and the pump light field are treated as one system. An atom in the dressed condensate can now spontaneously decay into a recoiling atom and a scattered photon which escapes. Inversion is maintained since the photons escape allowing, in principle, a complete transfer of the condensate atoms into the recoil mode.

The gain process can be explained in a semiclassical picture. The input matter wave of wave vector $\mathbf{K}_j$ interferes with the condensate at rest and forms a moving matter wave grating which diffracts the pump light with wave vector $\mathbf{k}_0$ into the momentum and energy conserving direction $\mathbf{k}_i$ = ($\mathbf{k}_0$-$\mathbf{K}_j$). The momentum imparted by the photon scattering is absorbed by the matter wave grating by coherently transferring an atom from the condensate into the recoil mode, which is the input mode. The rate of scattering, which is given by the square of the grating amplitude, is proportional to the number of atoms in the input mode N_j, implying an exponential growth of N_j (as long as one can neglect the depletion of the condensate at rest).

The amplification of atoms in a recoil mode j follows a gain equation[50,64]

$$\dot{N}_j = (G_j - L_j) N_j \tag{1}$$

with the gain coefficient

$$G_j = R N_0 \frac{\sin^2 \theta_j}{8\pi/3} \Omega_j \tag{2}$$

Here R is the rate for single-atom Rayleigh scattering, N_0 the number of atoms in the condensate at rest, θ_j the angle between the polarization of the incident light and the direction of the scattered light, and Ω_j the phase-matching solid angle for scattering into mode j. The loss term L_j describes the decoherence rate of the matter-wave grating and determines the threshold for exponential growth (see Ref. [50] for details).

Input matter waves with a well defined momentum were generated by exposing the condensate to a pulsed optical standing wave which transferred a small fraction of the atoms ($10^{-4} \sim 10^{-2}$) into a recoil mode by Bragg diffraction[51,44]. Both laser beams were red-detuned by 1.7 GHz from the $3S_{1/2}$, $F = 1 \rightarrow 3P_{3/2}$, $F = 0,1,2$ transition to suppress normal Rayleigh scattering.

The geometry of the light beams is shown in Fig. 12. The beam which was perpendicular to the long axis of the condensate (radial beam) was blue detuned by 50 kHz relative to the axial beam. This detuning fulfilled the Bragg resonance condition.

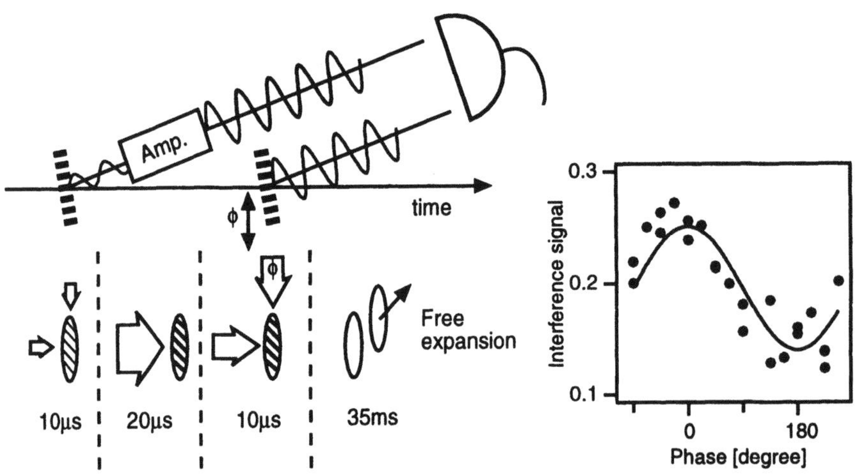

Figure 12. Experimental scheme for observing phase coherent matter wave amplification. A small-amplitude matter wave was split off the condensate by applying a pulse of two off-resonant laser beams (Bragg pulse). This input matter wave was amplified by passing it through the condensate pumped by a laser beam. The coherence of the amplified wave was verified by observing its interference with a reference matter wave, which was produced by applying a second (reference) Bragg pulse to the condensate. The interference signal was observed after 35 ms of ballistic expansion. The fringes on the right side show the interference between the amplified input and the reference matter wave. Reprinted by permission from Nature, Ref. [63], copyright 1999 Macmillan Magazines Ltd.

Amplification of the input matter wave was realized by applying an intense radial pump pulse for the next 20 μs, with a typical intensity of 40 mW/cm^2. The number of atoms in the recoil mode was determined by suddenly switching off the trap and observing the ballistically expanding atoms after 35 ms of time-of-flight using resonant absorption imaging. After the expansion, the condensate and the recoiling atoms were fully separated (Fig. 13c).

Fig. 13 shows the input-output characteristics of the amplifier. The number of input atoms was below the detection limit of our absorption imaging (Fig. 13a) and was determined from a calibration of the Bragg process at high laser powers, where the diffracted atoms were clearly visible in the images. The amplification pulse alone, although above the threshold for superradiance[50], did not generate a discernible signal of atoms in the recoil mode (Fig. 13b). When the weak input matter wave was added, the amplified signal was clearly visible (Fig. 13c). The gain was controlled by the intensity of the pump pulse (see Eq. (2)) and typically varied between 10 and 100. Fig. 13d shows the observed linear relationship between the atom numbers in the input and the amplified output with a number gain of 30.

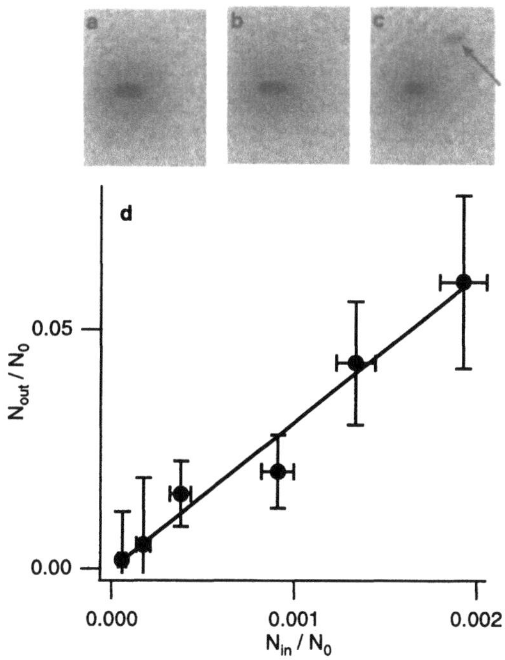

Figure 13. Input-output characteristic of the matter-wave amplifier. (**a-c**) Typical time-of-flight absorption images demonstrating matter wave amplification. The output of the seeded amplifier (**c**) is clearly visible, whereas no recoiling atoms are discernible in the case without amplification (**a**) or amplification without the input (**b**). The size of the images is 2.8 mm × 2.3 mm. (**d**) Output of the amplifier as a function of the number of atoms at the input. A straight line fit shows a number gain of 30. Reprinted by permission from Nature, Ref. [63], copyright 1999 Macmillan Magazines Ltd.

The phase of the amplified matter wave was determined with an interferometric technique (Fig. 12). For this, a reference matter wave was split off the condensate in the same way as the first (input) wave. The phase of the reference matter wave was scanned by shifting the phase of the radio-frequency signal that drove the acousto-optic modulator generating the axial Bragg beam. We then observed the interference between the reference and the amplified matter waves by measuring the number of atoms in the recoil mode.

The interference was observed by scanning the reference phase. When the input was comparable in intensity to the reference matter wave, high contrast fringes were observed even without amplification. Fringes were barely visible, when the input was about 40 times weaker in population. After amplification, we regained a large visibility (Fig. 12). This increase in visibility proved the coherent nature of the matter wave amplification process. The increase in visibility of the interference fringes was a factor of two, less than the expected square root of the total gain of thirty. This might be due to a distortion of the matter wave during the amplification, but this effect requires further study. A similar experiment with rubidium atoms was done at the University of Tokyo[65].

This experiment can be regarded as a demonstration of an active atom interferometer. It realizes a two-pulse atom interferometer with phase-coherent amplification in one of the arms. Such active interferometers may be advantageous for precise measurements of phase shifts in highly absorptive media, e.g. for measurements of the index of (matter wave) refraction when a condensate passes through a gas of atoms or molecules[66]. Since the most accurate optical gyroscopes are active interferometers[67], atom amplification might also play a role in future matter-wave gyroscopes[68].

8. CONDENSATE-CONDENSATE COLLISIONS

Bragg and Raman scattering has been used to realize output couplers for atom lasers. The realization of atom lasers with a large flux of atoms may require the use of much larger condensates. This raises the question of how the outcoupled atoms penetrate the condensate. Most theories on output couplers include only the coherent interactions between two modes, the condensate and the output mode. The coherent coupling between discrete modes leads for example to four-wave mixing[49].

However, when all other final states for elastic scattering are included, matter waves passing through a condensate are attenuated by elastic collisions with a cross section of $8\pi a^2$ and $4\pi a^2$ for atoms in the same or in different internal states, respectively, where a denotes the scattering length.

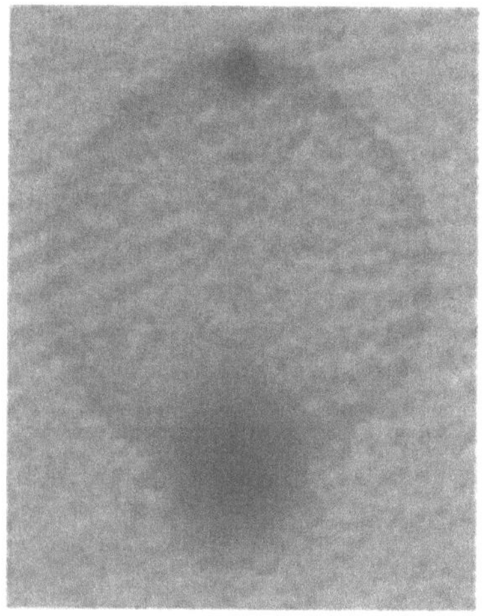

Figure 14. Observation of elastic collisions between the condensate (lower black dot) and Bragg diffracted atoms (upper spot). The image is taken after a time-of-flight of 30 ms, and shows the velocity distribution after the collision. The products of the collisions are distributed over a sphere in momentum space leading to the observed halo. The height of the image is 3.2 mm.

We have started to study collisions between out-coupled matter waves and the remaining condensate. We used counter-propagating laser beams and drove either a two-photon Bragg or Raman transition, thus coupling out either $m_F = 1$ or $m_F = 0$ atoms from the $m_F = -1$ condensate (see also Refs. [47,51]). A clear signature of collisions between the out-coupled atoms and the condensate is a halo in time-of-flight pictures. Due to momentum conservation, the final momenta are required to be distributed on the surface of a sphere, hence showing up as a halo in our two-dimensional images (Fig. 14). Typically, the collisional density along the radial and axial directions for our condensates is 1 and 10, respectively. A quantitative study should result in a determination of the scattering length which can be deduced directly from the number of out-coupled atoms which survive the passage through the condensate. Preliminary measurements show that the surviving fraction is higher for collisions between atoms in different hyperfine states, which is due to the absence of the exchange term in the scattering between "unlike" atoms.

9. A NEW WINDOW ON THE QUANTUM WORLD

Bose-Einstein condensation provides us with a new window on the quantum world, where wave properties of matter dominate on a macroscopic scale. The direct observations of the condensate's density distribution can be regarded as a direct visualization of the magnitude of the macroscopic wavefunction. This could be done even repeatedly and non-destructively, thus recording the real-time evolution of the squared wavefunction of a single condensate. A wavefunction is a probabilistic description of a system in the sense that it determines the distribution of measurements if many identical wavefunctions are repeatedly probed. In BEC, one simultaneously realizes millions of identical copies of the same wavefunction, and thus the wavefunction can be accurately determined while affecting only a small fraction of the condensed atoms by the measurement process. On the other hand, we have already observed quantum correlations which go beyond the simple single-particle picture[60].

The rapid pace of developments in atomic BEC during the last few years has taken the community by surprise. After decades of an elusive search nobody expected that condensates would be so robust and relatively easy to manipulate. Also, nobody imagined that such a simple system would pose so many challenges, not only to experimentalists, but also to our fundamental understanding of physics. The list of future challenges is long and includes the complete characterization of elastic and inelastic collisions at ultralow temperatures, the exploration of superfluidity, vortices, and second sound in Bose gases, the study of quantum-degenerate molecules and Fermi gases, the development of practical "high-power" atom lasers, and their application in atom optics and precision measurements.

ACKNOWLEDGMENTS

I gratefully acknowledge the contributions of my past and present collaborators M.R. Andrews, A.P. Chikkatur, K.B. Davis, D.S. Durfee, A. Görlitz, S. Gupta, Z. Hadzibabic, S. Inouye, M. Köhl, C.E. Kuklewicz, M.-O. Mewes, H.-J. Miesner, R. Onofrio, T. Pfau, D.E. Pritchard, C. Raman, D.M. Stamper-Kurn, J. Stenger, C.G. Townsend, N.J. van Druten, and J. Vogels. I thank A. Görlitz for valuable comments on the manuscript. This work was supported by the ONR, NSF, ARO, NASA, and the David and Lucile Packard Foundation.

REFERENCES

1. Einstein, A., 1925, Sitzungsber. Preuss. Akad. Wiss., Bericht 3, p. 18.
2. Anderson, M.H., Ensher, J.R., Matthews, M.R., Wieman, C.E., and Cornell, E.A., 1995, *Science* **269**: 198.

3. Davis, K.B., Mewes, M.-O., Andrews, M.R., van Druten, N.J., Durfee, D.S., Kurn, D.M., and Ketterle, W., 1995, *Phys. Rev. Lett.* **75**: 3969.
4. Bradley, C.C., Sackett, C.A., and Hulet, R.G., 1997, *Phys. Rev. Lett.* **78**: 985.
5. Fried, D.G., Killian, T.C., Willmann, L., Landhuis, D., Moss, S.C., Kleppner, D., and Greytak, T.J., 1998, *Phys. Rev. Lett.* **81**: 3811.
6. BEC home page of the Georgia Southern University, http://amo.phy.gasou.edu/bec.html.
7. Huang, K., 1964. In *Studies in Statistical Mechanics,* (J. de Boer and G.E. Uhlenbeck, eds.), North-Holland, Amsterdam, Vol. II, pp. 3-106.
8. Griffin, A., Snoke, D.W., and Stringari, S., 1995, *Bose-Einstein Condensation.* Cambridge University Press, Cambridge.
9. Ketterle, W., Durfee, D.S., and Stamper-Kurn, D.M., 1999. In *Bose-Einstein condensation in atomic gases,* Proceedings of the International School of Physics Enrico Fermi, Course CXL, (M. Inguscio, S. Stringari, and C.E. Wieman, eds.) IOS Press, Amsterdam, pp. 67-176.
10. Cornell, E.A., Ensher, J.R., and Wieman, C.E., 1999. In *Bose-Einstein condensation in atomic gases,* Proceedings of the International School of Physics Enrico Fermi, Course CXL, (M. Inguscio, S. Stringari, and C.E. Wieman, eds.) IOS Press, Amsterdam, p. 15-66.
11. Kleppner, D., Greytak, T.J., Killian, T.C., Fried, D.G., Willmann, L., Landhuis, D., and Moss, S.C., 1999. In *Bose-Einstein condensation in atomic gases,* Proceedings of the International School of Physics Enrico Fermi, Course CXL, (M. Inguscio, S. Stringari, and C.E. Wieman, eds.) IOS Press, Amsterdam, pp. 177-179.
12. Andrews, M.R., Mewes, M.-O., van Druten, N.J., Durfee, D.S., Kurn, D.M., and Ketterle, W., 1996, *Science* **273**, 84.
13. Mewes, M.-O., Andrews, M.R., van Druten, N.J., Kurn, D.M., Durfee, D.S., Townsend, C.G., and Ketterle, W.,1996, *Phys. Rev. Lett.* **77**: 988.
14. Jin, D.S., Ensher, J.R., Matthews, M.R., Wieman, C.E., and Cornell, E.A., 1996, *Phys. Rev. Lett.* **77**: 420.
15. Jin, D.S., Matthews, M.R., Ensher, J.R., Wieman, C.E., and Cornell, E.A., 1997, *Phys. Rev. Lett.* **78**: 764.
16. Stamper-Kurn, D.M., Miesner, H.-J., Inouye, S., Andrews, M.R., and Ketterle, W., 1998, *Phys. Rev. Lett.* **81**: 500.
17. Stringari, S., 1998, *Phys. Rev. A* **58**: 2365.
18. Andrews, M.R., Kurn, D.M., Miesner, H.-J., Durfee, D.S., Townsend, C.G., Inouye, S., and Ketterle, W., 1997, *Phys. Rev. Lett.* **79**: 553.
19. Dalfovo, F., Giorgini, S., Guilleumas, M., Pitaevskii, L.P., and Stringari, S.,1997, *Phys. Rev. A* **56**: 3840.
20. Al Khawaja, U., Pethick, C.J., and Smith, H., 1999, *Phys. Rev. A* **60**: 1507.
21. Onofrio, R., Durfee, D.S., Raman, C., Köhl, M., Kuklewicz, C.E., and Ketterle, W., 2000, *Phys. Rev. Lett.* **84**: 810.
22. Stringari, S., 1996, *Phys. Rev. Lett.* **77**, 2360.
23. Madison, K.W., Chevy, F., Wohlleben, W., and Dalibard, J., 2000, *Phys. Rev. Lett.* **84**: 806.
24. Myatt, C.J.,Burt, E.A., Ghrist, R.W., Cornell, E.A., and Wieman, C.E., 1997, *Phys. Rev. Lett.* **78**: 586.
25. Stamper-Kurn, D.M., Andrews, M.R., Chikkatur, A.P., Inouye, S., Miesner, H.-J., Stenger, J., and Ketterle, W., 1998, *Phys. Rev. Lett.* **80**: 2027.
26. Stenger, J., Inouye, S., Stamper-Kurn, D.M., Miesner, H.-J., Chikkatur, A.P., and Ketterle, W., 1998, *Nature* **396**: 345.

27. Ho, T.-L., 1998, *Phys. Rev. Lett.* **81**: 742.
28. Ohmi T., and Machida, K., 1998, *J. Phys. Soc. Jap.* **67**: 1822.
29. Law, C.K., Pu, H., and Bigelow, N.P.,1998, *Phys. Rev. Lett.* **81**: 5257.
30. Hall, D.S., Matthews, M.R., Ensher, J.R., Wieman, C.E., and Cornell, E.A, 1998, *Phys. Rev. Lett.* **81**: 4531.
31. Miesner, H.-J., Stamper-Kurn, D.M., Stenger, J., Inouye, S., Chikkatur, A.P., and Ketterle, W., 1999, *Phys. Rev. Lett.* **82**: 2228.
32. Guttman L., and Arnold, J.R., 1953, *Phys. Rev.* **92**: 547.
33. Ao, P., and Chui, S.T.,1998, *Phys. Rev. A* **58**: 4836.
34. Stamper-Kurn, D.M., Miesner, H.-J., Chikkatur, A.P., Inouye, S., Stenger, J., and Ketterle, W., 1999, *Phys. Rev. Lett.* **83**: 661.
35. Raman, C., Köhl, M., Onofrio, R., Durfee, D.S., Kuklewicz, C.E., Hadzibabic, Z., and Ketterle, W.,1999, *Phys. Rev. Lett.* **83**: 2502.
36. Huang, K., 1987, *Statistical Mechanics*. Wiley, New York.
37. Castelijns, C.A.M., Coates, K.F., Guénault, A.M., Mussett, S.G., and Pickett, G.R., 1985, *Phys. Rev. Lett.* **56**: 69.
38. Frisch, T., Pomeau, Y., and Rica, S., 1992, *Phys. Rev. Lett.* **69**: 1644.
39. Huepe, C., and Brachet, M.-E.,1997, C.R. Acad. Sci. Paris Série II **325**: 195.
40. Winiecki, T., McCann, J.F., and Adams, C.S., 1999, *Phys. Rev. Lett.* **82**: 5186.
41. Andrews, M.R., Townsend, C.G., Miesner, H.-J., Durfee, D.S., Kurn, D.M., and Ketterle, W., 1997, *Science* **275**: 637.
42. Anderson, B.P., and Kasevich, M.A., 1998, *Science* **282**: 1686.
43. Hall, D.S., Matthews, M.R., Wieman, C.E., and Cornell, E.A., 1998, *Phys. Rev. Lett.* **81**: 1543.
44. Stenger, J., Inouye, S., Chikkatur, A.P., Stamper-Kurn, D.M., Pritchard, D.E., and Ketterle, W., 1999, *Phys. Rev. Lett.* **82**: 4569.
45. Mewes, M.-O., Andrews, M.R., Kurn, D.M., Durfee, D.S., Townsend, C.G., and Ketterle, W., 1997, *Phys. Rev. Lett.* **78**: 582.
46. Bloch, I., Hänsch, T.W., and Esslinger, T., 1998, *Phys. Rev. Lett.* **82**: 3008.
47. Hagley, E.W., Deng, L., Kozuma, M., Wen, J., Helmerson, K., Rolston, S.L., and. Phillips, W.D., 1999, *Science* **283**: 1706.
48. Miesner, H.-J., Stamper-Kurn, D.M., Andrews, M.R., Durfee, D.S., Inouye, S., and Ketterle, W., 1998, *Science* **279**: 1005.
49. Deng, L., Hagley, E.W., Wen, J., Trippenbach, M., Band, Y., Julienne, P.S., Simsarian, J.E., Helmerson, K., Rolston, S.L., and Phillips, W.D., 1999, *Nature* **398**: 218.
50. Inouye, S., Chikkatur, A.P., Stamper-Kurn, D.M., Stenger, J., Pritchard, D.E., and Ketterle, W., 1999, *Science* **285**: 571.
51. Kozuma, M., Deng, L., Hagley, E.W., Wen, J., Lutwak, R., Helmerson, K., Rolston, S.L., and Phillips, W.D., 1999, *Phys. Rev. Lett.* **82**: 871.
52. Hau, L.V., Harris, S.E., Dutton, Z., and Behroozi, C.H., 1999, *Nature* **397**: 594.
53. Javanainen, J., 1994, *Phys. Rev. Lett.* **72**: 2375.
54. Politzer, H.D., 1991, *Phys. Rev. A* **43**: 6444.
55. Nozières, P., and Pines, D., 1990, *The Theory of Quantum Liquids*. AddisonWesley, Redwood City, CA.
56. Sokol, P.E., 1995. In *Bose-Einstein Condensation* (A. Griffin, D.W. Snoke, and S. Stringari, eds.), Cambridge University Press, Cambridge, pp. 51-85.
57. Hohenberg, P.C., and Platzman, P.M., 1966, *Phys. Rev.* **152**: 198.
58. Greytak, T.J., 1978. In *Quantum Liquids*, (J. Ruvalds and T. Regge, eds.), North-Holland, New York, pp. 121-165.

59. Hagley, E.W., Deng, L., Kozuma, M., Trippenbach, M., Band, Y.B., Edwards, M., Doery, M., Julienne, P.S., Helmerson, K., Rolston, S.L., and Phillips, W.D., 1999, *Phys. Rev. Lett.* **83**: 3112.
60. Stamper-Kurn, D.M., Chikkatur, A.P., Görlitz, A., Inouye, S., Gupta, S., Pritchard, D.E., and Ketterle, W., 1999, *Phys. Rev. Lett.* **83**, 2876.
61. Law, C.K., and Bigelow, N.P., 1998, *Phys. Rev. A* **58**: 4791.
62. Moore, M.G., and Meystre, P., 1999, *Phys. Rev. A* **59**: 1754.
63. Inouye, S., Pfau, T., Gupta, S., Chikkatur, A.P., Görlitz, A., Pritchard, D.E., and Ketterle, W., 1999, *Nature* **402**: 641.
64. Moore, M.G., and Meystre, P., 1999, *Phys. Rev. Lett.* **83**: 5202.
65. Kozuma, M., Suzuki, Y., Torii, Y., Sugiura, T., Kuga, T., Hagley, E.W., and Deng, L., 1999, *Science* **286**: 2309.
66. Schmiedmayer, J., Chapman, M.S., Ekstrom, C.R., Hammond, T.D., Wehinger, S., and Pritchard, D.E., 1995, *Phys. Rev. Lett.* **74**: 1043.
67. Stedman, G.E., 1997, *Rep. Prog. Phys.* **60**: 615.
68. Gustavson, T.L., Bouyer, P., and Kasevich, M.A., 1997, *Phys. Rev. Lett.* **78**: 2046.

Measurement Of The Relative Phase Between Two Bose-Einstein Condensates

D. S. HALL
JILA and department of Physics University of Colorado Boulder, CO 80309-0440 USA.
Currrent Address: Department of Physics, Amherst College, Amherst, MA 01002-5000 USA

1. INTRODUCTION

The experimental realisation of Bose-Einstein condensation in the dilute alkalis[1-3] and hydrogen[4] gases has made possible a variety of new and exciting experiments. As with other macroscopic quantum systems, such as superfluids and superconductors, the Bose-Einstein condensate (BEC) may be described by a macroscopic wavefunction: $\Phi = |\Phi| \exp(iS)$. Many of the interesting properties of a BEC can be traced back to the phase S of this wavefunction. Although the phase of an isolated condensate is not in itself meaningful, one can look for ways to measure the *relative* phase between two (or more) condensates.

The first (and perhaps most beautiful) of the experiments to examine the relative phase between two BECs was the 1997 experiment of Andrews et al.[5], in which a single condensate was cut into two pieces with a laser beam and then rejoined. A striking interference pattern, reminiscent of the interference pattern between two laser beams, was observed in the atomic density profile: fringes of enhanced and depleted density depending on the relative phase of the condensates at each position in space. This experiment established that one could indeed measure a relative phase between two condensates, but was unable to say anything about its time evolution due to mechanical instabilities.

At JILA, we adopted a different approach to looking at the relative phase between two quantum objects. In our system, the two condensates are in different *internal* states; these two states can only interconvert when they are coupled by a microwave field. The remainder of the time they behave like separate quantum fluids[6,7]. Using an interferometric technique[8], we can measure the time-evolution of the relative phase between our two condensates under well-controlled conditions[9]. In the first of these two Chapters I will present an introduction to our double-condensate system with an account of our first measurements of the time-evolution of the quantum fluids and their relative phase.

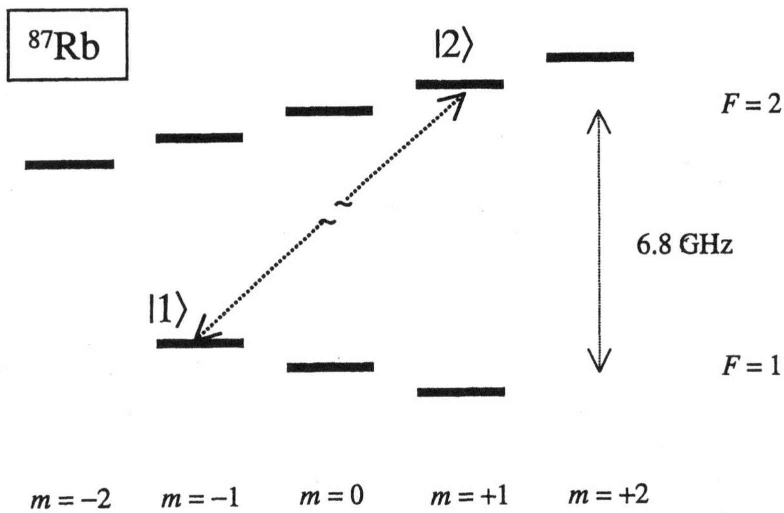

Figure 1. Energy level diagram of the ground states of ^{87}Rb. The atoms may be coherently transferred from state |1⟩ to state |2⟩, and back again, with microwave radiation.

Before embarking on these topics, I would like to mention some related experimental work which will no doubt be discussed further in other lectures at this School. The group of Ketterle at MIT has been working for some time with multicomponent BEC confined in optical traps, and has seen a variety of rich behaviours including metastability[10] and tunneling across spin domains[11]. Anderson and Kasevich at Yale have placed single-component condensates into optical lattices[12], effectively separating a single condensate into many smaller ones whose relative phases evolve at different rates; the interference between these condensates leads to coherent tunneling analogous to the Josephson effect in superconductors. Finally, a recently published experiment by Hagley et al. at NIST (Gaithersburg) uses a homodyne technique to extract information about the phase of a BEC[13].

2. OUR SYSTEM

The atom we use in our experiments is ^{87}Rb, the first of the dilute-gas alkalis to be Bose-condensed[1]. The atomic level structure of ^{87}Rb is shown in Fig. 1. As with the other alkalis, the ground state possesses two distinct hyperfine levels that are labeled by their total angular momentum $F = 1$ and $F = 2$; these levels are separated by 6.8 GHz. We shall be exclusively concerned with double condensates in the $|F = 1, m_f = -1\rangle$ and $|F = 2, m_f = 1\rangle$ states, labeled $|1\rangle$ and $|2\rangle$ in the Figure. These two states have nearly identical magnetic moments and may be coupled to one another via a two-photon transition. This two-level system thus constitutes a statistical spin-1/2 system.

It was not originally thought that condensates in both hyperfine states would simultaneously survive in a magnetic trap because of inelastic collisions. Each hyperfine-changing collision, for instance, releases 6.8 GHz of energy; this amount of energy is sufficient to drive all of the atoms out of a typical condensate, were it distributed evenly among them[7]. Fortunately, an early experiment showed that these hyperfine changing collisions are suppressed in ^{87}Rb[14]. This serendipitous circumstance results from an accidental degeneracy in the triplet and singlet *s*-wave scattering lengths [15,16] and makes ^{87}Rb unique (thus far) in its simultaneous accommodation of condensates in both hyperfine states.

The procedures we use to create a single condensate in ^{87}Rb are by now well established[1,17], and only a brief description will be given here. Atoms are first collected and precooled in a magneto-optical trap, optically pumped into the $|1\rangle$ state, and transferred into a time-averaged orbiting potential (TOP) magnetic trap[18], which confines the atoms by their magnetic moments. The sample is then evaporatively cooled by removing the most energetic atoms from the trap; this is accomplished by applying a radiofrequency drive that drives spatially-selective transitions between the magnetically trapped $|1\rangle$ state and the untrapped states $|1,0\rangle$ and $|1,+1\rangle$. The remaining atoms rethermalize at lower and lower temperatures until the critical temperature T_c is reached, at which point they begin to pile up in the ground state of the harmonic oscillator (condensate). The evaporation continues until most of the atoms are in the condensate and there is no discernible thermal component in the trap. Each condensate takes about a minute to prepare, and consists of $N \approx 5 \times 10^5$ atoms at a temperature below 50 nK.

We create the double condensate from the single $|1\rangle$ condensate by driving a two-photon transition from the $|1\rangle$ state to the $|2\rangle$ state, as suggested in Fig. 1[17]. The applied radiation induces coherent transitions (Rabi oscillations) between the states $|1\rangle$ and $|2\rangle$. We typically use field strengths such that the Rabi frequency Ω_R is of order 2π x 600 Hz. By varying the length of time that the drive is on, we can selectively put any fraction (from 0% to 100%) of the atoms into the state $|2\rangle$, with the remainder staying behind in state $|1\rangle$. For instance, a so-called $\pi/2$-pulse will create a double condensate, half of which is in state $|1\rangle$ (up to quantum fluctuations of order $\left(1/\sqrt{N}\right)$, with the remainder in state $|2\rangle$.

Once the (double) condensate is created, it may be imaged by one of a variety of techniques. In the experiments described in this first Chapter, we used resonant absorption imaging. The condensate is first released from the trap, so that it expands ballistically and drops under the influence of gravity. After 20 ms, a beam of light which is nearly resonant with one of the two internal states is passed through the condensate. The condensate atoms in that state absorb and scatter the light, impressing a shadow on the beam, which is then imaged on a charge-coupled device (CCD) array. The resulting image yields, after some processing, the density distribution of the atoms, their temperature, how much of the condensate is in each internal state, and so forth[1]. This type of imaging is destructive not only in that the condensate is released from the trap prior to the application of the probe beam, but that the condensate is irrevocably heated by the scattering of the probe beam.

Is the system we produce with a $\pi/2$-pulse best described as a system of two different condensates, or should it be treated as a single condensate in a superposition of two internal states? As with many questions in quantum mechanics, the answer we give depends on what experiment we later perform on the system. Consider the pair of experiments suggested in Fig. 2(a) and (b). In both experiments, we apply a $\pi/2$-pulse and then permit the system to evolve. In (a), we image the condensate components separately, as may be easily accomplished with appropriate choices of laser detunings. Since there can be no interconversion between the two species, and our imaging distinguishes one from the other, we may think of the system as being composed of two distinguishable BECs. The situation is quite analogous to that in which a laser beam is passed through a 50-50 beamsplitter; each of the two resulting beams is distinguishable as long as they remain spatially separated.

If the two laser beams were brought together and imaged on a screen, we would expect to see an interference pattern because of the coherence between the beams. Similarly, if we were to *recouple* the two condensates, we would expect them to interfere with one another, as suggested in Fig. 2(b). Thus, we can think of this system as a single condensate in a superposition of two internal states. Or we could say that the two condensates have a *predictable* relative phase. Which description we use depends largely on the details of the experiment we perform.

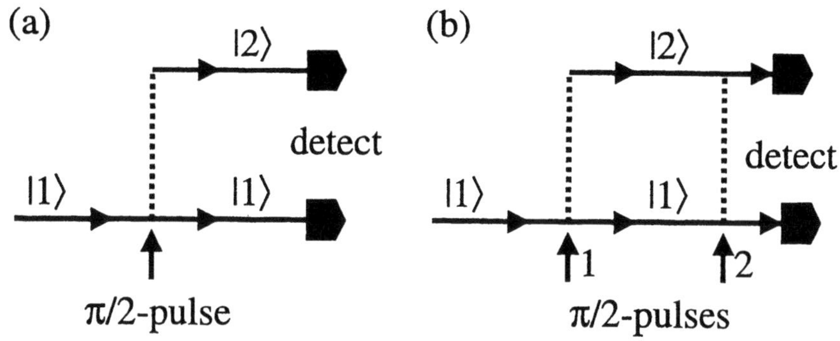

Figure 2. Our binary condensate system can be treated as two distinguishable quantum fluids (a) as long as the coupling drive remains off. We can observe interference effects related to the coherence between the condensates if we recouple the condensates at some later time (b).

3. DISTINGUISHABLE CONDENSATES

When created, our double condensate consists of two overlapping condensates with a well-defined relative phase[9]. The subsequent time-evolution of the two condensates is governed by a pair of Gross-Pitaevskii equations:

$$i\hbar\dot{\Phi}_1 = \left(-\frac{\hbar^2}{2m}\nabla^2 + V_1 + u_1|\Phi_1|^2 + u_{12}|\Phi_2|^2\right)\Phi_1 + \frac{\hbar\Omega(t)}{2}e^{i\omega_{rf}t}\Phi_2 \quad (1)$$

and

$$\begin{aligned} i\hbar\dot{\Phi}_2 = &\left(-\frac{\hbar^2}{2m}\nabla^2 + V_2 + V_{\mathrm{hf}} + u_2|\Phi_2|^2 + u_{21}|\Phi_1|^2\right)\Phi_2 + \\ &+\frac{\hbar\Omega(t)}{2}e^{-i\omega_{rf}t}\Phi_1, \end{aligned} \quad (2)$$

where m is the mass of the Rb atom, V_{hf} is the magnetic field-dependent hyperfine splitting between the two states in the absence of interactions, V is the trapping potential for state i, $u_i = 4\pi\hbar^2 a_i/m$ and $u_{ij} = 4\pi\hbar^2 a_{ij}/m$, $|\Phi_i|^2$ is the condensate density, and the intraspecies and interspecies scattering lengths are a_i and $a_{ij} = a_{ji}$, respectively. The coupling drive is characterized by its frequency ω_{rf} and Rabi frequency Ω_R, which is time-dependent in the sense that it can be turned on and off.

With both states in the same trapping potential, and the coupling drive off, the dominant asymmetry between the two states is the slight difference in scattering lengths, which are positive and in the ratio $a_1 : a_{12} : a_2 :: 1.03 : 1 : 0.97$, with a mean of 55(3) Å. Since $a_1 > a_2$, there is a tendency for the $|1\rangle$ condensate to be pushed outward in a shell around the $|2\rangle$ condensate, which itself draws inward[6]. The $|1\rangle$ condensate therefore has a positive "buoyancy" and "floats" on state $|2\rangle$. We have observed this separation into a "ball" and "shell," although some unknown and apparently uncontrollable asymmetry causes the condensates to separate asymmetrically at longer times.

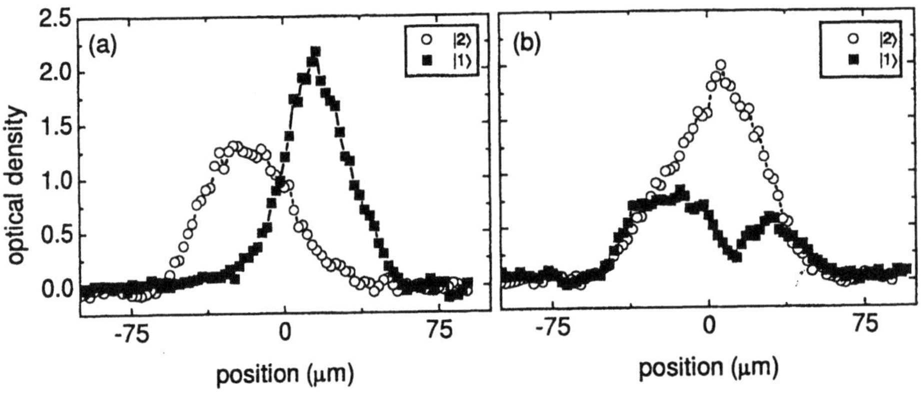

Figure 3. Cross-sections of the condense density profiles. The condensates reach a steady state after ~ 45 ms with a well-defined overlap region (a). After a second coupling pulse, we observe interference that results in phase-sensitive transfer from one condensate to the other (b). In the case shown, the interference is destructive for state $|1\rangle$ and constructive for state $|2\rangle$.

We can introduce a stronger and controllable asymmetry into the system by creating a slight offset in the vertical trapping centers for the two condensates[6]. This offset originates in the balance between gravity and the magnetic trapping potentials for the two states, and is further accentuated by subtleties associated with the rotating bias field of the TOP trap[19]. This fine control permits us to produce traps in which the centers overlap, or (alternatively) we can move one trap center up or down with respect to the other.

Working in an anisotropic trap with axial frequency $\omega_z/2\pi = 60$ Hz and radial frequency $\omega_r/2\pi = 21$ Hz, and a small vertical offset in the trap centers of 0.4 μm, we observed the time-evolution of a double condensate with roughly equal numbers of atoms in each of the two states. As a result of their mutual mean-field repulsions and the slight vertical offset in the trapping potentials, the two condensates separate spatially; we observe the ensuing time-evolution by taking a destructive image of each of the two states at each point in time. We find that the separation is quite violent and dramatic, with a distinctive and repeatable vertical structure forming as the system evolves. Remarkably, the sum of the individual density profiles, i.e., the total density profile, remains largely unperturbed as the individual components rearrange themselves.

After about 45 ms the system settles down, with the $|1\rangle$ condensate positioned above, and slightly overlapping, the $|2\rangle$ condensate; this is shown in cross-section in Fig. 3(a). Little subsequent evolution is observed. Clearly, the energy of the system has been damped, and is seen in an analysis of the relative center-of-mass motion of the two condensates[6]. The damping has been modeled, and while the qualitative features have been reproduced[20], the exact path that the condensate takes to its equilibrium state is not fully understood.

4. COHERENT FLUIDS

We now change our point of view, and consider our system as that of a single condensate in a superposition of two internal states (or, equivalently, two condensates with a well-defined relative phase). Of course, the relative phase itself is not an observable; we must come up with a way of turning the *phase* information into amplitude information, which will then appear as a variation in the *density of* the imaged condensates. That is, we need to interfere the condensates with one another.

The interference can result from a second microwave coupling pulse applied at some time T after the first. Since the interference can only occur in the region where the condensates overlap, we restrict our attention to the overlap region of Fig. 3(a). To see how the interference arises, consider the condensate in state $|2\rangle$ after *two* $\pi/2$ coupling pulses at times $t = 0$ and $t = T$. There are two possible paths by which the condensate could arrive in state $|2\rangle$. They are: (1) the condensate could have remained in state $|2\rangle$ after the first pulse, and been promoted to state $|2\rangle$ after the second; or (2) the condensate could have been promoted to state $|2\rangle$ by the first pulse and remained in state $|2\rangle$ after the second. It is the lack of information about "which path" the condensates took to reach state $|2\rangle$ that admits the possibility of interference between them.

Let us now consider the relative phase accumulated between these two paths. We may choose the initial relative phase between the condensates to be zero just after the first $\pi/2$-pulse at $t = 0$. For path (1), the condensate will accumulate a phase $\mu_1 T/\hbar$ (where μ_1, is its chemical potential) for the time spent in state $|1\rangle$, and then a discontinuous phase jump $\omega_{\text{rf}}T$ when it absorbs a microwave photon at time T. Its phase change is, therefore,

$$\Delta S_{(1)} = \frac{\mu_1}{\hbar}T + \omega_{\text{rf}}T\,. \tag{3}$$

For path (2), the condensate will accumulate a phase $\mu_2 T/\hbar$ for its time spent in state $|2\rangle$, leading to an amplitude

$$\Delta S_{(2)} = \frac{\mu_2}{\hbar}T\,. \tag{4}$$

The probability of arriving in state $|2\rangle$ at time T is proportional to the modulus squared of the sum of the two condensate amplitudes, giving rise to an interference term at the frequency $\Delta S = \mu_2/\hbar - \mu_1/\hbar - \omega_{\text{rf}}$. This is literally the difference in phase evolution frequencies $(\mu_2-\mu_1)/\hbar$ beat against a local oscillator (our microwave field at ω_{rf}). The resulting signal is essentially the *detuning* δ between the local oscillator and the rate of evolution of the relative phase[8].

Our analysis has perhaps been a bit simplistic, in that we have assumed a known evolution frequency $\mu_i/\hbar$ for a condensate in state $|i\rangle$. This needn't be so; in fact, couplings to the internal and external environment can lead to a diffusion of the relative phase. Some of these couplings are internal to the system and represent interesting physics: for instance, the couplings between the condensate and thermal atoms and its own internal excitations[21]. Given the violent rearrangement of the condensates after their creation, we might expect these contributions. to be quite large. Another category of couplings include phase variations that result from apparatus-related issues, such as changing magnetic fields or the effects of mechanical vibrations. In principle, the latter may be suppressed through increased experimental effort. Both categories do contribute to our lack of knowledge about the relative phase, which we can write as an additional phase term $\Delta\varphi_{\text{diff}}$.

In addition, there are quantum fluctuations in the number of atoms in each condensate; and, since the chemical potential of the condensate depends on the number of atoms in it, there are consequently fluctuations in the chemical potential. Such fluctuations are believed to lead to dispersive behavior in the relative phase[22,23] and phenomena such as collapses and revivals[24, 25]. It is believed at the present time that these fluctuations occur on time scales that are generally longer than the ~ 100 ms probed in our experiments[26], although they may in the future be accessible to measurements of this kind.

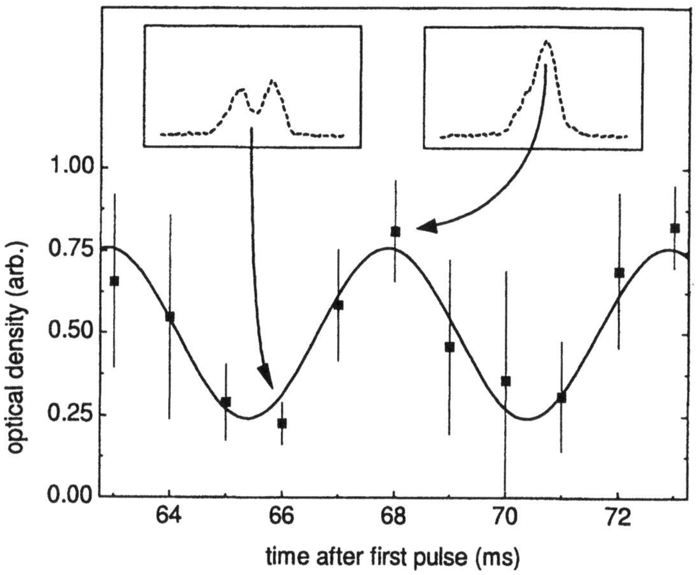

Figure 4. Time evolution of the relative phase, as determined by the condensate density in the overlap region after the second coupling pulse at time T (insets). Each data point represents the average of several individual measurements; the scatter in these measurements is given by the vertical bars. The emergence of a time-varying signal indicates that the memory of the initial relative phase remains despite the complicated rearrangement of the two components.

Given the potential for the relative phase to diffuse and disperse, we might well ask: will we see any interference at all in a particular run? And, if we do see an interference between the condensates, will the interference pattern be the same each time we repeat the measurement? We turn now to the experiment to provide the answers.

Our experiment begins in exactly the same manner as described above: the double condensate is formed and then allowed to evolve for some time T. For any time $T \geq 45$ ms the condensates have reached equilibrium [Fig. 3(a)]; we apply a second coupling pulse and release the condensate from the trap for imaging. A typical result is shown in Fig. 3(b). The image shows that there has been some transfer between the two states in their overlap region; this transfer is theresult of the interference between the two states. Were there no interference between the condensates there would be no enhancement of one condensate, nor depletion of the other, as a result of the coupling pulse.

Next, we consider the question of *reproducibility* of the relative phase from realization to realization of the experiment. In order to capture only the interference, we plot the density of the atoms at the center of the overlap region as a function of the time between the coupling pulses. The results in

Fig. 4 (solid points) show that the memory of the initial relative phase is preserved to about $\Delta\varphi_{\mathrm{diff}} \sim \pi/3$; these points are the averages of several individual measurements made at the various times T. The scatter in the individual measurements is represented by the thin vertical lines, and demonstrates that in any individual realization of the experiment we are capable of seeing almost complete contrast (i.e., all of the atoms in the overlapping region in either one state or the other) in the interference pattern. It is only upon averaging these individual results together, with the slight difference in phases that arise from realization to realization, that the contrast decreases.

The persistence of the phase coherence between these condensates was to us quite surprising. After all, our system entangles its external degrees of freedom with its internal degrees of freedom, and damping of one is generally expected to lead to decoherence in the other[27,28]. That we do not observe complete decoherence suggests that the relative phase between the condensates is robust against vagaries of its environment. We have seen the relative phase persist up to 100 ms after the first coupling pulse; in the future, we intend to examine the coherence at even longer times to quantify further the decoherence.

5. FUTURE DIRECTIONS

This Chapter has discussed experiments in which the coupling drive is mostly off, and the two internal spin states can interconvert only at distinct moments in time. A rich set of possibilities appears when the coupling drive is mostly on. One limit that can be explored is that of a weak coupling, in which a drive is tuned to be resonant with the overlap region between the two condensates. Atoms can then be locally transferred back and forth between the two states[29], giving rise to behavior reminiscent of Josephson junctions in superconductors. Indeed, such a system constitutes a physical realization of some Josephson junction thought experiments considered by Leggett and Sols[30].

Another limit is to put the condensates into dressed states, in which they are time-independent despite the presence of the coupling field. We have produced double condensates in which we reach the "equilibrium state" through a wholly coherent process, with little damping; it should be interesting to see what effect that has in our ability to look at phase diffusion at even longer times.

In my second Chapter, I will present results from experiments in which we produce topological excitations based on several of the techniques introduced here. In particular, we use spatially-dependent couplings to produce vortices, and identify their distinctive phase properties using the interferometric techniques described here.

ACKNOWLEDGMENTS

I would like especially to acknowledge my fellow experimenters at JILA for their work on this project: Eric Cornell, Jason Ensher, Michael Matthews, and Carl Wieman. These experiments are supported by the ONR, NIST, and NSF of the USA.

REFERENCES

1. Anderson, M. H., Ensher, J. R., Matthews, M. R., Wieman, C. E., and Cornell, E. A., 1995, Observation of Bose-Einstein condensation in a dilute atomic vapor, *Science* **269**: 198.
2. Davis, K. B., Mewes, M.-O., Andrews, M. R., van Druten, N. J., Durfee, D. S., Kurn, D. M., and Ketterle, W., 1995, Bose-Einstein condensation in a gas of sodium atoms, *Phys. Rev. Lett.* **75**: 3969.
3. Bradley, C. C., Sackett, C. A., and Hulet, R. G., 1997, Bose-Einstein condensation of lithium: Observation of limited condensate number, *Phys. Rev. Lett.* **78**: 985.
4. Fried, D. G., Killian, T. C., Willmann, L., Landhuis, D., Moss, S. C., Kleppner, D., and Greytak, T. J., 1998, Bose-Einstein condensation of atomic hydrogen, *Phys. Rev. Lett.* **81**: 3811.
5. Andrews, M. R., Townsend, C. G., Miesner, H.-J., Durfee, D. S., Kurn, D. M., and Ketterle, W., 1997, Observation of interference between two Bose-Einstein condensates, *Science* **275**: 637.
6. Hall, D. S., Matthews, M. R., Ensher, J. R., Wieman, C. E., and Cornell, E. A., 1998, Dynamics of component separation in a binary mixture of Bose-Einstein condensates, *Phys. Rev. Lett.* **81**: 1539.
7. Cornell, E. A., Hall, D. S., Matthews, M. R., and Wieman, C. E., 1998, Having it both ways: Distinguishable yet phase-coherent mixtures of Bose-Einstein condensates, J. Low *Temp. Phys.* **113**: 151.
8. Ramsey, N. F.,1956, *Molecular Beams,* Clarendon Press, Oxford.
9. Hall, D. S., Matthews, M. R., Wieman, C. E., and Cornell, E. A.,1998, Measurements of relative phase in two-component Bose-Einstein condensates, *Phys. Rev. Lett.* **81**: 1543.
10. Miesner, H.-J., Stamper-Kurn, D. M., Stenger, J., Inouye, S., Chikkatur, A. P., and Ketterle W., 1999, Observation of metastable states in spinor Bose-Einstein condensates, *Phys. Rev. Lett.* **82**: 2228.
11. Stamper-Kurn, D. M., Miesner, H.-J., Chikkatur, A. P., Inouye, S., Stenger, J., and Ketterle, W.,1999, Quantum tunneling across spin domains in a Bose-Einstein condensate, *Phys. Rev. Lett.* **83**: 661.
12. Anderson, B. P., and Kasevich, M. A., 1998, Macroscopic quantum interference from atomic tunnel arrays, *Science* **282**: 1686.
13. Hagley, E. W., Deng, L., Kozuma, M., Trippenbach, M., Band, Y. B., Edwards, M., Doery, M., Julienne, P. S., Helmerson, K., Rolston, S. L., and Phillips, W. D.,

Measurement of the coherence of a Bose-Einstein condensate, *to appear in Phys. Rev. Lett.*

14. Myatt, C. J., Burt, E. A., Ghrist, R. W., Cornell, E. A., and Wieman, C. E., 1997, Production of two overlapping Bose-Einstein condensates by sympathetic cooling, *Phys. Rev. Lett.* **78**: 586.
15. Burke, Jr., J. P., Bohn, J. L., Esry, B. D., and Greene, C. H., 1997, Impact of the ^{87}Rb singlet scattering length on suppressing inelastic collisions, *Phys. Rev. A* **55**: 112511.
16. Julienne, P. S., Mies, F. H., Tiesinga, E., and Williams C. J., 1997, Collisional stability of double Bose condensates, *Phys. Rev. Lett.* **78**: 1880.
17. Matthews, M. R., Hall, D. S., Jin, D. S., Ensher, J. R., Wieman, C. E., Cornell, E. A., Dalfovo, F., Minniti, C., and Stringari S., 1998, Dynamical response of a Bose-Einstein condensate to a discontinuous change in internal state, *Phys. Rev. Lett.* **81**: 243.
18. Petrich, W., Anderson, M. H., Ensher, J. R., and Cornell, E. A., 1995, Stable, tightly confining magnetic trap for evaporative cooling of neutral atoms, *Phys. Rev. Lett.* **74**: 3352.
19. Hall, D. S., Ensher, J. R., Jin, D. S., Matthews, M. R., Wieman, C. E., and Cornell, E. A., 1999, Recent experiments with Bose-condensed gases at JILA, Proc. SPIE 3270:98 (1998); e-print cond-mat/9903459.
20. Sinatra, A., Fedichev, P. O., Castin, Y., Dalibard, J., and Shlyapnikov, G. V., 1999, Dynamics of two interacting Bose-Einstein condensates, *Phys. Rev. Lett.* **82**: 251.
21. Graham, R.,1998, Decoherence of Bose-Einstein condensates in traps at finite temperatures, *Phys. Rev. Lett.* **81**: 5262.
22. Lewenstein, M., and You, L., 1996, Quantum phase diffusion of a Bose-Einstein condensate, *Phys. Rev. Lett.* **77**: 3489.
23. Javanainen, J., and Wilkens, M., 1997, Phase and phase diffusion of a split Bose-Einstein condensate, *Phys. Rev. Lett.* **78**: 4675.
24. Wright, E. M., Walls, D. F., and Garrison, J. C., 1996, Collapses and revivals of Bose-Einstein condensates formed in small atomic samples, *Phys. Rev. Lett.* **77**: 2158.
25. Castin Y., and Dalibard, J., 1997, Relative phase of two Bose-Einstein condensates, *Phys. Rev. A* **55**: 4330.
26. Sinatra A., and Castin, Y., 1999, Binary mixtures of Bose-Einstein condensates: Phase dynamics and spatial dynamics, e-print cond-mat/9904353.
27. Caldeira A. O., and Leggett, A. J., 1985, Influence of damping on quantum interference: An exactly soluble model, *Phys. Rev. A* **31**: 1059.
28. Walls D. F., and Milburn, G. J., 1985, Effect of dissipation on quantum coherence, *Phys. Rev. A* **31**: 2403.
29. Williams, J., Walser, R., Wieman, C., Cooper, J., and Holland, M., 1998, Achieving steady-state Bose-Einstein condensation, *Phys. Rev. A* **57**: 2030.
30. Leggett A. J., and Sols, F., 1991, On the concept of spontaneously broken gauge symmetry in condensed matter physics, *Found. Phys.* **21**: 353.

Intertwined Bose-Einstein Condensates

D. S. HALL
JILA and department of Physics University of Colorado Boulder, CO 80309-0440 USA.
Currrent Address: Department of Physics, Amherst College, Amherst, MA 01002-5000 USA

1. INTRODUCTION

In this second Chapter we'll look at the generation of topological states that we've excited in our double-condensate system. In particular, we have successfully produced and imaged vortices and other excitations in a trapped Bose-Einstein condensate (BEC). The key to generating these excitations is our ability to use the two internal states of our ^{87}Rb condensates to outmaneuver the topological constraints imposed upon condensates that possess only a single internal state.

2. SEEING DOUBLE, NONDESTRUCTIVELY

Our general approach to BEC was described in the first Chapter. The starting point for the experiments described here is a condensate of approximately 8×10^5 atoms in the $|1\rangle$ state of ^{87}Rb. We evaporatively cool the sample until there is no discernible thermal component remaining ($T \approx 0$).

Unlike the destructive imaging used in the previous experiments, we use here a state-selective form of phase-contrast microscopy to observe the condensates. As first developed by Zernike[1], and applied with great success to condensates by the group of Ketterle at MIT[2], the phase-contrast method relies upon the phase shift induced in an off-resonant laser beam that passes through the condensate. Since the imaging device (camera) is sensitive to light intensity rather than light phase, it is necessary to create an interference pattern by focusing the beam through a small "phase dot" that shifts the phase of the zeroth spatial order by $\pi/2$. After passing through the "phase dot," the probe beam intensity varies according to the magnitude of the phase shift which, in turn, is determined by the density of the condensate.

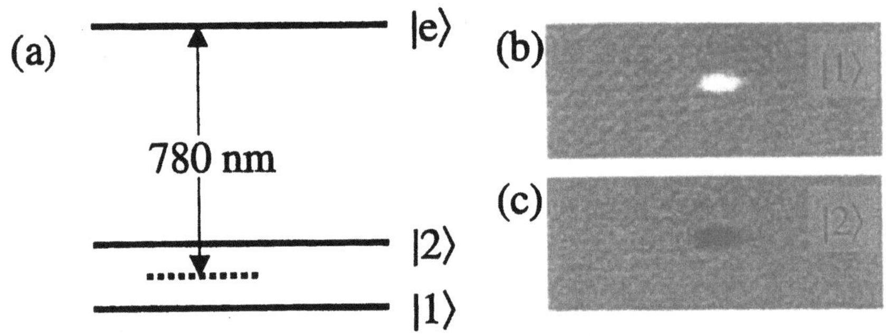

Figure 1. State-selective phase contrast imaging uses a probe laser that is detuned to the red of the $|1\rangle \rightarrow |e\rangle$ transition and to the blue of the $|2\rangle \rightarrow |e\rangle$ transition (a). The resulting phase shifts are positive and negative, respectively, meaning that a $|1\rangle$ condensate appears bright (b) and a $|2\rangle$ condensate dark (c) on a grey background.

The sign of the phase shift depends on whether the probe laser is detuned to the red (positive) or to the blue (negative) of the relevant atomic transition. With our two states $|1\rangle$ and $|2\rangle$, separated by 6.8 GHz, we can choose a probe frequency that is at once blue-detuned of the $|2\rangle \rightarrow |e\rangle$ transition and red-detuned of the $|1\rangle \rightarrow |e\rangle$ transition (where $|e\rangle$ represents the collection of $5P_{3/2}$ excited states on the D2 line). [See Fig. 1(a)]. Condensate atoms in $|1\rangle$ will thus induce a positive phase shift into the probe beam, whereas condensate atoms in the $|1\rangle$ state will induce a negative phase shift[3]. The resulting interference patterns appear as bright and dark images on a grey background for states $|1\rangle$ and $|2\rangle$, respectively, as shown in Fig. l(b) and (c).

The probe beam is detuned considerably from the atomic transitions to minimize heating due to spontaneous light scattering. Phase-contrast imaging is therefore non-destructive in the sense that the condensate density is largely unperturbed[2], which permits us to take multiple (individual) pictures of a single condensate while it remains in the magnetic trap. Alternatively, we can take "streak" images, in which the probe beam is left on while the "film" (actually a CCD array) is continuously scrolled. In Fig. 2 we show a streak image of a double condensate for which the two-photon microwave coupling drive is on. As the condensate oscillates back and forth between its two internal states (Rabi oscillations), the camera records an alternating pattern of bright and dark condensate images, just as we expect from the state-selectivity of the imaging. That we observe coherent Rabi oscillations even while the probe beam is on demonstrates that the phase-contrast imaging is largely nondestructive with respect to the relative phase between the condensates as well[3].

3. (UN)DOING THE TWIST

Let us associate the internal state at every point within the condensate with a classical spin vector (Bloch vector) **s**. If **s** points completely in the $\hat{\imath}$-direction in this spin space, then the condensate is entirely in state $|2\rangle$ at that point, whereas if **s** points in the $-\hat{\imath}$ direction it is in state $|1\rangle$. Of course, at each point the condensate may also be in a superposition of the two internal states, and this is represented by permitting **s** to trace out a circle in the *ij*-plane. A projection along $\hat{\jmath}$, therefore, represents an equal superposition of states $|1\rangle$ and $|2\rangle$ at that point in the condensate.

In the experiment described at the end of the previous section, the condensate atoms oscillated back and forth between the two internal states *in concert*. We expect this behavior whenever the effective Rabi frequency (i.e., the rate at which population is transferred from one internal state to the other) is approximately the same from point to point within the condensate. We can visualize this in terms of the spin vectors: the spin vectors at each point in the condensate are twisted at equal rates, and rotate synchronously in the *ij*-plane.

Figure 2. Streak images of the double condensate with the microwave coupling drive on. The observation of coherent Rabi oscillations indicates that the relative phase between the two condensates is largely unperturbed by the probe beam.

Now suppose that we introduce a *spatially-dependent* Rabi frequency. For instance, we can offset the trapping potentials slightly in the vertical direction[4], making the microwave transition frequency an explicit function of the vertical (spatial) position z. The definition of the Rabi frequency is $\Omega_{eff} = \sqrt{\Omega_R^2 + \delta^2}$, where Ω_R is the resonant Rabi frequency and δ is the detuning of the drive. A dependence of the detuning on spatial position, therefore, translates into a dependence on the spatial position of the effective Rabi frequency. Each part of the condensate will now oscillate back and forth between its two internal states at its own z-dependent rate. The spin vectors are differentially twisted, rotating at different rates and ultimately becoming out of phase with one another. Note that the twisting is continuous: there is no break in the helix described by the tips of the spin vectors as a function of the spatial position z. The helix appears as a series of alternating regions of $|1\rangle$ and $|2\rangle$ across the vertical extent of the condensate. Individual images of the condensates reveal this vertical structure; see Fig. 3(a)-(c). These pictures were taken of the condensate in an isotropic trap of frequency $\nu = 7.8(1)$ Hz. More and more bands appear as the condensate is increasingly twisted.

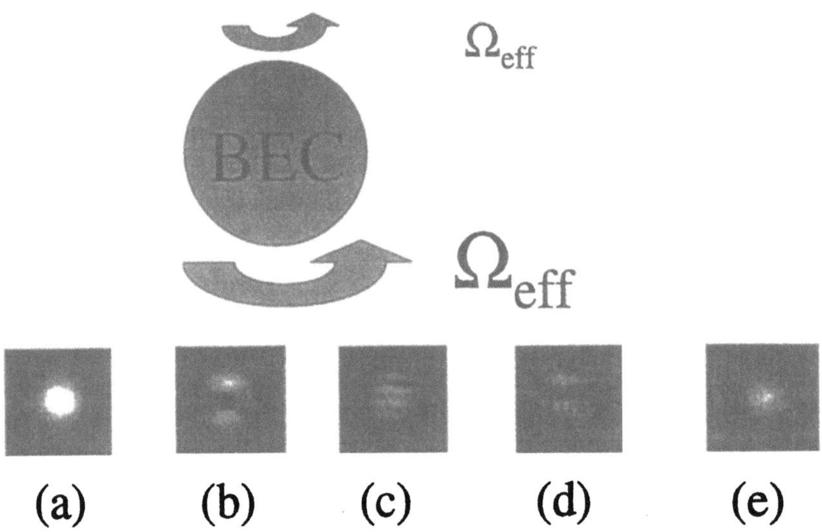

Figure 3. The effect of a spatially-dependent effective Rabi frequency is to introduce a twist to the condensate. In (a)-(c), the twisting introduces increasing numbers of bands of alternating condensate composition. Eventually the kinetic energy associated with the twists becomes too high for the condensate to support, and further twisting (c)-(e) actually ***untwists*** the condensate. State $|1\rangle$ is pictured. Each image block is 100-μm on a side.

4. SOLITONS

Consider again Fig. 3(b). In this picture, we see two regions of condensate predominantly in state $|1\rangle$, separated by a region of condensate predominantly in state $|2\rangle$ This situation is analogous to a dark soliton, in which the "notch" in the density of $|1\rangle$ is filled with the $|2\rangle$ condensate. The relative phase between the two $|1\rangle$ regions is π since the differential twisting has resulted in a 2πdifferential spin rotation. Had we turned off the coupling drive at this moment, and once again restricted the condensates to their separate spin spaces, we could conduct an experimental study of the time-evolution of these soliton-like excitations.

Nature's choice of excitation in this first set of experiments sparked our imagination, however, and led us to engineer the state of our choice in a second set of experiments. We now turn to the creation of a vortex state in a Bose-Einstein condensate.

5. VORTICES

Let us consider the definition of a vortex state and the origin of quantized vortices in a BEC[7]. Suppose we have a BEC described by a macroscopic wavefunction $\Phi=|\Phi|\exp(iS)$, where the phase S is the classical action. For our condensate, we take Φ and S to be single-valued functions of space and time. The condensate velocity, at any point, is related to the gradient of the phase S at that point:

$$\mathrm{v} = \frac{\hbar}{m_{Rb}} \nabla S \tag{1}$$

where m_{Rb} is the mass of a single Rubidium atom. The circulation k around any contour in the condensate is defined by

$$\kappa = \oint \mathrm{v} \cdot dl \tag{2}$$

which may also be written

$$\kappa = \frac{\hbar}{m_{Rb}} \oint (\nabla S \cdot dl) \tag{3}$$

$$\kappa = \frac{\hbar}{m_{Rb}}\left(\Delta S_{\text{contour}}\right). \tag{4}$$

Since the condensate wavefunction is single-valued, $\Delta S_{\text{contour}}$ must be zero in any simply-connected region with nonzero condensate density. Consider, however, a condensate in a multiply-connected space, such as a torus, in which the condensate is excluded from some region. The requirement that Φ be single-valued now means that ΔS can be an integer multiple of 2π for contours enclosing this region, and the circulation k, may be written

$$\kappa = n\frac{2\pi\hbar}{m_{Rb}} \tag{5}$$

where the number $n = \{..., -1, 0, 1, ...\}$ is the winding number of the vortex associated with the quantum of circulation $2\pi\hbar/m_{Rb}$. Vortices may also form in simply-connected regions about "cores" in which the condensate density is driven to zero by an angular momentum barrier. Once again, the circulation around the vortex core, and therefore the vortex angular momentum, is quantized.

In superfluid liquid helium, one produces vortices by rotating a container of helium while cooling the sample below the transition temperature. This method of production is not, however, readily generalized to the production of vortices in a dilute-gas condensate. It is difficult to apply a torque to a thermal gas of atoms to introduce the requisite angular momentum; a rotating field distortion in a magnetic trap is quite "slippery" to an atom. Even if the thermal gas were set spinning, the condensate forming at the (simply-connected) center of the trap would at first be smaller than its healing length and would therefore be unable to support a vortex core. Piercing the trap with a blue-detuned laser beam to generate a multiply-connected geometry might avoid this problem, but the time scale for coupling the angular momentum of the thermal gas to the condensate remains unknown.

Introducing a vortex into a single-component BEC that has already formed poses its own set of topological problems. In particular, there is no way to change the phase S of a condensate in a single internal state continuously from 0 to 2π without momentarily driving the condensate density to zero in some region of the ring. One might well wonder what dissipative processes are unleashed and how they affect the subsequent formation and evolution of the vortex state[8].

Our approach to vortex creation[9] avoids uncertainties and bypasses topological constraints by taking advantage of the internal (spin) degree of freedom of our condensates. With a spatially-dependent transition probability, we can transfer atoms from one internal state to another and, simultaneously, write in the phase of our choice to put them into a different external state. If that phase is simply the azimuthal angle, then the transferred condensate will be in a circulating state. Such state manipulation occurs under our precise control, and represents one path toward engineering general topological states in a BEC[10].

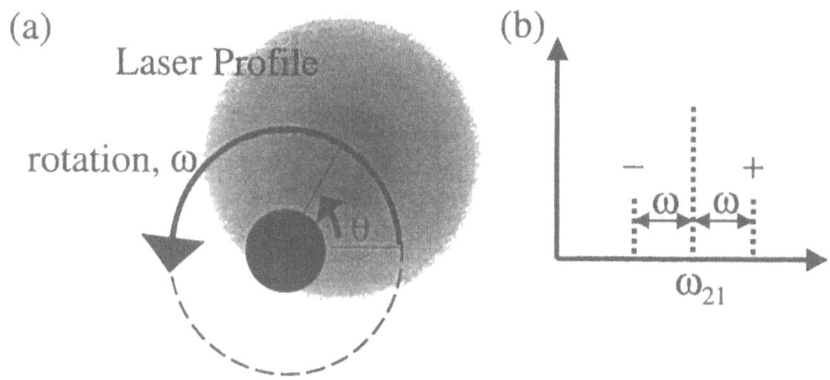

Figure 4. (a) Depiction of the rotating laser field that modulates the transition frequency. (b) The modulated spectrum, showing the first two sidebands ('+' and '-').

We have seen that soliton-like modes are produced when the two states see static trapping potentials that are offset slightly in the vertical direction and the condensate is twisted. The topological similarity between solitons and vortices prompted us to wonder whether we could engineer a vortex state in a similar fashion. Williams and Holland[10] considered the situation in which the offset in the trapping potentials were rotating rather than static. They found that, for appropriate choices of trap offset, rotation frequency, and microwave drive power and detuning, that one could indeed transfer atoms from one internal state to the other while simultaneously introducing the requisite vortex phase-winding[11-13]. For the remainder of this Chapter I will discuss the experimental realization of vortex states in a dilute-gas BEC.

The symmetry of the TOP trap is, generally, about an axis perpendicular to the plane of the rotating field; in our trap, this direction is parallel to gravity, and would seem to be naturally suited to be the axis of rotation for a vortex. Our imaging system was designed to look at the condensates from the side, however, as this is usually the more interesting view when condensates are released from the trap for imaging. Rather than redesign and rebuild the imaging system, we decided to create a spherical condensate with a vortex line along our traditional imaging axis. Reducing the magnetic field gradient introduces higher-order corrections to the trapping potential due to gravity and the "sag" of the condensate in the trap. We can thus vary the TOP trap to produce condensates with radial-vertical aspect ratios between[14] 0.94 and 2.8. The experiments described here were performed in an isotropic trap of frequency $\nu = 7.8 \pm 0.1$ Hz. An incidental benefit to working in this weaker trap is that the condensate expands until it is roughly 60 μm in diameter, easing the demands on our imaging system.

Next, we had to find a way to introduce a rotating offset between the trapping potentials. We chose to use the AC Stark shift introduced by a focused off-resonant laser beam that passes through the periphery of the condensate along the axis of observation. With an appropriate choice of laser detuning the two states experience different light-induced forces, giving us the necessary potential offset. The beam can then be rotated around the condensate at frequency ω to yield a trapping potential offset that rotates in time [Fig. 4(a)].

We use the microwave drive to transfer atoms from one internal state to the other with the phase appropriate to a vortex state. To understand this process, let us take the difference between the two energy levels of the internal states in the absence of the rotating potential offset to be $\omega_{21} = \omega_{|1\rangle} - \omega_{|2\rangle}$, where $\omega_{|1\rangle}$ and $\omega_{|2\rangle}$ are the energies for internal states $|1\rangle$ and $|2\rangle$, respectively. The introduction of the rotating laser field amounts to a frequency modulation of this interval, which we can approximate by

$$\Omega_{21} = \omega_{21}\left[1 + \varepsilon(r)\sin(\omega t)\right] \tag{6}$$

where $\varepsilon(r)$ is the "modulation index," a smoothly increasing function that reflects the extent to which the potential varies over a rotation of the laser beam a radial distance r from its rotation axis within the condensate. At $r = 0$, i.e., on-axis, there is no modulation of the potential, and thus $\varepsilon(r) = 0$.

Expanding $\exp[i\Omega_{21}t]$, we find that the single transition at frequency ω_{21} will now be accompanied by a series of sideband transition frequencies spaced by ω, the laser rotation frequency. Let's consider only the first sidebands ($n = \pm 1$).

$$\exp\left[i\Omega_{21}t\right] = \exp\left(i\omega_{21}t\right)\sum_{n=-\infty}^{+\infty}\exp(in\omega t)J_n\left[\varepsilon(r)\omega_{21}t\right] \tag{7}$$

$$
\begin{aligned}
\exp\left[i\Omega_{21}t\right] = \exp\left(i\omega_{21}t\right)J_0\left[\varepsilon(r)\omega_{21}t\right] + \\
+ \exp\left[\mathrm{i}\left(\omega_{21}+\omega\right)t\right]J_1\left[\varepsilon(r)\omega_{21}t\right] + \\
+ \exp\left[\mathrm{i}\left(\omega_{21}-\omega\right)t\right]J_{-1}\left[\varepsilon(r)\omega_{21}t\right] + \;\ldots\ldots
\end{aligned}
\tag{8}
$$

Note that the strength of the transitions also depends on $\varepsilon(r)$ through the Bessel functions: atoms at or near the center of the condensate will have correspondingly weaker (or vanishing) sidebands.

The microwave drive is now applied at a frequency ω_{rf}. Let us define $\delta = \omega_{21}-\omega_{rf}$. By Eq. (8), the microwave drive will induce transitions when $\delta = 0$ and $\delta = \pm\omega$. We ignore transitions for which $\delta = 0$ (i.e., those transitions resonant at the center of the condensate) since we are interested in changing the internal state of atoms at the periphery of the condensate only. Instead, we tune our drive such that it is on one of the sidebands, e.g., $\delta = \omega$. Atoms at or near the center of the condensate will be largely unaffected since $\varepsilon(r) \approx 0$ for $r \approx 0$, and atoms at the periphery will be most affected since they experience the largest modulation.

As the laser beam rotates around, different parts of the condensate will be transferred from state $|1\rangle$ to state $|2\rangle$ at different times. For instance, with $\delta = \omega$, atoms have a highest probability of transition at the azimuthal angle of the laser beam θ at any instant in time [Fig. 4(a)]. Let's take a little piece of condensate to be transferred from $|1\rangle$ to $|2\rangle$ at time $t = 0$, and let's call its phase $S = 0$. After some time T, the laser beam has rotated through some angle $\theta = \omega T$. The condensate piece transferred from $|1\rangle$ to $|2\rangle$ at time T will pick up a phase $\omega_{rf}T$ by absorbing the microwave photon, in addition to the phase $\omega_1 T$ it picked up since time $t = 0$ while remaining in state $|1\rangle$. On the other hand, the first piece of transferred condensate has picked up a phase $\omega_2 T$ since $t = 0$. The difference in phase between the two bits of condensate is, at T,

$$
\Delta S = \omega_2 T - \left(\omega_1 T + \omega_{rf} T\right) = \delta T = \omega T = \theta\,. \tag{9}
$$

That is, the phase difference is simply the azimuthal angle! As the beam rotates with the microwave drive on, we transfer from $|1\rangle$ to $|2\rangle$ and write the phase continuously around the periphery of the condensate from 0 to π — exactly what is required for a vortex with unit winding number. This analysis should make it clear that the circulation of state $|2\rangle$ is not simply due to a mechanical stirring effect, since driving the other sideband $\delta = -\omega$ results in a condensate that rotates in the opposite sense.

Our state-selective imaging is capable of resolving the density of the purported vortex state that is produced in the course of transferring atoms from one state to the other [Fig. 5(a)], but by itself it cannot tell us about the circulation of the condensate. In order to see the circulation, we take advantage of the interferometric technique we developed in Ref. [15] and interfere the rotating condensate in state $|2\rangle$ with the nonrotating condensate left behind in state $|1\rangle$. For this to work, the condensates must overlap one another slightly at their boundary, as we saw in the previous Chapter.

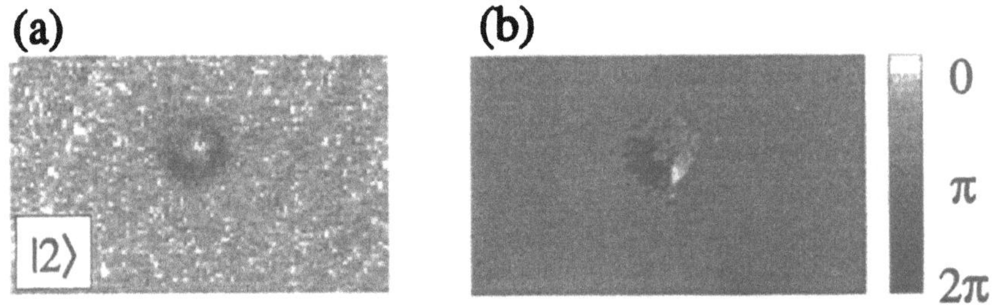

Figure 5. Pictures of a vortex. The density in the $|2\rangle$ state is shown in (a), and its phase (arbitrary offset) as extracted using the interferometric technique described in the text in (b). The vortices are formed with the assistance of a 10 nW, 780 nm laser beam with a waist of 180 μm, rotating at $\omega = 2\pi\times100$ Hz at a radius of 75 μm. A typical vortex forms in 70 ms.

The condensate in state $|1\rangle$ has phase $S_1 = 0$; the rotating condensate in state $|2\rangle$ has phase $S_2(\theta, t) = (\theta - \theta_o) + \omega_{21}t$, where θ_o *is* some initial (constant) phase offset, and ω_{21} is the relative rate of phase accumulation between the two condensates. After the vortex creation process has ceased we apply a resonant $\pi/2$ pulse of microwave radiation, $\omega_{rf} \approx \omega_{21}$, and subsequently image the atoms in the $|2\rangle$ state. The atoms arriving in state $|2\rangle$ from state $|1\rangle$ have a constant phase with respect to the phase gradient that varies as the azimuthal angle in state $|2\rangle$, as shown in Fig. 5(b).

The combination of density and phase information leads us to interpret these images as those of a vortex in state $|2\rangle$. At the core of the vortex is a condensate in state $|1\rangle$. Once the vortex is produced, the coupling drive can be turned off; no more interconversion between the states is possible, and the evolution of the system progresses according to individual Gross-Pitaevskii equations for each state. As noted in the previous Chapter, the $|1\rangle$ state prefers to "float" on state $|2\rangle$, and this instability drives the ensuing dynamics. Alternatively, a quick π-pulse can interchange the two states, resulting in a $|1\rangle$ vortex circulating about a $|2\rangle$ -filled core. In either case, the nonrotating condensate that fills the core can be selectively blown away by a pulse of laser radiation, leaving only the circulating condensate. A variety of experiments exploring the time-evolution these states can be envisioned.

ACKNOWLEDGMENTS

I would like especially to acknowledge the work of my fellow experimenters on this project: Brian Anderson, Eric Cornell, Paul Haljan, Michael Matthews, and Carl Wieman. Together, we have benefited enormously from the theoretical work of Murray Holland and Jamie Williams, and from conversations with Seamus Davis and Jason Ho. Dave Pritchard gave me the idea for the title of this talk. These experiments are supported by the ONR and NSF of the USA.

REFERENCES

1. Zernike, F., 1956, Nobel Prize Lecture.
2. Andrews, M. R., Mewes, M.-O., van Druten, N. J., Durfee, D. S., Kurn, D. M., and Ketterle, W., 1996, Direct, nondestructive observation of a Bose condensate, *Science* **273**: 84.
3. Matthews, M. R., Anderson, B. P., Haljan, P. C., Hall, D. S., Holland, M. J., Williams, J. E., Wieman, C. E., and Cornell, E. A., 1999, Watching a superfluid untwist itself: Recurrence of Rabi oscillations in a Bose-Einstein condensate, to appear in Phys. *Rev. Lett.;* e-print cond-mat/9906288.
4. Hall, D. S., Matthews, M. R., Ensher, J. R., Wieman, C. E., and Cornell, E. A., 1998, Dynamics of component separation in a binary mixture of Bose-Einstein condensates, *Phys. Rev. Lett.* **81**: 1539.
5. Ho, T.-L., and Shenoy, V. B., 1996, Local spin-gauge symmetry of the Bose-Einstein condensates in atomic gases, *Phys. Rev. Lett.* **77**: 2595.
6. Williams, J., Walser, R., Cooper, J., Cornell, E. A., and Holland, M., Excitation of an antisymmetric collective mode in a strongly coupled two-component Bose-Einstein condensate, e-print cond-mat/9904399.
7. Tilley D. R., and Tilley, J., 1990, *Superfluidity and Superconductivity, IOP* Publishing, Ltd., Philadelphia, 3rd ed..
8. Dobrek, L., Gajdal, M., Lewenstein, M., Sengstock, K., Birkl, G., and Ertmer, W., 1999, Optical generation of vortices in trapped Bose-Einstein condensates, e-print cond-mat/9907452.
9. Matthews, M. R., Anderson, B. P., Haljan, P. C., Hall, D. S., Wieman, C. E., and Cornell, E. A., 1999, Vortices in a Bose-Einstein condensate, *Phys. Rev. Lett.* **83**: 2498.
10. Williams, J., and Holland, M. J., 1999, Preparing topological states of a Bose-Einstein condensate, to appear in *Nature;* e-print cond-mat/9909163.
11. Marzlin, K.-P., Zhang, W., and Wright, E. M.,1997, Vortex coupler for atomic Bose-Einstein condensates, *Phys. Rev. Lett.* **79**: 4728.
12. Dum, R., Cirac, J. I., Lewenstein, M., and Zoller, P., 1998, Creation of dark solitons and vortices in Bose-Einstein condensates, *Phys. Rev. Lett.* **80**: 2972.
13. Bolda, E. L., and Walls, D. F., 1998, Creation of vortices in a Bose-Einstein condensate by a Raman technique, *Phys. Lett. A* **246**: 32.
14. Ensher, J. R., 1998, *The First Experiments with Bose-Einstein Condensation of* ^{87}*Rb,* Ph.D. thesis, University Of Colorado.

15. Hall, D. S., Matthews, M. R., Wieman, C. E., and Cornell, E. A., 1998, Measurements of relative phase in two-component Bose-Einstein condensates, *Phys. Rev. Lett.* **81**: 1543.

Coherent Atom Optics With Bose-Einstein Condensates

K. HELMERSON
Atomic Physics Division, Physics Laboratory National Institute of Standards and Technology, Gaithersburg, MD 20899-8424

1. INTRODUCTION

Atom optics, the manipulation of atoms with mirrors, beamsplitters and lenses in analogy to the manipulation of light, is a rapidly advancing field of research. Until recently, however, experiments have used thermal sources of atoms much as early experiments in optics used lamps. What was lacking was a coherent source of matter-waves similar to the laser for light.

The creation of a Bose-Einstein condensate (BEC) of a dilute atomic gas[1-4] has opened up the possibility of realizing a matter-wave source analogous to the optical laser. The macroscopic occupation of the ground state of a trap by a BEC is similar to the occupation of a single mode of an optical cavity by photons. The atoms forming the condensate all occupy the same wavefunction - both in terms of their internal and external degrees of freedom. This makes the BEC a highly coherent source of atoms.

Atoms released from a BEC should have coherence properties (1st-, 2nd- and higher-order) similar to those of an optical laser. First-order coherence has been observed by an analogue of the Young's double slit experiment[5]. Second and third-order coherence has been inferred from measurements of the mean field interaction[6] and three-body loss rates[7], respectively. In all cases, the results are consistent with the condensate having the statistical properties of a coherent state, similar to that of light from an optical laser.

In addition to their coherence properties, atoms from a BEC are nearly the ideal, monochromatic source for atom optics. Many atom optical

Bose-Einstein Condensates and Atom Lasers
Edited by Martellucci *et al.*, Kluwer Academic/Plenum Publishers, 2000

elements involve the interaction of the atoms with an optical field and the associated transfer of the photon momentum to the atoms. Because of the repulsive atom-atom interaction, which can be described by a mean field, the BEC swells to a size significantly larger than the ground state wavefunction of the harmonic trap confining the atoms[8]. The spatial extent of the resulting wavefunction can be several orders of magnitude larger than the optical wavelength. Hence the momentum width, given by the Heisenberg uncertainty principle, can be much less than the photon's momentum. Not all experiments will realise this reduced, intrinsic momentum width. The interaction energy may be converted to kinetic energy when the atoms are released from the trap. Nonetheless, the resulting additional momentum spread, due to the atom-atom interaction, can still be significantly less than the momentum of a single photon.

In this Chapter I will describe experiments, performed by the Laser Cooling and Trapping Group at the NIST in Gaithersburg, Maryland, on the use of a BEC for coherent atom optics.

2. BEC OF SODIUM IN A TRI-AXIAL TOP TRAP

We create a Bose-Einstein condensate of sodium atom in a time orbiting potential or TOP magnetic trap. Our TOP trap differs from the original design[9] in that our bias field rotates in a plane (x - y) that includes the quadrupole symmetry axis (x). This produces a potential for our BEC that is harmonic in the x, y and z directions with spring constants in the ratio of 4:2:1, respectively. Hence our trap is tri-axial; it has no rotational symmetry. The details of our trap and associated experimental apparatus for producing the condensate can be found in Ref. [10]. We routinely obtain a condensate of 1 to 3×10^6 sodium atoms with no discernible uncondensed fraction in a trap with harmonic frequencies of $\omega_x/2\pi$, $\omega_y/2\pi$ and $\omega_z/2\pi$ = 360, 250 and 180 Hz, respectively. The energy of the condensate is dominated by the mean field interaction and the wavefunction can be adequately described using a Thomas-Fermi approximation[8]. For many experiments it is advantageous to adiabatically expand the trap to reduce the mean field, decrease the momentum spread and increase the spatial extent of the BEC. Typical trap frequencies after expansion are, respectively, 25, 18, and 13 Hz with corresponding Thomas-Fermi radii of 30, 34 and 39 μm for 2×10^6 atoms.

The momentum distribution of the BEC is measured with probe absorption imaging after a variable time-of-flight period (typically 5 to 20 ms). The TOP trap is rapidly shut off and after the variable delay, a short laser pulse optically pumps the atoms from F = 1 to F = 2. Another short laser pulse resonant with the $3S_{1/2}$, F = 2 → $3P_{3/2}$, F' = 3 transition, is immediately applied to the atoms along the direction of gravity (the x direction). The light absorbed from this laser beam is imaged onto a CCD camera.

3. DIFFRACTION BY OPTICAL STANDING WAVES

When an atomic beam passes through a periodic optical potential formed by a standing light wave, it diffracts similar to the diffraction of light by a periodic grating. The diffraction can be divided into two regimes, normal and Bragg diffraction. Both diffraction processes can be thought of as arising from the simultaneous absorption of a photon from one laser beam of the optical standing wave, and stimulated emission of a photon due to the counter-propagating laser beam. This necessarily means that the momentum transfer to the atomic beam by the optical standing wave is quantized in units of $2\hbar k$, twice the momentum of a single photon.

The atoms in a BEC are, essentially, initially at rest. A situation similar to the passage of an atomic beam through the standing wave can be achieved by exposing the condensate to a pulsed, optical standing wave.

3.1 Normal Diffraction of a BEC

In normal diffraction, the condensate atoms are exposed to a non-adiabatic pulse of an optical standing wave. For short interaction times such that the atoms do not move appreciably along the direction of the standing wave, the Raman-Nath regime, the standing wave potential can be considered a thin phase grating that modifies the atomic de Broglie wave with a phase modulation that varies sinusoidally in space. For a square profile laser beam, this phase modulation is given by $\phi(x) = (U_0\tau / \hbar)\cos^2(kx)$, where U_0 is the maximum depth of the optical potential given by the AC Stark shift, and τ is the interaction time of the atomic beam with the standing wave. An atom, initially with zero momentum, is projected, by this phase modulation, onto states with momenta $p_n = n2\hbar k$, (n = 0, ±1, ±2,...) with populations $P_n = J_n^2(U_0\tau / 2\hbar)$ where $J_n(x)$ are Bessel functions of the first kind. Energy conservation is satisfied by the spread in energies associated with the non-adiabatic "turn-on" and "turn-off" of the standing wave.

Fig. 1 shows the result of the application of a 100 ns standing wave laser pulse to a BEC. The diffracted orders ($\pm 2n\hbar k$) are well separated from the BEC, which is the central dot in the picture. The degree of separation is related to the initial spatial coherence of the wavefunction compared to the wavelength of light. The number of diffracted orders is determined by the strength of the optical potential, U_0, which is proportional to the laser intensity.

Figure 1. Absorption images of a BEC, after 6 ms of time of flight, demonstrating diffraction by an optical standing wave pulse. The upper panel shows a BEC not exposed to the laser pulse. The lower panel shows the result of the application of a standing wave pulse of 100 ns duration, 1 W/cm^2 intensity, and −600 MHz detuning. Each dot is separated in momentum from an adjacent dot by $2\hbar k$.

3.2 Bragg Diffraction

In Bragg diffraction, the condensate atoms are exposed to an adiabatic pulse of an optical standing wave, and energy conservation must be explicitly satisfied in the interaction between the atoms and the light field. The energy difference of the atom after the change of momentum of $2\hbar k$ must come from the photon field. For an atomic beam, this is typically accomplished by choosing the angle of incidence such that the atoms see a differential Doppler shift between the two counter-propagating laser beams comprising the standing wave. In the case where we start with BEC essentially at rest, this differential Doppler shift can be created by moving the standing wave with respect to the atoms. We create our moving standing wave by having a frequency difference δ between the two counter-propagating waves that make up the standing wave. In the presence of this moving standing wave, an atom initially at rest will simultaneously absorb photons from the higher frequency laser beam and be stimulated to emit photons into lower frequency beam, acquiring a unidirectional momentum of $2n\hbar k$ in the process. In order to satisfy energy conservation, the detuning δ must be chosen such that $\hbar\delta = n4E_{rec}$, where E_{rec} is the recoil energy, the final energy that an atom initially at rest would have after absorption of a single photon.

We have performed Bragg diffraction on a stationary BEC, the details of which can be found in Ref. [11]. We have observed 1st-order Bragg diffraction with 100% diffraction efficiency and up to 6th-order, with a momentum transfer of ~$12\hbar k$ (corresponding to a velocity of 35 cm/s) with 15% efficiency. Because the Bragg diffraction process does not involve a change in the internal state of the atom, it can be applied to both trapped and untrapped atoms. We have also shown that Bragg diffraction is momentum selective. If the bandwidth of a Bragg pulse is narrower than the Doppler width associated with the velocity spread of a BEC, only the fraction of the BEC within the bandwidth of the Bragg pulse will diffract. This has been recently applied by the MIT group to measure the intrinsic momentum distribution of a trapped BEC[12].

3.3 Diffraction Beyond the Raman-Nath Regime

In the description of normal diffraction by an optical standing wave as a phase modulation of the atomic de Broglie wave, the pulse was considered short enough to be in the Raman-Nath regime, so that the phase modulation was essentially instantaneous. Alternatively, in the picture of diffraction as the simultaneous absorption and stimulated emission of photons, the bandwidth of the standing wave pulse was broad enough such that energy conservation can be satisfied for momentum transfer in both directions. If the laser pulse is left on for a longer time, we will violate the Raman-Nath approximation. The atoms can move along the periodic potential of the standing wave and the phase modulation will spatially average to a constant value. Alternatively, the pulse will have insufficient frequency width to satisfy energy conservation. This regime beyond Raman-Nath leads to periodic focusing and defocusing of the atoms and is relevant for atom lithography.

We have studied the behaviour of a BEC in a pulsed optical standing wave beyond the Raman-Nath regime, the details of which can be found in Ref. [13]. We observed oscillations in the intensity of the diffracted orders as a function of the laser pulse duration. Our results are in good agreement with a simple model where we project an incoming plane wave state onto a Bloch state basis, accumulate the differential phases due to the different energies of the Bloch bands, and then project back onto momentum eigenstates.

3.4 The Pulsed Talbot Effect

Periodic focusing and defocusing can also be studied within the thin diffraction grating (Raman-Nath) regime. This behaviour is known as the Talbot effect. In the optical Talbot effect, coherent light passing through a periodic grating will from an "image" of the grating at a characteristic distance known as the Talbot length. For a phase grating, this "image" corresponds to the initial intensity distribution of the light with the phase distribution of the grating. Unlike light, however, atoms can be initially at rest, and the "re-imaging" of the phase grating occurs at integer multiples of the Talbot time. Also, unlike light, atoms can be exposed to a pulsed, phase grating, which leads to a unique manifestation of the Talbot effect.

We have demonstrated a new manifestation of the Talbot effect using the diffraction of a BEC by pulsed optical standing waves. In our experiment, the details of which can be found in Ref. [14], we start with atoms at rest and apply a short pulse, optical standing wave to diffract the condensate atoms. A second identical diffraction pulse is applied after a variable delay to analyze the temporal evolution of the resulting condensate wavefunction. We observe that the initial phase distribution reimages itself at integer multiples of 10 μs, the Talbot time for our parameters. When the second pulse is applied at odd multiples of half the Talbot time, self imaging of the condensate in momentum space is observed. Intermediate delays produce more complicated momentum-space patterns that are in excellent agreement with theory. The coherent property of the condensate provides signals of very high contrast. In addition, we observe that the dynamics of the short pulse is different from that of a static grating because it has a broad frequency spectrum and hence can add energy to the system. It is the dispersion relation of matter waves, not the path length difference as in the case of static gratings, that results in this new manifestation of the Talbot effect.

4. A COLLIMATED, DIRECTIONAL ATOM LASER

A BEC, with its high degree of coherence, is an ideal starting point in order to realise an atom laser[15], the matter wave equivalent of an optical laser. All that is required is to coherently extract the condensate atoms from a BEC; that is, an atom output coupler is needed. The first demonstration of an output coupler for BEC was reported by Ketterle's group at MIT in 1997[16]. They used coherent, rf-induced transitions to change the internal state of the atoms from a trapped state to an untrapped one. This method, however, did not allow the direction of the output coupled atoms to be chosen. The extracted atoms fell under the influence of gravity and expanded because of the intrinsic repulsion of the atoms.

We have developed a highly directional method to couple out a variable fraction of a condensate, the details of which can be found in Ref. [17]. We use stimulated Raman transitions to coherently transfer trapped condensate atoms in the $F = 1$, $m = -1$ magnetic sublevel to the untrapped $F = 1$, $m = 0$ sublevel, while giving them a momentum kick. This is similar to the process of Bragg diffraction of atoms discussed earlier; however, output coupling involves a stimulated Raman transition between different internal, as well as external, states. The frequency difference between the lasers includes the

Zeeman energy between the two magnetic sublevels, in addition to the change in kinetic energy of the atoms.

The first demonstration of an output coupler for BEC[16] was pulsed due to fluctuating magnetic fields. Subsequent experiments[18,19] with stable magnetic fields have demonstrated continuous extraction. Our output coupling was also pulsed, because the condensate atoms were displaced by gravity away from the zero of the quadrupole field such that the local magnetic field was modulated by the rotating TOP bias field. In order to avoid changes in the Raman resonance frequency between different magnetic sublevels we synchronised the application of the Raman pulses to our rotating TOP field. By repeatedly applying the Raman pulses at a fast enough rate, we were able to effectively produce a continuous beam of atoms extracted from the condensate.

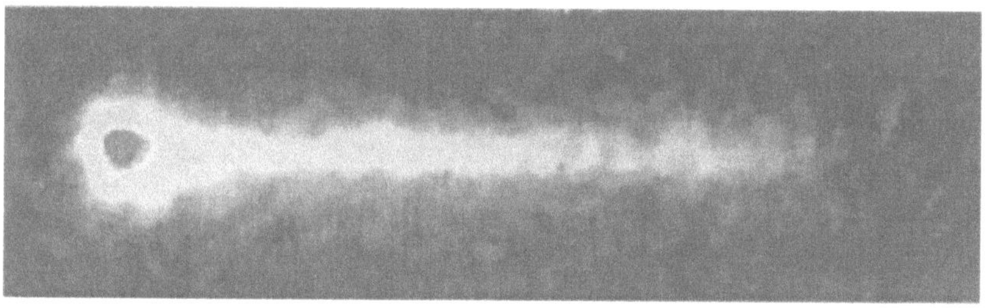

Figure 2. Demonstration of a highly directional, well collimated atom laser. Bose-Einstein condensed atoms in the $m = -1$ magnetic sublevel are confined in a magnetic trap. A stimulated optical Raman transition is used to couple atoms to the $m = 0$ magnetic sublevel which is not affected by the magnetic fields. The two photon Raman process imparts a $2\hbar k$ momentum kick to the transferred atoms causing them to leave the region of the condensate. The orientation of the laser beams determines the direction of the momentum transfer, which in our case is perpendicular to gravity (into the page). The optical Raman laser beams are pulsed, with a duration of 1 μs. When the pulse rate is sufficiently fast (20 kHz in our case), the cloud of atoms transferred have only moved a small distance (2.9 μm) between pulses compared to their spatial extent (~50 μm). Hence there is a substantial overlap of the pulses of transferred atoms and the atom laser beam appears continuous.

Fig. 2 is an image of the distribution of atoms 1.6 ms after applying 140 Raman output coupling pulses at a rate of 20 kHz, the frequency of the TOP rotating bias field. In the time between two subsequent Raman pulses each output coupled wavepacket, with a velocity kick of 6 cm/s, moves only 2.9 μm. These pulses strongly overlap because this spatial separation is much smaller than the ~50 μm size of the condensate, therefore the output coupled atoms form a continuous matter wave.

The momentum kick from the Raman output coupling process produces a highly directional beam of atoms. Unlike the other output couplers demonstrated[16,18,19], which rely on gravity to determine the direction of the beam of atoms, the direction of our beam can be chosen by a suitable orientation of the Raman laser beams. In fact, the beam of extracted atom shown in Fig. 2 is perpendicular to gravity.

Our Raman output-coupling scheme dramatically reduces the transverse momentum width of the extracted atoms compared to other methods such as rf output coupling. This dramatic reduction occurs because the output coupled atoms have received a substantial momentum kick from the Raman process. If the atoms were simply released from the trap with no momentum transfer, they would undergo a burst of expansion due to the repulsive interactions with the other condensate atoms. In our output coupling scheme, however, this additional expansion energy is primarily channelled into the forward direction. The increase in the transverse momentum width due to the interaction between the atoms is reduced by roughly the ratio of the characteristic time it takes the output coupled atoms to leave the still trapped condensate, divided by the time-scale over which the mean field repulsion acts on the freely expanding condensate. In our case, the reduction ratio is about a factor of ~20 which results in a well collimated beam of atoms.

5. SPATIAL PHASE VARIATIONS OF A BEC

We have used an unequal arm length interferometer, based on normal diffraction by pulsed optical standing waves, to study the spatial coherence of a BEC, the details of which can be found in Ref. [20]. Two optical standing wave pulses of duration 100 ns and separation time Δt are applied to the condensate. Each standing-wave phase grating diffracts the condensate, making small "copies" of the condensate displaced in momentum space by twice the momentum of a single photon. As the first copy moves away from the condensate its phase is evolving at $4E_R/\hbar$, where E_R is the single photon recoil energy. (For sodium atoms with an excitation wavelength of 589 nm, E_R/h is 25 kHz.) After the second copy is created at a time Δt later, the phase of both copies then evolve at nominally the same rate. The quantum mechanical amplitudes of each copy interfere, and the total number of atoms coupled out of the condensate by the two pulses is measured. The resulting interferogram oscillates at the expected 100 kHz phase evolution of the first copy with respect to the second copy. The decay of the envelope of the interferogram is due to both the spatial overlap of the two copies (since the first copy has moved during Δt due to the momentum kick) and on the initial spatial phase variations across the condensate.

When the coherence measurement is made on a condensate held in the trap, we obtain an interferogram whose envelope decays essentially as the spatial overlap of the two coupled out copies. The results are consistent with the trapped condensate having a uniform spatial phase. Hence we have experimentally verified that the trapped BEC, despite being spatially expanded due to the mean-field interaction between the atoms, has a momentum spread that is determined by the Heisenberg uncertainty principle. This result, which we measured in the time domain, was also obtained earlier, from measurements in the frequency domain, by Ketterle's group at MIT using Bragg spectroscopy[12]. Alternatively, a released BEC exhibits large phase variations across the condensate as the mean-field interaction is converted into kinetic energy. This is apparent in our measurements where we obtain an interferogram with an envelope that decays much faster than the spatial overlap of the two copies. Our measurements also confirm that the successive, Raman output coupled pulses of atoms in our atom laser are fully coherent.

ACKNOWLEDGMENTS

Much of the experimental work described here was carried out at NIST by Lu Deng, Johannes Denschlag, Ed Hagley, Mikio Kozuma, Robert Lutwak, Yuri Ovchinnikov, Jesse Simsarian, Jesse Wen, KH, Steve Rolston, and Bill Phillips. We have benefited greatly from discussions with our theoretical colleagues, Yehuda Band, Charles Clark, Marya Doery, Mark Edwards, David Feder, Paul Julienne, and Marek Trippenbach. This work was partially supported by the U. S. Office of Naval Research and NASA.

REFERENCES

1. Anderson, M.H., Ensher, J.R., Matthews, M.R., Wieman, C.E., and Cornell, E.A., 1995, Observation of Bose-Einstein condensation in a dilute atomic vapor. *Science* **269**: 198-201.
2. Davis, K.B., Mewes, M.-O., Andrews, M.R., van Druten, N.J., Durfee, D.S., Kurn, D.M., and Ketterle, W., 1995, Bose-Einstein Condensation in a gas of sodium atoms. *Phys. Rev. Lett.* **75**: 3969-3972.
3. Bradley, C.C., Sackett, C.A., and Hulet, R.G., 1997, Bose-Einstein condensation of lithium: Observation of limited condensate number. *Phys. Rev. Lett.* **78**: 985-988.
4. Fried, D., Killian, T.C., Willmann, L., Landhuis, D., Moss, S.C., Kleppner, D., and Greytak, T.J., 1999, Bose-Einstein condensation of atomic hydrogen. *Phys. Rev. Lett.* **81**: 3807-3810.
5. Andrews, M.R., Townsend, C.G., Miesner, H.-J., Durfee, D.S., Kurn, D.M., and Ketterle, W., 1997, Observation of interference between two Bose-Einstein condensates. *Science* **275**: 637-641.

6. Mewes, M.-O., Andrews, M.R., van Druten, N.J., Kurn, D.M., Durfee, D.S., and Ketterle, W., 1996, Bose-Einstein condensation in a tightly confining dc magnetic trap. *Phys. Rev. Lett.* **77**: 416-419.
7. Burt, E.A., Ghrist, R.W., Myatt, C.J., Holland, M.J., Cornell, E.A., Wieman, C.E., 1997, Correlations and collisions: What one learns about Bose-Einstein condensates from their decay. *Phys. Rev. Lett.* **79**: 337-340.
8. Dalfovo, F., Giorgini, S., Pitaevskii, L.P., and Stringari, S., 1999, Theory of Bose-Einstein condensation in trapped gases. *Rev. Mod. Phys.* **71**: 463-512.
9. Petrich, W., Anderson, M.H., Ensher, J.R., and Cornell, E.A., 1995, Stable, tightly confining magnetic trap for evaporatively cooling of neutral atoms. *Phys. Rev. Lett.* **74**: 3352-3361.
10. Helmerson, K., and Phillips, W.D., 1999, Cooling, trapping and manipulation of neutral atoms and BEC by electromagnetic fields. In *Proceedings of the International School of Physics "Enrico Fermi" Course CXL* (M. Inguscio, S. Stringari and C. Wieman, eds.) IOS Press, Amsterdam, pp. 391-438.
11. Kozuma, M., Deng, L., Hagley, E.W., Wen, J., Lutwak, R., Helmerson, K., Rolston, S.L., and Phillips, W.D., 1999, Optically-induced Bragg diffraction of a Bose-Einstein condensate. *Phys. Rev. Lett.* **82**: 871-874.
12. Stenger, J., Inouye, S. Chikkatur, A.P., Stamper-Kurn, D.M., Pritchard, D.E., and Ketterle, W., 1999, Bragg spectroscopy of a Bose-Einstein condensate. *Phys. Rev. Lett.* **82**: 4569-4573.
13. Ovchinnikov, Yu.B., Müller, J.-H., Doery, M.R., Vredenbregt, E.J.D., Helmerson, K., Rolston, S.L., and Phillips, W.D., 1999, Diffraction of a released Bose-Einstein condensate by a pulsed standing light wave. *Phys. Rev. Lett.* **83**: 284-287.
14. Deng, L., Hagley, E.W., Denschlag, J., Simsarian, J.E., Edwards, M.A., Clark, C.W., Helmerson, K., Rolston, S.L., and Phillips, W.D., 1999, Temporal matter-wave dispersion Talbot effect. *Phys. Rev. Lett.*, accepted for publication.
15. Helmerson, K., Hutchinson, D., Burnett, K., and Phillips, W.D., 1999, Atom Lasers. *Physics World*, August: 31-35.
16. Mewes, M.-O., Andrews, M.R., Kurn, D.M., Durfee, D.S., Townsend, C.G., and Ketterle, W., 1997, Output coupler for Bose-Einstein condensed atoms. *Phys. Rev. Lett.* **78**: 582-585.
17. Hagley, E.W., Deng., L., Kozuma, M., Wen, J., Helmerson, K., Rolston, S.L., and Phillips, W.D., 1999, A well collimated, quasi-continuous atom laser. *Science* **283**: 1706-1709.
18. Bloch, I., Hänsch, T.W., and Esslinger T., 1999, Atom laser with a cw output coupler. *Phys. Rev. Lett.* **82**: 3008-3011.
19. Ertmer, W., *private communication.*
20. Hagley, E.W., Deng., L., Kozuma, M., Trippenbach, M., Band, Y.B., Edwards, M., Doery, M., Julienne, P.S., Helmerson, K., Rolston, S.L., and Phillips, W.D., 1999, Measurement of the coherence of a Bose-Einstein condensate. *Phys. Rev. Lett.* **83**: 3112-3115.

Non-Linear Atom Optics With Bose-Einstein Condensates

K. HELMERSON
Atomic Physics Division, Physics Laboratory National Institute of Standards and Technology, Gaithersburg, MD 20899-8424

1. INTRODUCTION

The advent of the laser as an intense, coherent light source enabled the field of non-linear optics to flourish. The interaction of light, mediated by materials whose index of refraction depends on intensity, has led to effects such as multi-wave mixing of optical fields to produce coherent light of a new frequency, and optical solitons, pulses of light that propagate without dispersion. Non-linear optics now plays an important role in many areas of science and technology. With the experimental realisation of Bose-Einstein condensation[1-4] (many atoms in a single quantum state) and the matter-wave or atom "laser"[5-8] (atoms coherently extracted from a condensate), we now have an intense source of matter-waves analogous to the source of light from an optical laser. This has led us to the threshold of a new field of physics: non-linear atom optics[9].

The analogy between non-linear optics with lasers and non-linear atom optics with Bose-Einstein condensates (BECs) can be seen in the similarities between the equations that govern each system. For a condensate of interacting bosons, in a trapping potential V, the macroscopic wave function Ψ satisfies a non-linear Schrödinger equation[10],

$$i\hbar\frac{\partial\Psi}{\partial t}=\left(-\frac{\hbar^2}{2M}\nabla^2+V+U_0|\Psi|^2\right)\Psi \qquad (1)$$

where M is the atomic mass, g describes the strength of the atom-atom interaction ($g > 0$ for sodium atoms), and $|\Psi|^2$ is proportional to atomic number density.

This Chapter describes the experimental effort of the Laser Cooling and Trapping Group, at NIST in Gaithersburg, Maryland, to observe non-linear matter wave phenomena with Bose-Einstein condensates. In particular: the demonstration of four-wave mixing of matter waves, and the creation of dark solitons.

2. FOUR-WAVE MIXING OF MATTER WAVES

The non-linear term in Eq. (1) is similar to the third-order susceptibility term, $\chi^{(3)}$, in the wave equation for the electric field describing optical four-wave mixing. We therefore expect that if three coherent matter waves are spatially overlapped with the appropriate momentum, a fourth matter wave will be produced due to the non-linear interaction; a process analogous to optical four-wave mixing. In contrast to optical four-wave mixing, the non-linearity in matter wave four-wave mixing comes from atom-atom interactions, described by a mean-field; there is no need for an external non-linear medium.

Using the atoms from a BEC, we have observed such four-wave mixing of matter waves. This work is described in detail in Ref. [11]. In our four-wave mixing experiment, we used optically induced Bragg diffraction[12] to create three overlapping wavepackets with appropriately chosen momentum. When the three wavepackets spatially separated, a fourth wavepacket, due to the wave-mixing process, was observed (see Fig. 1).

The process of four-wave mixing of matter waves (and also optical waves), can be thought of as Bragg diffraction off of a matter grating. In this picture, two of the matter waves interfere to form a standing wave grating. The third wave can Bragg diffract off of this grating, giving rise to the fourth wave. An alternative picture of four-wave mixing is in terms of stimulated emission. In this picture it is helpful to view the four-wave mixing process in a reference frame where the process looks like degenerate four-wave mixing; that is, all of the waves have the same energy.

In four-wave mixing, both energy and momentum (corresponding to phase matching) must be conserved. Since atoms, unlike photons, can not be created out of the vacuum we have the additional requirement, for matter waves, of particle number conservation. (If we included the rest mass of the atom, particle number conservation is contained in energy conservation.) Given these three conditions, one can show that the only four-wave mixing configurations possible with matter waves are those that can be viewed in some frame of reference as degenerate four-wave mixing. This is also the geometry of phase conjugation. Fig. 1 shows the four-wave mixing geometry for matter waves viewed in the degenerate or phase conjugation frame.

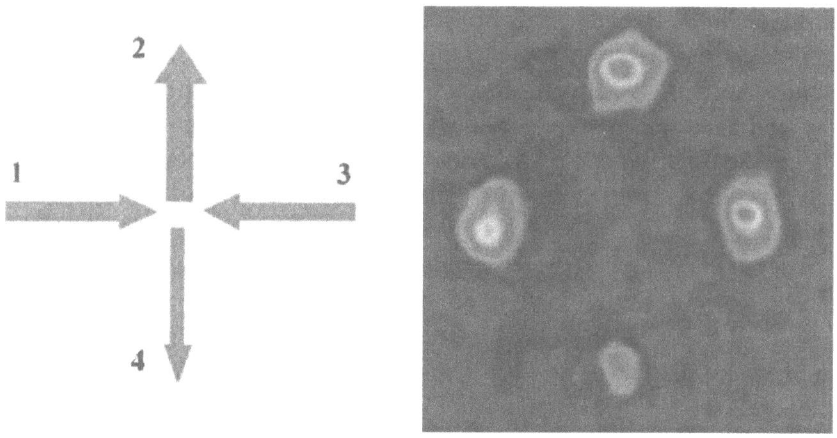

Figure 1. The process of four-wave of mixing of matter waves can always be transformed to a reference frame where the mixing process is degenerate (all of the waves have the same energy left). The non-linear term describing the mean-field, s-wave interaction of the atoms is responsible for the four-wave mixing. Atoms from waves 1 and 3 scatter off each other and go off back-to-back. The scattering process can be stimulated by wave 2, so that it is more likely that one of the scattering pairs goes into this wave. By momentum conservation, wave 4 is created. The small cloud of atoms in the image on the right is the fourth wave generated by four-wave mixing of matter waves.

In the picture of four-wave mixing as arising from stimulated emission, atoms in waves 1 and 3 can be considered as undergoing an elastic collision. The scattering process results in atoms going off back-to-back in order to conserve momentum, but at some arbitrary angle with respect to the incident direction. (The scattering process is typically s-wave and the outgoing waves can be considered spherical). In the presence of wave 2, however, this scattering process can be stimulated. There is an enhanced probability that one of the atoms from the collision of waves 1 and 3 will scatter into wave 2. (This probability is enhanced by the number of atoms in wave 2). Because of momentum conservation, the enhanced scattering of atoms into wave 2 results in an enhanced number of atoms in wave 4. In this picture, it is obvious that the four-wave mixing process removes atoms from waves 1 and 3 and puts them into waves 2 and 4. This may have some interesting consequences in terms of quantum correlations between the waves.

3. QUANTUM PHASE ENGINEERING

A three-dimensional image of an arbitrarily complex object can be constructed by sending light, with sufficient spatial coherence, through the appropriate phase and/or amplitude mask. This is the basic principle behind physical optics, which includes wave phenomena like diffraction and holography. Diffraction can be achieved with a periodic phase and/or amplitude mask while a more complicated mask is needed to construct a complex holographic image. In each case, the mask modifies the incoming wave and subsequent propagation produces the desired pattern of light. This idea can be readily adapted to atom optics, especially when the "incoming" matter wave is from a highly coherent source such as a Bose-Einstein condensate.

We have developed a technique to optically imprint complex phase patterns onto a Bose-Einstein condensate in order to create interesting topological states. This technique is analogous to sending a wave through a thin phase mask. The basic idea is to expose the condensate atoms to a short pulse of laser light with a spatially varying intensity pattern. The laser detuning is chosen such that spontaneous emission is negligible. (The phase mask can also serve as an amplitude mask by tuning closer to resonance, so that spontaneous emission is significant.) The pulse duration is sufficiently short such that the atoms do not move an appreciable distance (i.e. the wavelength of light) during the pulse. This is sometimes referred to as the Raman-Nath regime. During the laser pulse, the AC Stark effect shifts the energy of the atoms by $U(r,t)$. Hence the effect on the atomic wavefunction is to "instantaneous" change its phase. This effect can be represented by multiplying the wavefunction by the phase factor exp $(i\phi(r))$, where $\phi(r) = -\int U(r, t)\, dt/\hbar$. Since the AC Stark or light shift is proportional to the intensity of the light, any spatial intensity variation in the light field will be written onto the BEC wavefunction as a spatial variation in its phase.

Optically induced phase imprinting is a tool for "quantum phase engineering" the wavefunction to create a wide variety of states. For example, the application of a short pulse of standing wave light will imprint a sinusoidal phase onto the condensate. The subsequent evolution of the wavefunction produces wavepackets of atoms with momentum $\pm 2n\hbar k$ (n = 0, 1, 2, ...); the wavefunction appears to have diffracted off of the sinusoidal potential[13]. It should be possible to use quantum phase engineering to produce collective states of excitation of the interacting BEC, such as solitons and vortices. The application of a uniform intensity light field to half of the BEC imprints a relative phase difference between the two halves. This phase step is expected to give rise dark solitons (see following section). Such solitons will propagate with a speed related to the phase difference[14], which can be adjusted by the intensity of the laser pulse. It should also be possible to produce one or more vortices by applying a laser pulse, which has a linearly varying azimuthal intensity dependence[15]. This will produce a topological winding of the BEC phase, which if large enough (i.e. 2π) should produce a vortex. Numerical solutions to a 3-D Gross-Pitaevskii equation[16] show that this is the case; and also show that such a vortex, although unstable because it is created in a non-rotating trap, will live for a sufficient time to be observable. Increasing the phase winding will generate multiple vortices (vortices with more than $\hbar$ of angular momentum are not stable and will immediately split into multiple vortices each with angular momentum $\hbar$). Quantum phase engineering can generate arbitrary phase patterns, and perhaps other interesting quantum states. In this sense, it is a form of atom holography[17]. The technological challenge is mostly one of imaging. Any complicated pattern must be imaged to the size of the BEC, typically of order 50 μm.

4. SOLITONS

Solitons are stable, localised waves that propagate in a non-linear medium without spreading. They may be either bright or dark, depending on the details of the governing non-linear wave equation. A bright soliton is a peak in the amplitude while a dark soliton is a notch with a characteristic phase step across it. Eq. (1), which describes the weakly interacting, zero-temperature BEC also supports solitons. The solitons propagate without spreading (dispersing) because the nonlinearity balances the dispersion; for Eq. (1) the corresponding terms are the non-linear interaction $g|\Psi|^2$, and the kinetic energy $-\left(\hbar^2/2M\right)\nabla^2$, respectively. Our sodium condensate only supports dark solitons because the atom-atom interactions are repulsive[14,18] $(g > 0)$.

A distinguishing characteristic of a dark soliton is that its velocity is less than the Bogoliubov speed of sound[14,18] $\upsilon_0 = \left(gn/M\right)^{1/2}$ where n is the unperturbed condensate density. The soliton speed υ_s can be expressed either in terms of the phase step δ $(0 < \delta \le \pi)$, or the soliton "depth" n_d, which is the difference between n and the density at the bottom of the notch[14,18]:

$$\frac{\upsilon_S}{\upsilon_0} = \cos(\delta/2) = \sqrt{1 - n_d/n} \tag{2}$$

For $\delta = \pi$ the soliton has zero velocity, zero density at its centre, a width on the order of the healing length[18], and a discontinuous phase step. As δ decreases the velocity increases, approaching the speed of sound. The solitons are shallower and wider, with a more gradual phase step. They travel opposite to the direction of the phase gradient. Because a soliton has a characteristic phase step, optically imprinting a phase step on the BEC wavefunction should be a way to create a soliton.

4.1 Optically Imprinting the Soliton Phase

We modified the phase distribution of the BEC by employing the technique of quantum phase engineering discussed in Sect. 3.

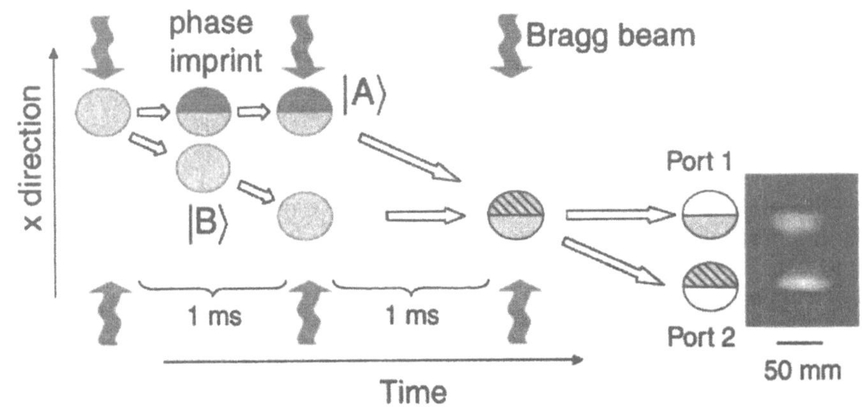

Figure 2. Space-time diagram of the matter wave interferometer used to measure the spatial phase distribution imposed on the BEC. Three optically induced Bragg diffraction[12] pulses formed the interferometer. Each pulse consisted of two counter-propagating laser beams detuned by about −2 GHz from atomic resonance (so that spontaneous emission is negligible) with their frequencies differing by 100 kHz. The first pulse had a duration of 8 μs and coherently split the condensate into two components, $|A\rangle$ and $|B\rangle$, with equal number of atoms. $|A\rangle$ remained at rest and $|B\rangle$ received two photon recoils of momentum. When they were completely separated, we exposed the top half of $|A\rangle$ to a phase imprinting pulse of π, which changed the phase distribution of $|A\rangle$ while $|B\rangle$ served as a phase reference. 1 ms after the first Bragg pulse, a second Bragg pulse of 16 μs duration brought $|B\rangle$ to rest and imparted two photon momenta to $|A\rangle$. When they overlapped again, 1 ms later, a third pulse of 8 μs duration converted their phase differences into density distributions at ports 1 and 2, which appears in the images.

The condensate atoms were exposed to a pulsed, off-resonant laser beam, propagating coaxial with the absorption probe beam, with a spatial intensity profile such that only half of the BEC was illuminated. This was accomplished by blocking half of the laser beam with a razor blade and imaging this razor blade onto the condensate. The intensity pattern at the condensate, as observed by our absorption imaging system, had a light to dark (90% to 10%) transition region of 7 μm. The intensity required to imprint a phase of π was checked by atom interferometry.

We constructed a Mach-Zehnder atom interferometer based on optically induced Bragg diffraction[19,20], to directly measure the spatial phase variation across a BEC. Our Bragg interferometer differs from previous ones in that we can independently manipulate atoms in the two arms (because of their large separation) and can resolve the output ports to reveal the spatial distribution of the condensate phase. In our interferometer a Bragg pulse splits the initial condensate into two states, $|A\rangle$ and $|B\rangle$, differing only in their momenta (Fig. 2). After they spatially separate, the phase step is imprinted on $|A\rangle$, while $|B\rangle$ is unaffected and serves as a phase reference. When recombined, they interfere according to their local phase difference. Where this phase difference is 0, atoms appear in port 1, and where it is π, atoms appear in port 2. Imaging the density distributions of ports 1 and 2 displays the spatially varying phase. Fig. 2 shows the output of the interferometer when a phase of π was imprinted on the upper half of $|A\rangle$. The high-contrast "half-moons" are direct evidence that we can imprint the phase step appropriate for a soliton.

4.2 Soliton Propagation

To observe the creation and propagation of solitons, we do not use interferometry, but instead measure BEC density distributions with absorption imaging after imprinting a phase step. Fig. 3 shows the evolution of the condensate after the top half was phase imprinted with $\phi = 1.5\ \pi$, a phase for which we observe a single deep soliton (the reason for imprinting a phase step larger than π is discussed below). Immediately after the phase imprint, there is a steep phase gradient across the middle of the condensate such that this portion has a large velocity in the $+x$ direction. This velocity can be understood as arising from the impulse imparted by the optical dipole force, and results in a positive density disturbance that travels at or above the speed of sound. A dark notch is left behind, which is a soliton moving slowly in the $-x$ direction (opposite to the direction of the applied force).

A striking feature of the images is the curvature of the soliton. This curvature is due to the 3-D geometry of the trapped condensate, and occurs for two reasons. First, the speed of sound υ_0 is largest at the trap centre where the density is greatest, and decreases towards the condensate edge. Second, as the soliton moves into regions of lower condensate density, we find numerically that the density at its centre, $n - n_d$, approaches zero, δ approaches π, and υ_s decreases to zero before reaching the edge. This is because the soliton depth n_d, rather than its phase offset δ, appears to be a conserved quantity in a non-uniform medium.

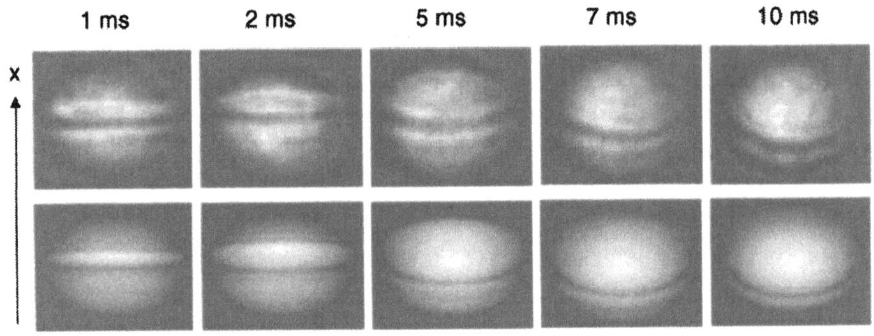

Figure 3. Experimental (upper) and theoretical (lower) images of the integrated BEC density for various times after we imprint a phase of about 1.5π on the top half of the condensate with a 1 μs pulse. The measure number of atoms in the condensate was $1.7(3)\times10^6$, and this value was used in the calculations. A positive density disturbance moved rapidly in the $+x$ direction and a dark soliton moves oppositely at significantly less than the speed of sound. Because the imaging is destructive, each image shows a different BEC. The width of the images is 70 μm.

A clear indication that the notches seen in Fig. 3 are solitons, rather than simply sound waves, is their subsonic propagation velocity. To determine this velocity, we measure the distance after propagation between the notch and the position of the imprinted phase step along the x direction. Because the position of our condensate varies randomly from shot-to-shot (presumably due to stray, time varying fields) we cannot always apply the phase step at the centre. A marker for the location of the initial phase step is the intersection of the soliton with the condensate edge, because at this point the soliton has zero velocity. Using images taken 5 ms after the imprint, at which time the soliton has not traveled far from the BEC centre, we obtain a mean soliton velocity of 1.8(4) mm/s. This speed is significantly less than the mean Bogoliubov speed of sound υ_0 = 2.8(1) mm/s. From the propagation of the notch in the numerical solutions (Fig. 3, lower images) we obtain a mean soliton velocity, υ_S = 1.6 mm/s, in agreement with the experimental value. The experimental uncertainty is mainly due to the difficulty in determining the position of the initial phase step.

From the lower image of Fig. 3 at 5 ms, we can extract the theoretical density and phase profile along the x-axis through the centre of the condensate. The dark soliton notch and its phase step are centred at $x = -8$ μm. This phase step, δ = 0.58π, is less than the imprinted phase of 1.5π. The difference is caused by the mismatch between the phase imprint and the phase and depth of the soliton solution of Eq. (1): Our imprinting resolution of 7 μm is larger than the soliton width, which is of the order of the healing length (~0.7 μm), and we do not control the amplitude of the wave function.

In order to improve our measurement of the soliton velocity, we avoid the uncertainty in the position of the initial phase step by replacing the razor blade mask with a thin slit. This produces a stripe of light with a Gaussian profile ($1/e^2$ full width ≈ 15 μm). With this stripe in the centre of the condensate, numerical simulations predict the generation of solitons that propagate symmetrically outwards. We select experimental images with solitons symmetrically located about the middle of the condensate, and measure the distance between them. For a small phase imprint of $\phi \approx 0.5\pi$ (at Gaussian maximum), we observe solitons moving at the Bogoliubov speed of sound, within experimental uncertainty. For a larger phase imprint of $\phi \approx 1.5\pi$, we observe much slower soliton propagation, in agreement with numerical simulations. An even larger phase imprint generates many solitons. The results of these experiments on the creation and propagation of solitons can be found in Ref. [21]. Solitons in a BEC have also been observed by a group in Hannover[22].

ACKNOWLEDGMENTS

The experimental work described here was carried out at NIST by Lu Deng, Johannes Denschlag, Ed Hagley, Mikio Kozuma, Jesse Simsarian, KH, Steve Rolston, and Bill Phillips. We have benefited greatly from discussions with our theoretical colleagues, Yehuda Band, Charles Clark, Marya Doery, Mark Edwards, David Feder, Paul Julienne, and Marek Trippenbach. This work was partially supported by the U.S. Office of Naval Research and NASA.

REFERENCES

1. Anderson, M.H., Ensher, J.R., Matthews, M.R., Wieman, C.E., and Cornell, E.A., 1995, Observation of Bose-Einstein condensation in a dilute atomic vapor. *Science* **269**: 198-201.
2. Davis, K.B., Mewes, M.-O., Andrews, M.R., van Druten, N.J., Durfee, D.S., Kurn, D.M., and Ketterle, W., 1995, Bose-Einstein Condensation in a gas of sodium atoms. *Phys. Rev. Lett.* **75**: 3969-3972.
3. Bradley, C.C., Sackett, C.A., and Hulet, R.G., 1997, Bose-Einstein condensation of lithium: Observation of limited condensate number. *Phys. Rev. Lett.* **78**: 985-988.
4. Fried, D., Killian, T.C., Willmann, L., Landhuis, D., Moss, S.C., Kleppner, D., and Greytak, T.J., 1999, Bose-Einstein condensation of atomic hydrogen. *Phys. Rev. Lett.* **81**: 3807-3810.

5. Mewes, M.-O., Andrews, M.R., Kurn, D.M., Durfee, D.S., Townsend, C.G., and Ketterle, W., 1997, Output coupler for Bose-Einstein condensed atoms. *Phys. Rev. Lett.* **78**: 582-585.
6. Anderson, B.P., and Kasevich, M.A., 1998, Macroscopic quantum interference from atomic tunnel arrays. *Science* **282**: 1686-1689.
7. Hagley, E.W., Deng., L., Kozuma, M., Wen, J., Helmerson, K., Rolston, S.L., and Phillips, W.D., 1999, A well collimated, quasi-continuous atom laser. *Science* **283**: 1706-1709.
8. Bloch, I., Hänsch, T.W., and Esslinger T., 1999, Atom laser with a cw output coupler. *Phys. Rev. Lett.* **82**: 3008-3011.
9. Lens, G., Meystre, P., and Wright, E.W., 1993, Nonlinear atom optics. *Phys. Rev. Lett.* **71**: 3271-3274.
10. Dalfovo, F., Giorgini, S., Pitaevskii, L.P., and Stringari, S., 1999, Theory of Bose-Einstein condensation in trapped gases. *Rev. Mod. Phys.* **71**: 463-512.
11. Deng, L., Hagley, E.W., Wen, J., Trippenbach, M., Band, Y., Julienne, P.S., Simsarian, J.E., Helmerson, K., Rolston, S.L., and Phillips, W.D., 1999, Four-wave mixing with matter waves. *Nature* **398**: 218-220.
12. Kozuma, M., Deng, L., Hagley, E.W., Wen, J., Lutwak, R., Helmerson, K., Rolston, S.L., and Phillips, W.D., 1999, Optically-induced Bragg diffraction of a Bose-Einstein condensate. *Phys. Rev. Lett.* **82**: 871-874.
13. Ovchinnikov, Yu.B., Müller, J.H., Doery, M.R., Vredenbregt, E.J.D., Helmerson, K., Rolston, S.L., and Phillips, W.D., 1999, Diffraction of a released Bose-Einstein condensate by a pulsed standing light wave. *Phys. Rev. Lett.* **83**: 284-287; see also, Helmerson, K., Coherent atom optics with Bose-Einstein condensates. *these proceedings.*
14. Reinhardt, W.P., and Clark, C.W., 1997, Soliton dynamics in the collisions of Bose-Einstein condensates: an analogue of the Josephson effect. *J. Phys. B: At. Mol. Opt. Phys.* **30**: L785-L789.
15. Dobrek, L., Gajda, M., Lewenstein, M., Sengstock, K., Birkl, G., and Ertmer, W., 1999, Optical generation of vortices in trapped Bose-Einstein condensates. *Phys. Rev. A.* **60**: R3381-3384.
16. Feder, D.L., Clark, C.W., and Schneider, B.I., 1996, Vortex stability of interacting Bose-Einstein condensates confined in anisotropic harmonic traps. *Phys. Rev. Lett.* **82**: 4956-4959.
17. Fujita, J., Morinaga, M., Kishimoto, T., Yasuda, M., Matsui, S., and Shimizu, F., 1996, Manipulation of an atomic beam by a computer-generated hologram. *Nature.* **380**: 691-694.
18. Jackson, A.D., Kavoulakis, G.M., and Pethick, C.J., 1998, Solitary waves in clouds of Bose-Einstein condensed atoms. *Phys. Rev. A* **58**: 2417-2422.
19. Torii, Y., Suzuki, Y., Kozuma, M., Kuga, T., Deng, L., and Hagley, E.W., 1999, Mach-Zehnder Bragg interferometer for a Bose-Einstein condensate. *cond-mat/9908160.*
20. Giltner, D.M., Mc Gowan, R.W., and Lee, S.A., 1995, Atom interferometer based on Bragg scattering from standing light waves. *Phys. Rev. Lett.* **75**: 2638-2641.
21. Denschlag, J., Simsarian, J.E., Feder, D.L., Clark, C.W., Collins, L.A., Cubizolles, J., Deng, L., Hagley, E.W., Helmerson, K., Reinhardt, W.P., Rolston, S.L., Schneider, B.I., and Phillips, W.D., 1999, Generating solitons by phase engineering a Bose-Einstein condensate. *Science*, accepted for publication.

22. Burger, S., Bongs, K., Dettmer, S., Ertmer, W., Sengstock, K., Sanpera, A., Shlyapnikov, G.V., Lewenstein, M., 1999, Dark solitons in Bose-Einstein condensates. *Phys. Rev. Lett.*, accepted for publication.

Momentum Distribution Of A Bose Condensed Trapped Gas

[1]S. STRINGARI, [1,2]L. PITAEVSKII, [3]D.M. STAMPER-KURN, AND [1]F. ZAMBELLI
[1]Dipartimento di Fisica, Università di Trento, and Istituto Nazionale per la Fisica della Materia, I-38050 Povo, Italy; [2]Kapitza Institute for Physical Problems, 117334 Moscow, Russia; [3]Norman Bridge Laboratory of Physics, California Institute of Technology 12-33, Pasadena, CA 91125.

1. INTRODUCTION

A peculiarity of trapped atomic gases is that Bose-Einstein condensation (BEC) shows up not only in momentum space, where it is usually discussed in traditional textbooks, but also in coordinate space. This is the consequence of the inhomogeneity of these systems, which makes it possible to separate the condensate from the thermal component also in coordinate space. Actually, most of measurements in trapped gases have been so far limited to the study of density profiles and of the effects of BEC in coordinate space. The possibility of a direct measurement of the momentum distribution, as emerged from recent experiments based on two-photon Bragg scattering[1], is highly appealing. In fact, despite the inhomogeneity of the gas, BEC shows up more deeply in momentum than in coordinate space. In current experiments on harmonically-confined Bose gases, the sizes of the condensate R and of the thermal cloud R_T are in fact typically comparable. In the Thomas-Fermi regime[2], the ratio between the two radii is given as

$$\frac{R}{R_T} \approx \sqrt{\frac{\mu}{k_B T}} = \sqrt{\frac{\hbar\omega_{ho}}{k_B T}}\left(15\frac{N_0 a}{a_{ho}}\right)^{1/5} \tag{1}$$

Bose-Einstein Condensates and Atom Lasers
Edited by Martellucci *et al.*, Kluwer Academic/Plenum Publishers, 2000

where N_0 is the number of atoms in the condensate, μ is the chemical potential, $a_{ho} = \sqrt{\hbar/m\omega_{ho}}$ is the oscillator length, and a is the s-wave scattering length. Due to the large value of the Thomas-Fermi parameter $N_0 a/a_{ho}$, this ratio is typically close to unity. Eq. (1) also provides an estimate for the ratio of sizes of the two components measured in time-of-flight experiments, in which the trap is suddenly switched off and the gas allowed to freely expand. While the expansion of the thermal cloud is indicative of the non-condensate momentum distribution before release from the trap, the expansion of the condensate in the Thomas-Fermi regime is dominated by the release of interaction energy and does not reveal its initial momentum distribution.

In contrast, the distinction between the condensate and the thermal cloud in momentum space is stark. A confined condensate of finite size has a momentum distribution of width $\Delta p_c \approx \hbar/R$ fixed by the inverse of the size R of the condensate. The momentum width of the thermal cloud is instead given by the temperature of the gas as $\Delta p_T \approx \sqrt{mk_B T}$. For harmonic confinement in the Thomas-Fermi regime, one then finds

$$\frac{\Delta p_c}{\Delta p_T} \approx \sqrt{\frac{\hbar\omega_{ho}}{k_B T}} \left(15 \frac{N_0 a}{a_{ho}}\right)^{-1/5} \tag{2}$$

In contrast with the comparison of the condensate and the thermal cloud in coordinate space, the distinction between the two components in momentum space is strongly *enhanced* by two body interactions as the Thomas-Fermi parameter $N_0 a/a_{ho}$ increases. The investigation of the momentum distribution consequently provides a deeper understanding of the phenomenon of BEC. In particular the smallness of the width Δp_c reflects the presence of long-range coherence.

The purpose of this chapter is to summarize some of the key features exhibited by a trapped Bose-Einstein condensed gas in momentum space.

2. DYNAMIC STRUCTURE FACTOR AND MOMENTUM DISTRIBUTION

The momentum distribution of a many-body system can be investigated through the analysis of the dynamic structure factor

$$S(\vec{q},E)=\frac{1}{Z}\sum_{mn} e^{-\beta E_m}\left|\langle m\,|\,\rho_{\vec{q}}\,|\,n\rangle\right|^2\delta(E-E_m+E_n) \tag{3}$$

measurable in inelastic scattering reactions, where a scattering probe transfers to the sample momentum $\vec{q}$ and energy E. In Eq. (3) $|n\rangle$ and E_n are the eigenstates and eigenvalues of the Hamiltonian of the system, $e^{-\beta E_n}$ is the usual Boltzmann factor, $\rho_{\vec{q}}=\sum_j e^{-i\vec{q}\cdot\vec{r}_j/\hbar}$ is the Fourier transform of the one-body density operator, and Z is the usual canonical partition function.

The quantity $S(\vec{q},E)$ has been recently measured in Bose-Einstein condensates via two-photon Bragg scattering using two detuned laser beams both in the high q [1] and in the phonon regime[2]. Actually these experiments measure the difference $S(\vec{q},E)-S(-\vec{q},-E)$ which corresponds to the imaginary part of the response function. A consequence of this difference is that thermal effects tend to cancel out, and these measurements essentially provide the $T=0$ value of the dynamic structure factor.

When the momentum transfer q is very high the scattering process involves essentially single particles and the system in the final state can be described as $(N-1)$ atoms remaining in the unperturbed configuration, plus the scattered atom moving with an extra momentum $\vec{q}$. In this limit the dynamic structure factor can be expressed in terms of the momentum distribution by using the so called impulse approximation (IA)[4]

$$S_{IA}(\vec{q},E)=\int d\,\vec{p}\,\delta\left(E-\frac{(\vec{p}+\vec{q})^2}{2m}+\frac{p^2}{2m}\right)n(\vec{p}) \tag{4}$$

In Eq. (4) $n(\vec{p})$ is the momentum distribution of the system, defined by the relation

$$n(\vec{p})=\left\langle\hat{\psi}^+(\vec{p})\hat{\psi}(\vec{p})\right\rangle, \tag{5}$$

where

$$\hat{\psi}(\vec{p})=\frac{1}{(2\pi\hbar)^{3/2}}\int d\,\vec{r}\,\hat{\psi}(\vec{r})e^{i\vec{p}\cdot\vec{r}/\hbar} \tag{6}$$

is the Fourier transform of the usual field operator. The momentum distribution is also related to the off diagonal one-body density matrix $\rho^{(1)}(\vec{s}) = \langle \hat{\psi}^+(\vec{s})\hat{\psi}(\vec{s}) \rangle$ by the relation

$$n(\vec{p}) = \frac{1}{(2\pi\hbar)^3} \int d\vec{s}\, \rho^{(1)}(\vec{s}) e^{i\vec{p}\cdot\vec{s}/\hbar} \tag{7}$$

The validity of Eq. (4) is not restricted to ideal gases, but holds also for interacting and non uniform systems, independently of quantum statistics. Of course for interacting systems the momentum distribution will differ significantly from that of the ideal gas and interactions show up in the form of $n(\vec{p})$. Actually most of the information on the momentum distribution of superfluid helium, including the estimate of the condensate fraction[5], comes from the measurements of $S(\vec{q}, E)$. In the case of ^{4}He, however, final state interactions are important also for the largest available values of q, and corrections to Eq. (4) must be included for a safe analysis of experimental data.

Before discussing more in details the IA, let us introduce the systems which we are going to study: in the following we will limit ourselves to the $T = 0$ case focusing on dilute Bose gases where Bogoliubov theory is applicable. This restricts the range of momenta q to the "macroscopic" regime $qa << \hbar$, where a is the s-wave scattering length. For larger values of q short range correlations become important and Bogoliubov theory is no longer adequate. In the conditions of the experiment of Ref. [1], carried out on a gas of sodium atoms, the Bogoliubov approach is well applicable since $q/a \approx 20\,\mu m^{-1}$ and $qa \approx 0.06\hbar$. Moreover we will always make the harmonic axially symmetric choice for the external potential, which reads as

$$V_{ext}(\vec{r}) = \frac{1}{2} m\omega_\perp^2 (x^2 + y^2) + \frac{1}{2} m\omega_z^2 z^2 \tag{8}$$

The Eq. (4) for the impulse approximation can be also written in the form

$$S_{IA}(\vec{q}, E) = \frac{m}{q} \int dp_y dp_z n(p_x, p_y, p_z) \tag{9}$$

where we have assumed that the vector $\vec{q}$ is oriented along the x axis, and $p_x = m(E - E_r)/q$, with

$$E_r = \frac{q^2}{2m} \tag{10}$$

The integral $\int dp_y dp_z n(p_x, p_y, p_z)$ is also called the longitudinal momentum distribution.

In a dilute Bose gas at zero temperature the momentum distribution is given by $n(\vec{p}) = |\Phi(\vec{p})|^2$ where

$$\Phi(\vec{p}) = \frac{1}{(2\pi\hbar)^{3/2}} \int d\vec{r}\, \psi(\vec{r}) e^{i\vec{p}\cdot\vec{r}/\hbar} \tag{11}$$

is the Fourier transform of the order parameter $\psi(\vec{r})$. The form of $n(\vec{p})$ for a trapped condensate has been discussed previously[6,7]. In the Thomas-Fermi limit $Na/a_{ho} >> 1$ one finds the simple analytic result

$$n_{TF}(\vec{p}) = N \frac{15}{16\lambda} \left(\frac{R_\perp}{\hbar}\right)^3 \left[\frac{J_2(\tilde{p})}{\tilde{p}^2}\right]^2 \tag{12}$$

where $J_2(z)$ is the usual Bessel function of order 2,

$$R_\perp = \left(15 \frac{Na}{a_{ho}}\right)^{1/5} \lambda^{1/3} a_{ho} \tag{13}$$

is the Thomas-Fermi radius of the condensate in the $x - y$ plane, and $\tilde{p} = \sqrt{p_x^2 + p_y^2 + (p_z/\lambda)^2}\, R_\perp/\hbar$ is a dimensionless variable, with the parameter $\lambda = \omega_z/\omega_\perp$ fixing the anisotropy of the external potential. In Eq. (13) $a_{ho} = \sqrt{\hbar/m\omega_{ho}}$ is the oscillator length calculated using the geometrical average $\omega_{ho} = \left(\omega_z \omega_\perp^2\right)^{1/3}$ of the oscillator frequencies.

Eq. (12) explicitly shows that the momentum distribution scales as $1/R_\perp$ and is consequently much narrower than that of the non-interacting gas

$$n_{IBG}(\vec{p}) = N\left(\frac{a_{ho}}{\hbar\sqrt{\pi}}\right)^3 \exp\left[-\frac{a_{ho}^2}{\hbar^2}\lambda^{1/3}\left(p_x^2 + p_y^2 + \frac{p_z^2}{\lambda}\right)\right] \tag{14}$$

since $R_\perp >> a_{ho}$.

In the impulse approximation the peak of $S(\vec{q},E)$ coincides with the recoil energy defined in Eq. (10), while the curve is broadened due to the Doppler effect in the momentum distribution. A useful estimate of the broadening can be obtained carrying out a Gaussian expansion in the dynamic structure factor of Eq. (9) near the peak value $E = E_r$. One finds:

$$S_{IA}(\vec{q},E) \cong S_{IA}(\vec{q},E_r)\exp\left[-\frac{(E-E_r)^2}{2\Delta_{IA}^2}\right] \tag{15}$$

with $\Delta_{IA}^2 = -\left[S_{IA}(\vec{q},E)/\partial_E^2 S_{IA}(\vec{q},E)\right]\Big|_{E=E_r}$. By calculating the second derivative of Eq. (9) with the Thomas-Fermi profile of Eq. (12) for the momentum distribution, we obtain, after some straightforward algebra

$$\Delta_{IA} = \sqrt{\frac{8}{3}}\frac{q\hbar}{mR_\perp} \tag{16}$$

The Gaussian profile (Eq. (15)) reproduces very well the exact curve, so that the Doppler width (Eq. (16)) can be usefully compared with experiments, where the widths are usually extracted through Gaussian fits to the measured signal.

The investigation of the dynamic structure factor also provides information on the coherence effects exhibited by the system and in particular on the behaviour of the off-diagonal one-body density (see Eq. (7)). By taking the Fourier transform of Eq. (9) with respect to p_x, one finds the result

$$\rho^{(1)}(s_x,0,0) = \int dE\, S_{IA}(\vec{q},E)\exp\left[-i\frac{ms_x}{\hbar q}(E-E_r)\right] \tag{17}$$

which shows that the one-body density is a measurable quantity if one works at high q where $S(\vec{q},E) \cong S_{IA}(\vec{q},E)$. In a uniform Bose-Einstein gas $\rho^{(1)}(s_x,0,0)$ tends to a constant value when s_x is large. In a finite system

$\rho^{(1)}(s_x,0,0)$ always tends to zero when $s_x \to \infty$. The typical length over which $\rho^{(1)}$ decreases can be of the order of the size of the sample or smaller depending on the degree of coherence. Using, for example, the Gaussian profile of Eq. (15) for $S(\vec{q},E)$ one finds

$$\rho^{(1)}(s_x,0,0) \approx N \exp\left[-\frac{s_x^2}{2\chi_x^2}\right] \tag{18}$$

with $\chi_x = R_\perp \sqrt{3/8}$. One can see from Eq. (18) that χ_x plays the role of a coherence length, which turns out to be of the order of the size of the system. This result reflects the fact that in a Bose-Einstein condensate the Heisenberg inequality $\Delta R \Delta p_c \geq \hbar/2$ is close to an identity. Note that the coherence length should not be confused with the healing length ξ (see Eq. (25) in Sect. 4) which, differently from χ_x, becomes small as the density of the sample increases.

3. ROLE OF THE MEAN FIELD INTERACTIONS

The impulse approximation discussed in the previous section ignores the mean field effects predicted by Bogoliubov theory, and works better for large values of momentum transfer. In the experiments of Refs. [1, 3] the value of q was kept fixed, while the density of the sample was varied within a wide range of values $(n(0) \approx 5\,10^{13} \div 5\,10^{14}\ \mathrm{cm}^{-3})$. Depending on the density of the cloud the mean field interactions can play a relevant role, which results in a deviation of the dynamic structure factor from the IA predictions (Eqs. (4), (15)).

In Ref. [8] we have developed an eikonal expansion to evaluate the high energy solutions of the Bogoliubov equations. In particular it was proven that the importance of the mean field interactions is determined by the so called Born parameter b

$$b = \frac{\mu}{E_r}\frac{qR_\perp}{\hbar} \tag{19}$$

where E_r is the free recoil energy (Eq. (10)), $R_\perp$ is the Thomas-Fermi radius (Eq. (13)), and μ is the chemical potential, given by

$$\mu = g\, n(0) = \frac{1}{2}\left(15\frac{Na}{a_{ho}}\right)^{2/5} \hbar\omega_{ho} \tag{20}$$

which is fixed by the central density, $g = 4\pi\hbar^2 a/m$ being the coupling constant proportional to the s-wave scattering length

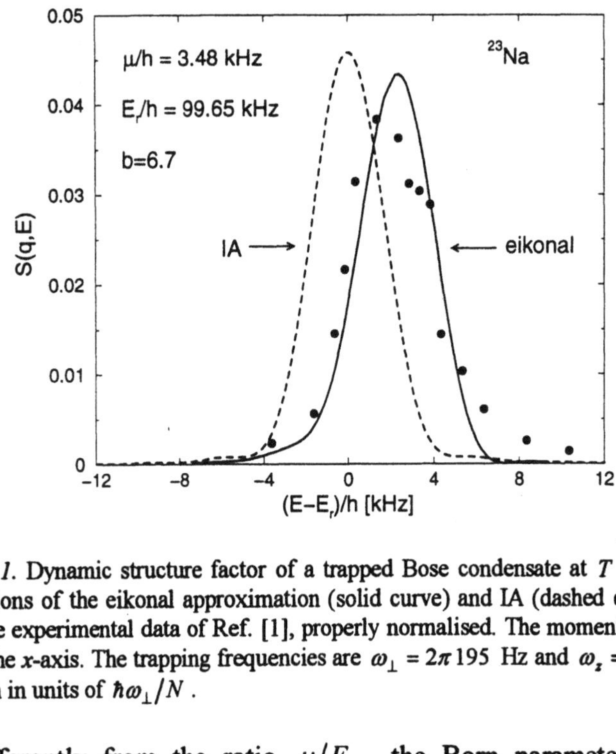

Figure 1. Dynamic structure factor of a trapped Bose condensate at $T = 0$. The numerical predictions of the eikonal approximation (solid curve) and IA (dashed curve) are compared with the experimental data of Ref. [1], properly normalised. The momentum transfer is taken along the x-axis. The trapping frequencies are $\omega_\perp = 2\pi\, 195$ Hz and $\omega_z = 2\pi\, 17$ Hz. $S(q,E)$ is given in units of $\hbar\omega_\perp/N$.

Differently from the ratio μ/E_r the Born parameter depends more explicitly on the size of the atomic cloud. One can show, using the eikonal expansion, that the dynamic structure factor in Thomas-Fermi regime takes the form

$$S(\vec{q},E) = \int d\,\vec{p}\,|M|^2 \delta\left(E - \frac{q^2}{2m} - \frac{qp_x}{m}\right) \tag{21}$$

with

$$M = \frac{1}{(2\pi\hbar)^{3/2}} \int d\vec{r}\, \psi_0(\vec{r}) \exp\left[i \frac{m}{\hbar q} \int d\tilde{x} V_{eff}(\tilde{x}, y, z) \right] \cdot \left(1 - \frac{g(\psi_0(\vec{r}))^2}{q^2/m} \right) \exp\left[-i \left(\frac{\vec{p}\cdot\vec{r}}{\hbar} \right) \right] \tag{22}$$

where $\psi_0(\vec{r}) = \sqrt{(\mu - V_{ext}(\vec{r}))/g}$ is the order parameter calculated in the Thomas-Fermi approximation, and $V_{eff}(\vec{r})$ is an effective potential equal to $g(\psi_0(\vec{r}))^2$ inside and to $V_{ext}(r) - \mu$ outside the condensate. One can show that in the limit $b \to 0$ the eikonal expansion approaches the IA result of Eq. (4).

Fig. 1 shows the dynamic structure factor of a trapped Bose gas for a typical configuration characterized by $b = 6.7$ and $n(0) \cong 2.29\,10^{14}$ cm^{-3}. The predictions of Eq. (4) (dashed line) and of Eq. (21) (solid line) are compared with the experimental results of Ref. [1], which turn out to be in good agreement with theory.

4. VORTICES

In this Section we show that the measurement of the dynamic structure factor in the IA regime would represent a powerful tool to analyse the structure of vortices in a trapped Bose gas. In fact a vortex strongly affects the momentum distribution of the system. This can be easily understood by noting that the kinetic energy of a trapped condensate is roughly doubled by the addition of a vortex[9].

The study of vortices in trapped Bose gases is presently a challenging topics of both theoretical and experimental investigation. First experimental evidence of vortices has been recently reported[10,11].

On the theoretical side the structure of vortices, the corresponding stability conditions, as well as their consequences on the dynamic behaviour of the condensate have already attracted the attention of many physicists. The identification of suitable methods of detection has also been the object of theoretical investigation. These include the expansion of the condensate[12], the shift of the collective excitation[13] and the occurrence of dislocations in interference patterns[14].

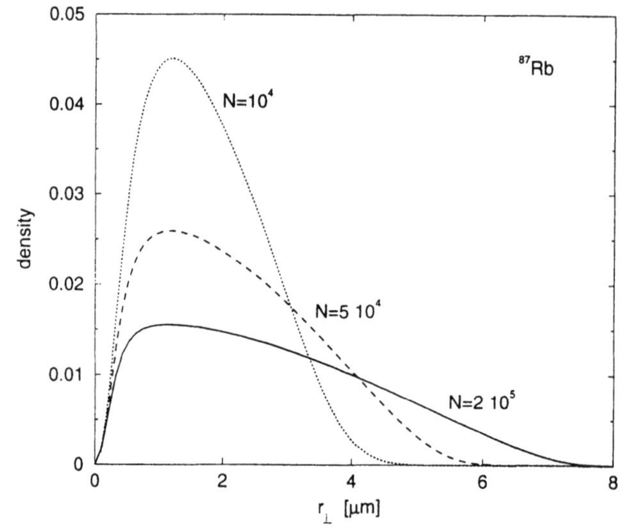

Figure 2. Density profiles of trapped Bose condensates with different number of atoms as a function of the radial variables $r_\perp$, integrated along z. The ^{87}Rb atoms are trapped in an isotropic harmonic potential with frequency $\omega_z = \omega_\perp = 2\pi\, 100$ Hz.

In the presence of a quantized vortex aligned along the z-axis the wave function of the condensate takes the form

$$\psi(\vec{r}) = e^{i\varphi}\psi_0(r_\perp, z) \tag{23}$$

where $\psi_0(r_\perp, z)$ is the solution of the Gross-Pitaevskii equation[16,17]

$$\left[-\frac{\hbar^2\nabla^2}{2m} + \frac{\hbar^2}{2mr_\perp^2} + \frac{m}{2}\omega_\perp^2 r_\perp^2 + \frac{m}{2}\omega_z^2 z^2 + g(\psi_0(r_\perp, z))^2\right]\psi_0(r_\perp, z) = \mu\psi_0(r_\perp, z), \tag{24}$$

which contains the additional centrifugal term $\hbar^2/2mr_\perp^2$. Solutions of Eq. (24) have been obtained numerically in Ref. [17]. The density distribution $|\psi_0(r_\perp, z)|$ exhibits a hole whose size is of the order of the healing length of the gas

$$\xi = \frac{1}{\sqrt{8\pi\, na}} \tag{25}$$

where n is the central density of the cloud. It is worth noticing that in the Thomas-Fermi limit ξ is much smaller than the size of the condensate (see Fig. 2). Also in momentum space the distribution exhibits a hole as shown in Fig. 3. This is the consequence of the phase in Eq. (23), which gives a vanishing value to the integral of Eq. (11) at $p_\perp = 0$ where $p_\perp$ is the radial component of the momentum vector $\vec{p}$ .The size of the hole is of the order of $\hbar/R$, and consequently comparable to the total size of the condensate in momentum space (see Fig. 3).

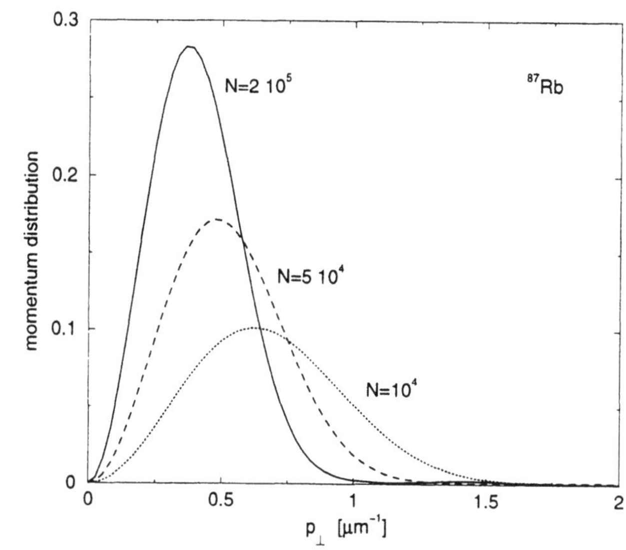

Figure 3. Momentum distribution of trapped Bose condensates with different number of atoms as a function of the variable $p_\perp$, integrated along p_z. The ^{87}Rb atoms are trapped in an isotropic harmonic potential characterised by the same parameters as in Fig. 2.

This can be easily seen calculating the momentum distribution in the Thomas-Fermi limit. In this limit the main effect of the vortex on the momentum distribution arises from the phase $e^{i\varphi}$, and one can safely use for ψ_0 the Thomas-Fermi expression $\sqrt{(\mu - V_{ext}(\vec{r}))/g}$, holding in the absence of the vortex. The result for $n(\vec{p})$ can be written in the form

$$n_{TF}(\vec{p}) = N\frac{R_\perp^3}{\hbar}\frac{15}{16\lambda}\frac{1}{\tilde{p}_z^2}\left|\int_0^1 dr_\perp r_\perp^2 J_1\left(\tilde{p}_\perp\sqrt{1-r_\perp^2}\right)J_1\left(|\tilde{p}_z|r_\perp\right)\right|^2 \tag{26}$$

where $\vec{\tilde{p}}$ is the scaled momentum vector $\tilde{p} \equiv (p_x, p_y, p_z/\lambda)R_\perp/\hbar$, already introduced in Section 2. Notice that in the Thomas-Fermi limit the effect of the vortex is factorized through a dimensionless integral.

In Fig. 6 we report the dynamic structure factor calculated in the IA (see Eq. (4)) with and without the vortex (the corresponding density profile and momentum distribution are shown in Figs. 4, 5). The calculation was carried out for a gas of $N = 10^4$ ^{87}Rb atoms trapped in a disk type geometry ($\lambda = \sqrt{8}$). For this low density sample the IA is very accurate. The double peak structure in $S(\vec{q}, E)$ reflects the occurrence of a peculiar Doppler effect. In fact the vortex generates a velocity field in the condensate with significant components both parallel and antiparallel to the momentum transfer $\vec{q}$.

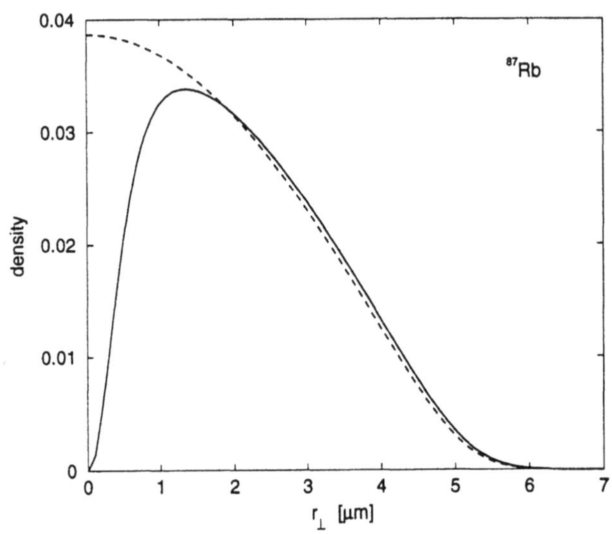

Figure 4. Density profile of a trapped Bose condensate as a function of the radial variable $r_\perp$, integrated along z, in the absence (dashed line) and in the presence (full line) of a quantized vortex. These profiles correspond, in the absence of the vortex, to a central density $n(0) \cong 10^{14}$ cm^{-3}. The trap parameters are $\omega_z = 2\pi\, 220$ Hz, and $\lambda = \sqrt{8}$.

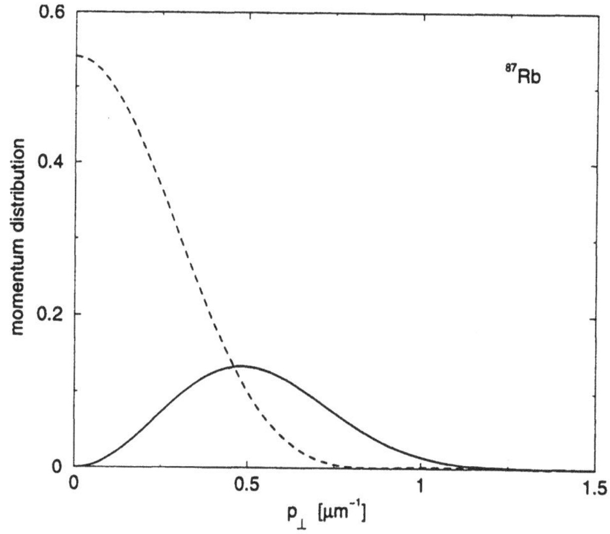

Figure 5. Momentum distribution of a trapped Bose condensate as a function of the variable $p_\perp$, integrated along p_z, in the absence (dashed line) and in the presence (full line) of a quantized vortex. Number of atoms and trap parameters are the same as in Fig. 4.

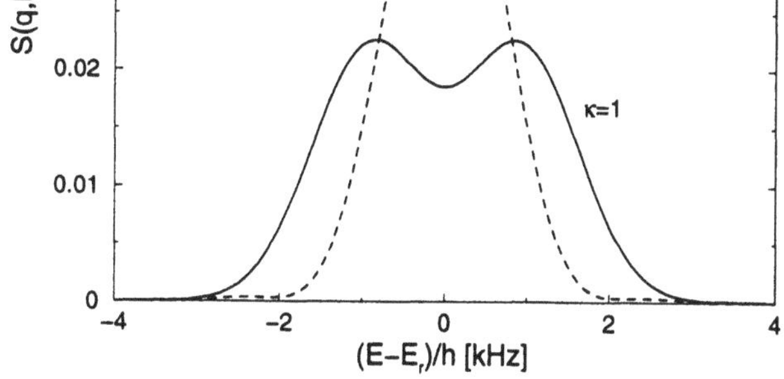

Figure 6. Dynamic structure factor of a trapped Bose condensate at $T = 0$ in the presence (solid line) and in the absence (dashed line) of the vortex, calculated using the impulse approximation (4). The momentum transfer is taken along the x-axis. The trap parameters and the central density are the same as in Fig. 4. $S(q,E)$ is given in units of $\hbar\omega_\perp/N$.

The eikonal expansion described in the previous section can be easily extended to explore the effect of the mean field interactions on the dynamic structure factor in the presence of a vortex. This has been done using the Thomas-Fermi ansatz for the order parameter, and the numerical results are shown in Figs. 7-8. The main effect of the mean field interaction is a deformation of the lineshape of $S(\vec{q},E)$ near the hole: in particular the strengths of the two peaks are different, reflecting the same asymmetry of the curve which characterises the dynamic structure factor in the absence of the vortex.

5. INTERFERENCE EFFECTS IN MOMENTUM SPACE

Finally, let us discuss the dynamic structure factor in terms of the occurrence of interference phenomena in momentum space. Interference has been so far investigated in coordinate space by imaging two overlapping Bose Einstein condensates[18]. However, even if the two condensates do not overlap in space, they can interfere in momentum space[19].

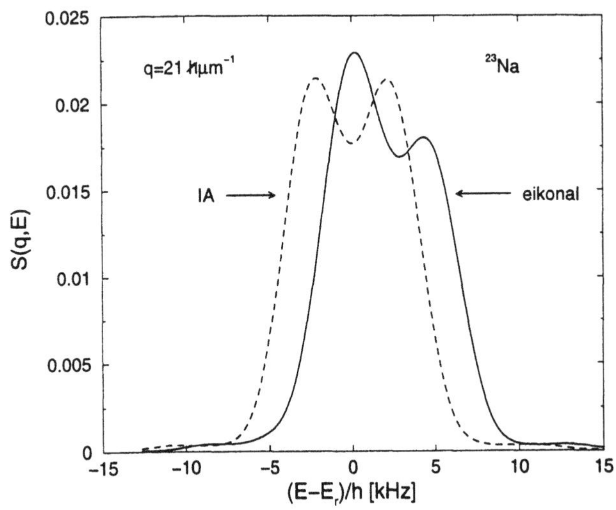

Figure 7. Dynamic structure factor of a trapped Bose condensate at $T=0$ in the presence of a vortex. The numerical predictions of the eikonal approximation (solid curve) and IA (dashed curve) are shown. The momentum transfer is taken along the x-axis. The trapping frequencies are the same as in Fig. 1. In the absence of the vortex this configuration would correspond to a central density $n(0) \cong 2.29 \cdot 10^{14}$ cm^{-3}. $S(q,E)$ is given in units of $\hbar\omega_\perp/N$.

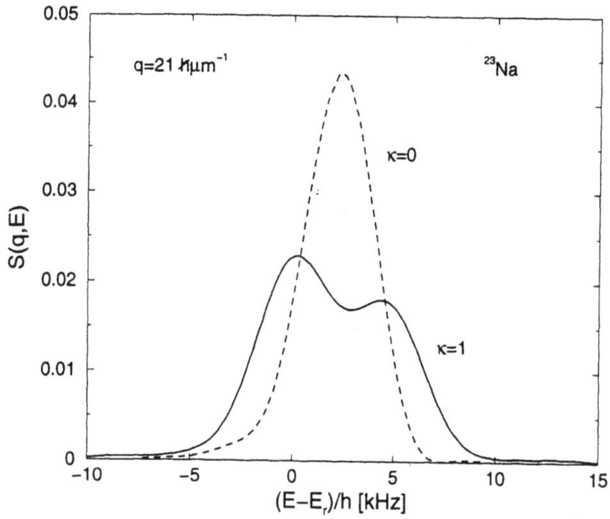

Figure 8. Dynamic structure factor of a trapped Bose condensate at $T=0$ in the presence (solid line) and in the absence (dashed line) of the vortex, calculated using the eikonal expansion (21). The momentum transfer is taken along the x-axis. The trap parameters are the same as in Fig. 1, and the central density in the absence of the vortex is $n(0)\cong 2.29\cdot 10^{14}$ cm^{-3}. $S(q,E)$ is given in units of $\hbar\omega_\perp/N$.

This opens the possibility of investigating interference while avoiding any interaction between the two condensates.

Consider, for example, a double well potential and let us assume, for simplicity, that the condensates in the two wells (condensates a and b respectively) have the same number of atoms (see Fig. 9).

If the distance d between the two wells is large enough to avoid overlapping, and if the potential acting on the condensates a and b can be obtained by a simple translation, then their wave functions can be written as:

$$\psi_{a,b}(\vec{r})=\psi_0\left(\vec{r}\pm\frac{\vec{d}}{2}\right) \tag{27}$$

where ψ_0 is the solution of the Gross Pitaevskii equation for each condensate. The Fourier transforms of Eq. (27) hence read:

$$\Phi_{a,b}(\vec{p})=e^{\pm ip_x d/2\hbar}\Phi_0(\vec{p}) \tag{28}$$

having taken the displacement between the two wells along the x axis.

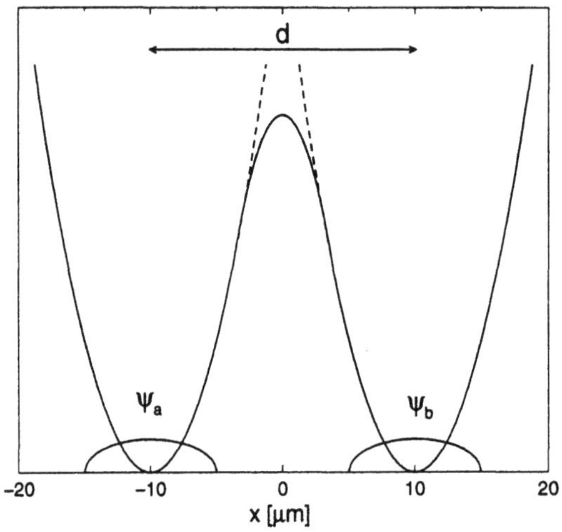

Figure 9. Typical configuration to explore interference of two Bose condensates in momentum space: two harmonic wells are separated by a distance $d = 20\,\mu m$; each well confines a condensate of sodium atoms with radial size $R_\perp = 5\,\mu m$ and central density. $n(0) \cong 0.7 \cdot 10^{14}$ cm^{-3}.

Under the above conditions any linear combination

$$\psi_C(\vec{r}) = \psi_a(\vec{r}) + e^{i\varphi}\psi_b(\vec{r}) \tag{29}$$

of the wavefunctions (27) corresponds to a solution of the Gross Pitaevskii equation. These combinations represent coherent configurations which exhibit interference patterns in the momentum distribution:

$$n(\vec{p}) = 2n_0(\vec{p})\left[1 + \cos\left(\frac{p_x d}{\hbar} + \varphi\right)\right] \tag{30}$$

where $n_0(\vec{p}) = |\Phi_0(\vec{p})|^2$. These patterns have interesting consequences on the shape of the dynamic structure factor which, in the IA, takes the form

$$S(\vec{q}, E) = 2S_0(\vec{q}, E)\left[1 + \cos\left(\frac{Yd}{\hbar} + \varphi\right)\right] \tag{31}$$

where Y is the given by

$$Y = \frac{m}{q}(E - E_r) \tag{32}$$

with E_r defined in Eq. (10). The dynamic structure factor (Eq. (31)) exhibits fringes with frequency period

$$\Delta\nu = \frac{\Delta E}{h} = \frac{q}{md} \tag{33}$$

In Fig. 10 we show a typical result for $S(\vec{q},E)$ corresponding to a distance between the two condensates four times larger than their radial width. The position of the fringes depends crucially on the value of the relative phase between the two condensates.

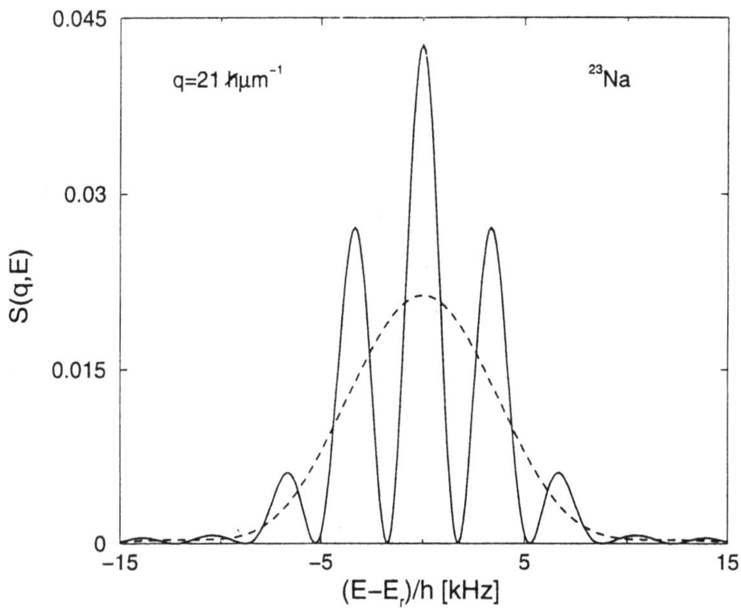

Figure 10. Dynamic structure factor relative to the configuration reported in Fig. 9 without (dashed line) and with (solid line) coherence between the two condensates (Eq. (31)). The momentum transfer q is taken along the x-axis. The relative phase between the two condensates is chosen equal to zero. The trap parameters are the same as in Fig. 1. $S(q,E)$ is given in units of $\hbar\omega_\perp/N$.

ACKNOWLEDGMENTS

We are very grateful to W. Ketterle and A.P. Chikkatur for many fruitful discussions. D.M. Stamper-Kurn acknowledges support of Millikan Prize Postdoctoral Fellowship. This work has been supported by the Istituto Nazionale per la Fisica della Materia (INFM) through the Advanced Research Project on BEC, and by Ministero dell'Università e della Ricerca Scientifica e Tecnologica (MURST).

REFERENCES

1. Stenger, J., Inouye, S., Chikkatur, A.P., Stamper-Kurn, D.M., Pritchard, D., and Ketterle, W., 1999, Bragg Spectroscopy of a Bose-Einstein Condensate, *Phys. Rev. Lett.* **82**: 4569.
2. Dalfovo, F., Giorgini, S., Pitaevskii , L., and Stringari, S., 1999,Theory of Bose-Einstein Condensation in trapped gases, *Rev. Mod. Phys.* **71**: 463.
3. Stamper-Kurn, D.M., Chikkatur, A.P., Görlitz, A., Inouye, S., Gupta, S., Pritchard, D.E., and Ketterle, W., 1999, Excitation of Phonons in a Bose-Einstein Condensate by Light Scattering, *Phys. Rev. Lett.* **83**: 2876.
4. Hohenberg, P.C., and Platzman, P.M., 1966, High-Energy Neutron Scattering from liquid He4, *Phys. Rev.* **152**: 198.
5. Sokol, P., 1995, Bose-Einstein Condensation in Liquid Helium. *In Bose-Einstein Condensation* (A. Griffin, D.W. Snooke, and S. Stringari, eds.), Cambridge University Press, Cambridge, pp. 51-85.
6. Baym, G., and Pethick, C., 1996, Ground-State properties of Magnetically Trapped Bose-Condensed Rubidium gas, *Phys. Rev. Lett.* **76**: 6.
7. Dalfovo, F., Pitaevskii, L., and Stringari, S., 1996, Bosons in a Magnetic Trap: The Condensate Wave Function, *Physica Scripta* **T66**: 234.
8. Zambelli, F., Pitaevskii, L., Stamper-Kurn, D.M., and Stringari, S., e-print cond-mat/9912089, accepted for publication in *Phys. Rev. A*.
9. Lundh, E., Pethick, C., and Smith, H., 1997, Zero-Temperature properties of a trapped Bose-condensed gas: Beyond the Thomas-Fermi approximation, *Phys. Rev. A* **55**: 2126.
10. Matthews, M.R., Anderson, B.P., Haljan, P.C., Hall, D.H., Wieman, C.E., and Cornell, E.A., 1999, Vortices in a Bose-Einstein Condensate, *Phys. Rev. Lett.* **83**: 2498-2501.
11. Madison, K.W., Chevy, F., Wohlleben, W., and Dalibard, J., 2000, Vortex Formation in a Stirred Bose-Einstein Condensate, *Phys. Rev. Lett.* **84**: 806.
12. Lundh, E., Pethick, C.J., and Smith, H., 1998, Vortices in Bose-Einstein-condensed atomic clouds, *Phys. Rev. A* **58**: 4816; Dalfovo, F., and Modugno, M., 2000, Free expansion of Bose-Einstein condensates with quantized vortices, *Phys. Rev. A* **61**: 023605.
13. Zambelli, F., and Stringari, S., 1998, Quantized vortices and Collective Oscillations of a Trapped Bose-Einstein Condensate, *Phys. Rev. Lett.* **81**: 1754; Svidzinsky, A.A., and Fetter, A., 1998, Normal modes of a vortex in a trapped Bose-Einstein condensate, *Phys. Rev. A* **58**: 3168.
14. Bolda, E.L., and Walls, D.F., 1998, Detection of Vorticity in Bose-Einstein Condensed Gases by Matter-Wave Interference, Phys. *Rev. Lett.* **81**: 5477.
15. Pitaevskii, L.P., 1961, Vortex lines in an imperfect Bose Gas, *Sov. Phys.* JETP **13**: 451.

16. Gross, E.P., 1961, Structure of a Quantized Vortex in Boson systems, *Nuovo Cimento* **20**: 454.
17. Dalfovo, F., and Stringari, S., 1996, Bosons in anisotropic traps: Ground state and vortices, *Phys. Rev. A* **53**: 2477.
18. Andrews, M.R., Townsend, C.G., Miesner, H.-J., Durfee, D.S., Kurn, D.M., and, Ketterle, W., 1997, Observation of Interference Between Two Bose Condensates, *Science* **275**: 637.
19. L.P. Pitaevskii and S. Stringari, 1999, Interference of Bose-Einstein Condensates in Momentum Space, *Phys. Rev. Lett.* **83**: 4237.

Atom Optics With Bose-Einstein Condensates

S. BURGER, K. BONGS, K. SENGSTOCK, AND W. ERTMER
Institut für Quantenoptik, Universität Hannover, 30167 Hannover, Germany

1. INTRODUCTION

The purpose of this Chapter is to give an insight into coherent atom optics by discussing recent experiments using Bose-Einstein condensates in atom optical experiments.

Since the first experimental realisation of Bose-Einstein condensation (BEC) in weakly interacting atomic systems[1-4] many fundamental experiments with BECs have been performed[5,6]. One of the most interesting future prospects for Bose-Einstein condensates is their application as a source of coherent matter waves[7-10], e.g., in atom optics and atom interferometry. This offers a significant advance similar to the introduction of lasers in light optics.

The application of coherent matter waves in phase sensitive experiments, like interferometers, demands for the understanding of their evolution when being manipulated by atom optical elements like mirrors and beamsplitters. The dynamics of coherent matter waves during and after the interaction with these elements is in comparison to single-atom optics much more complex and may easily lead to 'non-linear atom optics' , e.g., to four wave mixing[11].

In the first part of this Chapter we will focus on the design of typical atom optical elements. One of the key elements are atom mirrors[12] which we have applied to Bose-Einstein condensates[13]. Another very important element for future applications are atom waveguides[14]. They may be used to confine the motion of Bose-Einstein condensates or the output of an atom laser to one dimension. In addition, they allow for new interferometric geometries by holding the condensate against gravity. Together with atom

Bose-Einstein Condensates and Atom Lasers
Edited by Martellucci *et al.*, Kluwer Academic/Plenum Publishers, 2000

mirrors they can be used as resonators for atomic de Broglie waves. In combination with partially transmitting mirrors this would be the reservoir of a directed atom laser. Finally, waveguides could even be used to explore superfluid behaviour of BECs. We will discuss these aspects in the fourth part.

Atom mirrors and waveguides are 'classical' elements in the sense that they are represented by a conservative potential directly influencing the density distribution of the atomic ensemble. Recently, holographic methods which only have an effect on the phase of an atom or the collective phase of a BEC have been developed[15]. Imprinting a spatially varying phase on Bose-Einstein condensates is a powerful tool of atom holography. This technique and its application in converting ground state Bose-Einstein condensates to excited states like vortex or soliton states will be discussed in the last part of the Chapter.

2. DIPOLE POTENTIALS

Due to the interaction between an atom and a far detuned light field (described by the ac Stark shift), atoms are either attracted or repelled from regions of high intensity depending on the sign of the detuning of the light. The coherent manipulation of atomic motion by these dipole forces is widely used in atom optics[16].

In first order approximation, the potential energy corresponding to the dipole force acting on an atom in a far detuned light field is given by

$$U(r) = \frac{\hbar\Omega^2(r)}{4\delta_c} \tag{1}$$

Here, $\Omega(r)$ is the Rabi frequency , and $\delta_c = \omega_l - \omega_0$ is the detuning of the light field, where ω_l is the laser frequency, and ω_0 is the frequency of the relevant atomic transition.

For a large laser detuning, spontaneous processes can often be neglected. This allows for a coherent manipulation of the atomic motion. For a dipole potential, then acting as a *coherent* atom optical element, the appropriate laser frequency has to be chosen. As can be seen from Eq. (1), blue detuned light ($\omega_l >> \omega_0$) repels the atoms whereas red detuned light ($\omega_l << \omega_0$) attracts the atoms to regions of high laser intensity. Thus it is possible to trap or guide atoms in appropriate geometries of dipole potentials. Compared to red detuned light potentials, blue detuned light potentials have the advantage of confining the atoms in a region of low laser intensity, thereby light scattering which leads to decoherence is suppressed even more.

In the subsequent sections we describe experiments in which optical dipole potentials are created by blue detuned, far off-resonant laser light.

3. MIRRORS FOR BEC

Here we discuss experiments demonstrating the bouncing of atomic BECs off a mirror formed by a repulsive dipole potential. Condensates released from a magnetic trap fall under the influence of gravity and interact with a blue-detuned far-off-resonant 'sheet' of light.

In the experimental setup[17], condensates typically containing 10^5 ^{87}Rb atoms in the (F=2, m_F=+2)-state are produced every 20s. Less than 10 % of the atoms of the cloud are in the non-condensate fraction, this corresponds to a temperature range of $\cong$ 200 nK. The fundamental frequencies of the magnetic trap (a 'cloverleaf' trap[18]) are $\omega_{\|} = 2\pi \times 17$Hz and $\omega_{\perp} = 2\pi \times 340$Hz along the axial and radial directions, respectively. Therefore, the condensates are pencil-shaped with the long axis (length $\approx$ 100μm) oriented horizontally. The trap can be switched off within 200μs, and - after a variable time delay - the density distribution of the atomic sample can be detected using absorption imaging, dark field imaging, or phase-contrast imaging. The 2D-imageplane contains the weak trap axis as well as the strong trap axis along the direction of gravity.

The atom mirror is created by a Gaussian laser beam from a frequency doubled Nd:Vanadate laser (λ = 532nm), focused to a waist of about 10 μm and spatially modulated with an acousto-optic deflector in the horizontal plane. The modulation period of typically 10 μs is much shorter than the time the atoms are spending inside the dipole potential. For the atoms this results in a time-averaged static dipole potential. In contrast to magnetic mirrors or evanescent wave mirrors, changing the modulation waveform gives the flexibility to externally define the intensity profile. For the experiments presented here, a flat light sheet oriented nearly perpendicular to the direction of gravity is used. The time-averaged beam profile has a spatial extent of $\cong$ 200μm in the horizontal direction. The interaction of the atoms with the light field is dominated by the 780nm and 795nm dipole transitions, leading to a repulsive potential barrier with a detuning large enough that spontaneous emission becomes negligible. Ultra cold atomic clouds dropped from a height of up to 300 μm can thus totally be reflected.

In the experiments described here, the clouds were reflected off the mirror up to three times before they laterally move out of the field of view due to a slight slope in the orientation of the light sheet. The behaviour of the reflected atomic samples can be modified by changing the drop height, as well as the power and waist of the light sheet. This allows for the observation of two regimes in the wave packet dynamics: the dispersive reflection off weak dipole potentials ('soft' mirror) and the nearly non-dispersive reflection off strong dipole potentials ('hard' mirror).

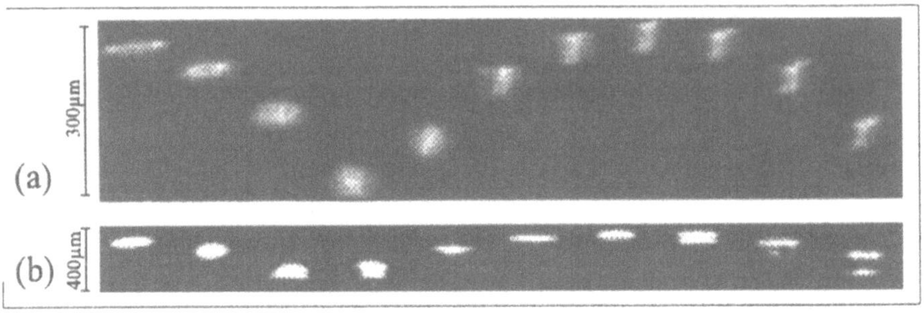

Figure 1. (a) Series of dark-field images for BECs of ^{87}Rb bouncing off a light sheet 270 μm below the magnetic trap. Each image was taken with a new condensate and with an additional time delay of 2ms. The density of the condensate during the first few ms of expansion causes a phaseshift in the detection light of more than 2π, which explains the dark area in the middle of the first two images. (b) A thermal cloud bouncing off a light sheet placed 230μm below the magnetic trap splits into two parts.

Fig. 1(a) shows a time-of-flight series of Bose-Einstein condensates bouncing off a soft mirror. Each frame is recorded with a different condensate, created under identical experimental conditions. The light sheet is positioned 270μm below the magnetic trap and shows up as the sharp lower edge in the fourth frame. The high kinetic energy accumulated before hitting the atom mirror causes the atomic cloud to penetrate deep into the dipole potential before being reflected.

The corresponding classical turning point is situated close to the maximum of the Gaussian intensity profile, in a region with a weak gradient of the repulsive potential. Further increasing the drop height or reducing the laser power results in a partial transmission through the mirror.

As the condensate reapproaches the initial altitude, it develops self interference structures (frames 6 to 11 in Fig. 1(a) and Fig. 2(a)) which do not occur for temperatures above the critical temperature for BEC, T_c. The self-interference structures prove the persistence of matter-wave coherence for BECs reflected off the dipole potential atom mirror[13].

In another set of measurements, a nearly non-dispersive mirror for Bose-Einstein condensates was created by placing an intense light sheet closer to the magnetic trap (155μm drop height). This resembles reflection off a much steeper potential step. Close to the upper turning point, the atom cloud is refocused to a narrow distribution along the direction of gravity, and develops into a double-peak structure shortly after the upper turning point. No interference structures such as those presented in Fig. 1(a) are observed here.

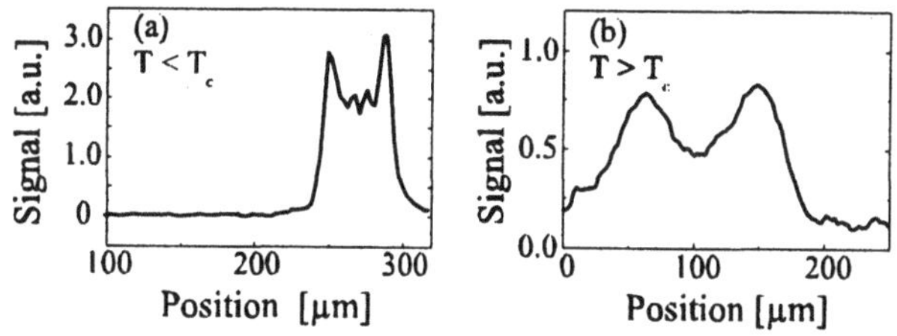

Figure 2. Cross sections of images of a bouncing BEC (a) and a bouncing thermal cloud (b). Graph (a) corresponds frame 9 of Fig. 1(a). Graph (b) is a cross section of the last frame in Fig. 1(b).

In order to compare the bouncing of condensates to single atom optical effects, bouncing experiments with atom clouds cooled to a temperature just above the critical temperature, T_c, have been performed. Surprisingly, thermal clouds of ultra cold atoms also reveal splitting after reflection off a hard atom mirror. Typically, this splitting is more pronounced than the double peaked structure for BECs (see Fig. 1(a) and 1(b)).

From simulations of the classical Liouville equation for the flow of density in phase space it can be seen that, indeed, classical dynamics leads to splitting of the cloud right after passing through the upper turning point.

To understand how the interference structures in bouncing BECs arise, numerical simulations of the Gross-Pitaevskii equation have been performed to mimic the experimental behaviour of the condensate. The trapped condensate has a radial width of $\Delta z \sim$ few µm. The velocity spread is thus of order $\Delta v \cong \hbar / M\Delta z \cong 10^{-1}$ µm/ms, with atom mass M. A classical cloud with such parameters does not show any splitting or specific structure. Similarly, the wave packet of a single atom evolving according to the linear Schrödinger equation does not split. In the condensate, however, the potential energy of the atom-atom interaction is transferred into kinetic energy within few ms of ballistic expansion, allowing thus for the splitting.

As an important result from the simulations, splitting and self-interference structures of the condensate are very sensitive to the value of the initial mean field energy - therefore, its appearance allows to estimate the number of atoms in the condensate fraction. For the value of the nonlinear coupling given above, splitting is observed only for atom numbers $N \geq 5 \times 10^4$. This is in good agreement with the experimental results, as the splitting vanished for smaller condensates.

In agreement with the experimental results described above, the numerical results also show additional structures before and after the upper turning point when a soft mirror is placed 270 μm below the trap (Fig. 3). Evidently, when the initial gravitational energy is comparable to the mirror height, the softness of the mirror causes velocity dependent dispersion for the reflected matter waves, leading to interference and density modulation.

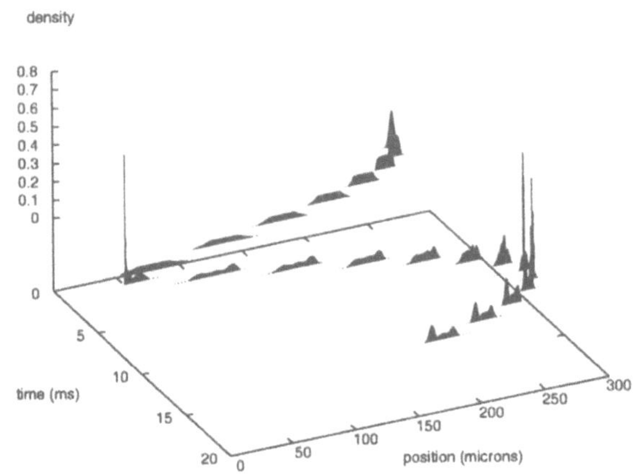

Figure 3. Numerical results showing the time evolution of the condensate corresponding to the parameters of Fig. 1. The 'soft' mirror is situated 270 microns below the magnetic trap, the number of atoms is N= 8 × 10^4.

The interference structure can be explained by the fact that the splitted parts of the wave packet overlap in the central region, where the particles have positions and velocities very close to the mean ($\Delta v \sim 0.5$mm/s, which corresponds to a fringe separation of $h/M\Delta v \sim 10\mu$m, as observed). Each of the two parts has a different spatial phase dependence, and interference is observed. This effect is enhanced for soft mirrors. Indeed, the analysis indicates that in the quantum regime splitting cannot be regarded as a purely classical effect.

The numerical results agree well with the experimental observations and clearly explain the appearance of the different splitting behaviour for non-condensed samples and BECs as well as the self-interference structure for bouncing off a 'soft' mirror.

The observation of splitting and of interference can be used to characterise and determine mirror properties such as roughness and steepness, and coherence properties of the condensate.

In addition to creating an atom mirror with reflectivity close to unity, partially reflecting mirrors and a phase shifter can be created by reducing the intensity of the light sheet. Then, the optical potential delays the atoms but does not cause reflection (see Fig. 4).

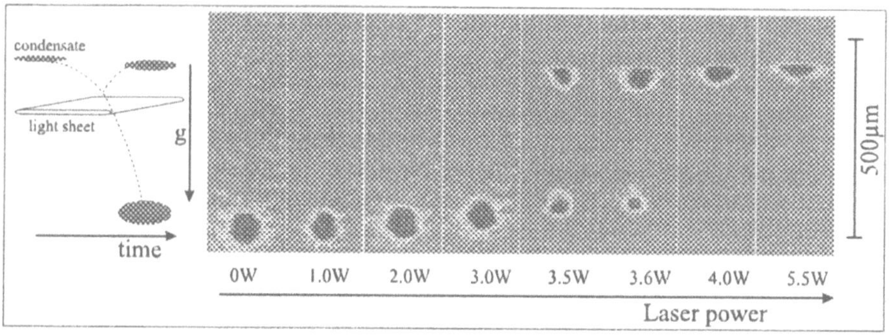

Figure 4. Absorption images of BECs after 10 ms time-of-flight and after interaction with the dipole potential of a light sheet placed about 200μm below the magnetic trap for different laser powers of the light sheet. For different potential heights the light sheet acts as phase shifter, delaying the atoms (0W,...,3W), as dispersive beam splitter (3.5W, 3.6W), or as atom mirror (4W, 5.5W).

These elements can in the future be applied to develop atom interferometers for Bose-Einstein condensates. The observed coherent splitting itself may also be applied to realise an atom interferometer in a pulsed scheme, e.g., by the application of additional light fields acting as mirrors and phase shifters for the individual partial waves. Furthermore, mirrors based on optical potentials can serve as detection scheme for matter wave coherence, i.e., the onset of BEC or the output properties of an atom laser. They may even be used to systematically characterise the coherence properties of these sources, or of coherent matter waves being manipulated by other techniques.

4. LOADING BECs INTO A DE BROGLIE WAVEGUIDE

Guiding of atoms has been demonstrated in the last years in experiments using hollow fiber-guides[19-22] or freely propagating light beams[23,24]. In these experiments, however, the temperature ranges have been orders of magnitude higher than the temperatures achievable in a BEC experiment. With an ultracold atom cloud slowly moving in a waveguide the population of the lowest transverse modes of the wave-guide becomes feasible.

This section deals with the loading of a Bose-Einstein condensate to a linear waveguide formed by a far off resonant, hollow laser beam. Due to the repulsive dipole force of the beam the atomic ensemble could be confined to the low intensity region with a loading rate of up to 100%. Trapping times are so far mainly limited by the longitudinal movement inside the waveguide. Heating effects within the waveguide have been observed.

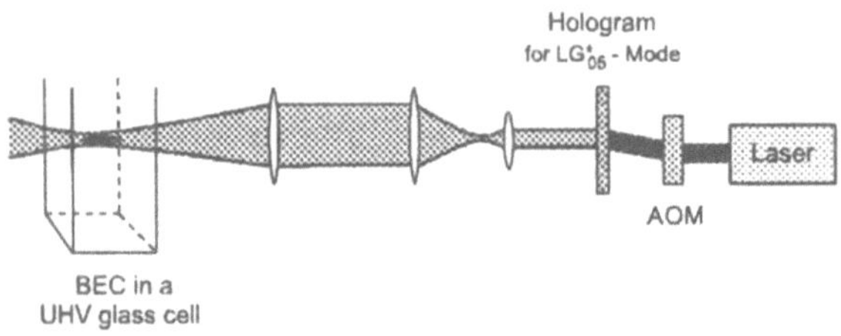

Figure 5. Experimental setup: the light from a frequency doubled Nd:Vanadate laser passes an acousto-optical modulator, is converted to a Laguerre-Gaussian beam, and is directed to the magnetically trapped BEC inside a UHV glass cell.

We use a Laguerre-Gaussian beam of fifth order, LG_{05} ('donut-mode' of 5th order), to hold and guide the atoms loaded from Bose-Einstein condensates. The donut beam is generated from a TEM_{00} mode by a blazed mode converting hologram[25] . The intensity distribution in the donut mode, $I(r)$, is given by

$$I(r) = \frac{8\mathrm{P}r^{10}}{15\pi r_0^{12}} e^{-2\frac{r^2}{r_0^2}} \tag{2}$$

with the laser power, P, and the radial parameter of the LG_{05} beam, r_0.

After transmission through the hologram, the laser beam is expanded by a telescope, and finally focused to a size of $r_0 = 12\mu m$ (see Fig. 5).

Using Eq. (1) with a laser wavelength of $\lambda = 532$nm, and a power of P = 1.2W, the resulting dipole potential at the focal plane has a maximum value of $U_{max} \approx 8 \times 10^{-28}$ J, corresponding to a temperature of $T \approx 60\mu K$.

After a BEC has formed, we instantaneously switch from the magnetic trap to the waveguide potential. The donut beam is adjusted such that the Bose-Einstein condensate is trapped in two dimensions in the dark inner region of the light field. The beam axis is aligned with the long axis of the pencil-shaped condensate, slightly tilted which allows for a longitudinal movement of the atoms inside the waveguide due to gravity.

The evolution of the BEC inside the donut mode is governed by gravity, expansion due to the initial mean field energy and by heating effects. Freely propagating dipole potential waveguides allow to easily monitor the evolution of BECs inside the donut by taking absorption and dark field images of the atomic cloud after a variable evolution time.

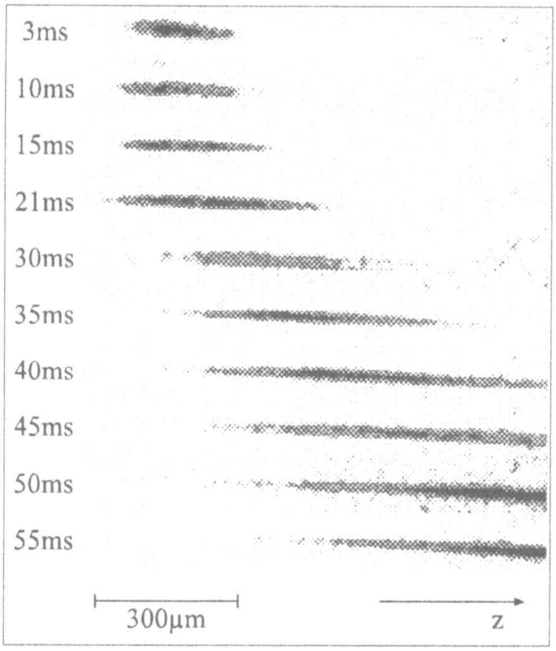

Figure 6. Atoms from a BEC loaded to a donut mode waveguide. Broadening of the cloud due to heating effects and to expansion of the BEC is observed. With ultra-cold atomic clouds it has nicely become possible to monitor the atomic dynamics within the waveguide.

Fig. 6 shows the evolution of a BEC loaded into the waveguide. The loading efficiency η, i.e., the number of atoms trapped in the waveguide over the total number of atoms, depends on the potential height as well as on the spatial overlap of the BEC in the magnetic trap and the waveguide. In this measurement, the overlap between condensate and waveguide resulted in a loading efficiency to the waveguide of $\eta = (70 \pm 10)\%$. The waveguide is tilted by an angle α relative to the horizontal direction, such that the atoms captured in the dipole potential are accelerated downwards with an acceleration $a = g \sin \alpha$, where g is the gravitational acceleration. The centre of the atomic distributions follows a parabola from which - in comparison to free fall - the slope of the donut mode is determined to be (4.2 ± 0.3)degrees. In this measurement the longitudinal expansion of the atomic cloud is dominated by heating due to beam pointing and intensity fluctuations of the donut laser beam. For a stabilised beam it should be possible to transfer the BEC to the lowest transverse vibrational level of the dipole potential, which will enable coherent transport in future experiments.

A main advantage of dipole traps over magnetic traps is that they allow to trap atoms independent of their magnetic substate. Using a dipole trap as a reservoir for an atom laser then allows, e.g., to generate an atom laser beam of any magnetic substate.

In order to form an analogue to an optical laser cavity for an atom laser it is possible to introduce two additional dipole potential mirrors closing the waveguide (see Fig. 7). As has been shown in the last section, the reflectivity of these mirrors is velocity-dependent. Therefore by applying Bragg-pulses transferring an appropriate momentum to the condensate it should be possible to couple out parts of the condensate. A main advantage of this scheme would be the directed output of coherent matter waves into the donut waveguide.

Other interesting aspects for future work are guiding in different potential shapes using Laguerre-Gaussian modes of different orders as waveguides, one dimensional expansion due to the mean field energy of the condensate in a tightly confining LG_{01} mode, or the use of wave-guides in atom interferometers.

5. ATOM HOLOGRAPHY: ENGINEERING THE PHASE OF A BEC

In the first part of this section, the phase imprinting method in connection to the creation of vortices is discussed. The second part describes experimental results on the creation of dark solitons in BECs with the phase-imprinting method.

5.1 Generation Of Vortices In BECs By The Phase Imprinting Method

One of the challenges of the physics of trapped Bose-Einstein condensates concerns the demonstration of their superfluid behaviour. Superfluidity is inevitably related to the existence of vortices and persistent currents in BEC.

Several methods were proposed to generate vortices in non-rotating traps: stirring the condensate using a blue detuned laser beam[26,27], or several laser beams[28], adiabatic passage[29] or Raman transitions[30] in bi-condensate systems. Such vortices are typically not stable, and can exhibit dynamical or energetic instability. In the first case vortices decay rapidly, in the second the vortices are stable within the framework of the mean field theory, and their corresponding decay requires to take into account interactions of the BEC with the thermal cloud. In the latter case the vortex dynamics is expected to be sufficiently slow, and thus experimentally accessible[31].

In rotating traps vortices appear in a natural way as thermodynamic ground states with quantized angular momentum[32,33]. Stability and other properties of vortices in rotating traps have been thoroughly discussed in Refs. [34-36].

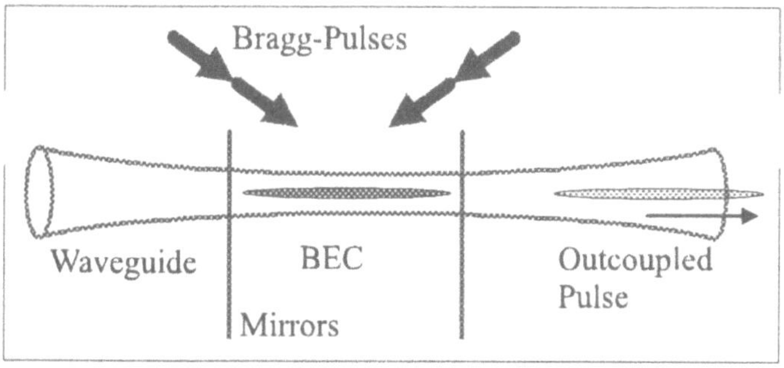

Figure 7. Experimental set-up for a 'donut-guided' atom laser.

Recently, vortices were successfully generated in a rotating trap by Madison *et al.*[52]. Vortices in a two component BEC have been observed by Matthews *et al.*[37]. Raman *et al.*[38] measured a critical onset of the heating effects of perturbations of Bose-Einstein condensates which is also closely connected to superfluidity and the existence of vortices.

Here, we refer to a procedure of vortex generation using 'phase imprinting'[15]. This method consists of i) passing a far off resonant laser pulse through an absorption plate whose absorption coefficient depends on the rotation angle φ around the propagation axis and ii) creating the corresponding Stark shift potential inside a BEC by imaging this laser pulse onto the condensate which leads to a φ dependent phase shift in the condensate wave function. This method is very efficient and robust, and allows for engineering of a variety of excited states of BECs containing vortices. It is expected that in the ideal case the method allows to generate genuine vortices with integer angular momenta. The presence of imperfections should typically result in more complex vortex patterns. The dynamical generation of vortices differs from the case of rotating traps, in which a pure vortex state with angular momentum $L_z = 1$ (in units of $\hbar$) is selected in the process of reaching the equilibrium. For the phase imprinting method, generation of pure vortices requires a fine tuning of parameters which is hard to achieve in experiments. This method is suitable for creation of generic states with vorticity[39], i.e., states with several vortex lines, around which the circulation of velocity does not vanish[40].

It can be shown that a short pulse of light with a typical duration of the order of fractions of microseconds with properly modulated intensity profile creates vorticity in a Bose-Einstein condensate initially in its ground state[15]. If the incident light is detuned far from the atomic transition frequency its main effect on the atoms is to induce a Stark shift of the internal energy levels. As the intensity of light depends on the position, the Stark shift will also be position dependent as will be the phase of the condensate.

The main feature characterising a vortex is related to the particular behaviour of the phase of the wave-function at the vortex line: the phase 'winds up' around this line, i.e., it changes by an integer multiple m of 2π on a path surrounding the vortex, where m is the vortex charge. The light beam, before impinging on the atomic system, is shaped by an absorption plate with an absorption coefficient that varies linearly around the plate axis with rotation angle φ (see Fig. 1). As a result, V_l(r, t) depends on the distance ρ from the propagation axis, and the azimuthal angle φ.

In the ideal case this absorption plate causes a real jump of the potential at, say $\varphi = 0$. Due to a finite imaging resolution, in the real case the potential corresponds to

$$V_l(\rho,\varphi) = \begin{cases} \hbar I \varphi sin^2(\pi\rho/2L) & \text{for } \rho \le L \\ \hbar I \varphi & \text{for } \rho > L \end{cases} \tag{3}$$

during the short laser pulse. Here, I denotes the characteristic Stark shift, L is the characteristic length scale on which the absorption profile is smoothed.

One of the most important questions concerning the investigation of vortices is an efficient method for their detection. Experimentally, monitoring density profiles with the necessary resolution is difficult, since the vortex core is very small. The best way is to monitor the phase of the wave function in an interference measurement. Such interference measurements are routinely done in non-linear optics[41]. In the context of vortex detection in BEC, they were proposed by Bolda and Walls[39],similar methods were proposed by Tempere *et al.*[42]. If both condensates are in the ground state (no vortices), one expects interference fringes as those observed by Andrews *et al.*[43]. In the case of interference of one condensate in the ground state with the second one in the $m = 1$ vortex state, a fork-like dislocation in the interference pattern appears. The distance between the interference fringes is determined by the relative velocity of the condensates, which can be controlled experimentally. This is a very efficient and clear method of vorticity detection. It requires, however, the use of two independent condensates.

As has been suggested[15], the detection can be simplified by combining the interference method with the recently developed Bragg diffraction technique[44,45]. The idea is to transfer part of the atoms coherently to another momentum state using one or several stimulated two-photon Raman scattering processes. The procedure is as follows:
i) First a vortex or vorticity state is created in the trap.
ii) The trap is opened and the condensate expands.
iii) Bragg pulses are applied in order to transfer a small momentum to part of the wavefunction. The resulting wave function is the superposition of two vortex (or vorticity) states moving apart from each other, with a velocity that can be easily controlled by choice of the angle between the Bragg beams. In simulations using the split operator method in 2D, typical velocities were of the order of 1mm/s, which allow for efficient detection after 4-5 ms when the vortices are about 5μm apart.
iv) Detection consists of optical imaging that is accomplished within a few μs. The interference patterns have a characteristic length scale of a few μm.

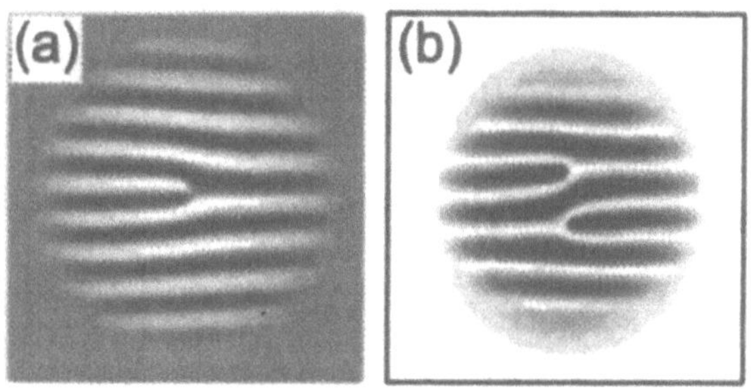

Figure 8. Interference pattern of the vortex state (5 ms after phase imprinting): (a) superimposed with a plane wave of $k = 1.8\ \mu m^{-1}$ moving vertically in the plane of the figure; (b) 5 ms after applying Bragg pulses transferring momentum to a part of the condensate (along the vertical axis in the plane of the figure). The ideal case of equal splitting of the whole condensate has been assumed. The incident pulse was focused at the trap center and had a sharp step in the intensity profile.

Fig. 8(a) shows a typical result of the numerical simulations obtained a few ms after the creation of a vortex state using the detection method [39] based on the interference of a condensate in the vortex state with a condensate in the plane wave state. The characteristic fork-like pattern reflects the fact that the phase winds up by 2π as one circulates around the vortex line. Fig. 8(b) shows results of the simulation of the interference pattern for the method of detection proposed by Dobtrek *et al.*[15]. A few milliseconds after applying the Bragg pulses a double fork structure can clearly be seen. The forks are oriented in opposite directions, because the condensates have the same helicity, but opposite velocities.

In future experiments it should be possible to study the creation and dynamics of vorticity using the phase imprinting and interference detection methods.

5.2 Generating Dark Solitons In Condensates Of ^{87}Rb

Beside vortices, an important class of excited states of BECs are the so called dark solitons or 'kink-wise' states. Their properties and ways to create them have been discussed extensively in the literature[15,29,46-48].

The macroscopic wavefunction of a dark soliton in a cylindrical harmonic trap has a nodal plane perpendicular to the symmetry axis of the confining potential. Thus, the corresponding density distribution shows a minimum around the nodal plane with a width of the order of the correlation length $l_0 = (4\pi a n_0)^{-1/2}$ of the condensate. A dark soliton state in a homogeneous BEC of density n_0 is described by the wavefunction[46]

$$\Psi_0(x) = \sqrt{n_0}\left[i\frac{v_k}{v_s} + \sqrt{1-\frac{v_k^2}{v_s^2}}\tanh\left(\frac{x-x_k}{l_0}\sqrt{1-\frac{v_k^2}{v_s^2}}\right)\right] \qquad (4)$$

with the position and velocity of the nodal plane x_k and v_k, and the speed of sound $v_s = \sqrt{4\pi a n_0}\,\hbar/m$, where a and m are the s-wave scattering length and the mass of the atomic species, respectively.

For $T = 0$ in 1D, dark solitons are stable. In this case, only solitons with zero velocity in the trap center do not move; otherwise they oscillate along the trap axis[49]. However, in 3D at finite T, dark solitons exhibit thermodynamic and dynamical instabilities. The interaction of the soliton with the thermal cloud causes dissipation which accelerates the soliton. Ultimately, it reaches the speed of sound and disappears[46].

The dynamical instability originates from the transfer of the (axial) soliton energy to the radial degrees of freedom and leads to the undulation of the DS-plane, and ultimately to the destruction of the soliton. This instability is essentially suppressed for solitons in cigar-shaped traps with a strong radial confinement[47].

Here we present experiments using the phase imprinting method to create soliton states[50] of BECs of ^{87}Rb. In these experiments, condensates containing $\approx 10^5$ atoms in the (F=2, m_F =+2)-state in a nearly 1D geometry are produced every 20s. The fundamental frequencies of our magnetic trap are in this case $\omega_{\|} = 2\pi \times 14$ Hz and $\omega_{\perp} = 2\pi \times 425$Hz along the axial and radial directions, respectively. Therefore, the condensates are pencil-shaped with the long axis oriented horizontally. Typical dimensions of the condensates along this axis are 100 – 130μm. The density distribution of the atomic sample is detected using absorption imaging.

The creation of dark solitons consists again of the imprinting of an appropriate phase and a successive evolution in the magnetic trap. The phase distribution is imprinted by temporarily exposing the half space ($x < 0$) to an additional homogenous potential (where the center of the condensate wavefunction is located at $x = 0$). This is done with the of a defocused laser beam (λ = 532nm) which is partly blocked with a razor blade (see Fig. 9).

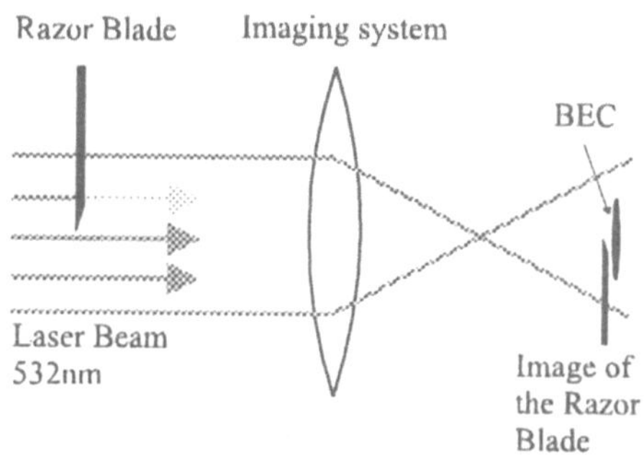

Figure 9. Setup for a phase-imprinting experiment.

The transition from dark to bright at the position of the BEC can be as sharp as the resolution of the optical system which images the razor blade to the condensate. In our experiment this resolution is < 3μm.

The time the dipole potential is applied, t_p, is short compared to the correlation time of the condensate, $t_c = \hbar/\mu$ with the chemical potential μ. This ensures that the effect of the light pulse is mainly a change of the topological phase of the BEC, and the change of the condensate density during the pulse due to the dipole force can be neglected.

The intensity of the applied laser field of $I \approx 50W/mm^2$ results in a dipole potential of $U \approx 5 \times 10^{-29}$ J. Thus a pulse of ten microseconds results in a phase shift $\Delta\Phi = U \times \tau / \hbar$ of the order of π.

After applying the dipole potential profile we let the atoms evolve in the magnetic trap for a variable time t_{ev}. In order to image the BEC it is then released from the trap and an absorption image is taken after a time-of-flight of 4ms.

As can be seen in Fig. 10(a), for a short evolution time t_{ev} the resulting density profile of the BEC shows a pronounced minimum at the centre of the cloud. As t_{ev} is increased this minimum moves along the cloud's axis and after a time of typically $t_{ev} \approx$ 1.5ms, a second density minimum can be monitored (Fig. 10(b)). This minimum thereafter moves in the opposite direction. Fig. 11 shows the evolution of these minima over a time of 8ms, after which they fade away.

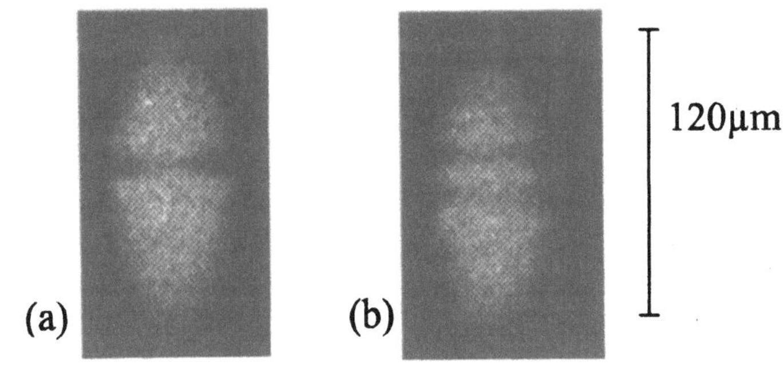

Figure 10. Absorption images of two BECs with kink-wise structures propagating in the direction of the long condensate axis. The evolution time in the magnetic trap was 0.2 ms in image (a) and 3 ms in image (b). The images were taken after 4 ms of time-of flight.

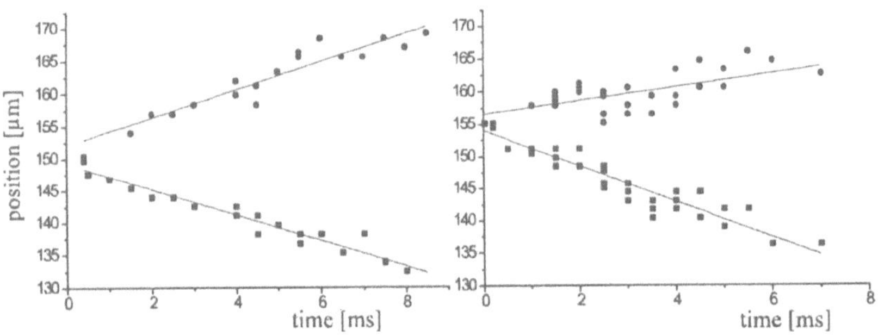

Figure 11. Evolution of the density minima in the magnetic trap. Position of the observed density minima (see Fig. 10) in dependence of the evolution time in the magnetic trap. The two sets are taken with different phase-imprinting times τ. The lines are linear fits to guide the eye; they correspond to constant velocities. The structures move with different velocities and velocities smaller than the speed of sound $v_s \approx 3.7$mm/s.

It is remarkable that the structures move with velocities which are much smaller than the speed of sound which is about $v_s \approx 3.7$mm/s for our experimental parameters.

The velocities depend on the imprinted intensity of the laser field, for properly chosen parameter sets (Fig. 11(b)) the two density minima apparently move with highly differing velocities. The comparison to numerical simulations of the Gross-Pitaevskii equation for our experimental conditions confirms this behaviour[50].

Dark solitons as demonstrated here are a new type of excitations of non-linear matter waves and are closely related to non-linear optics[51] and superfluidity. Further experiments can therefore lead to a better understanding of the connections between these fields and the physics of matter waves.

6. CONCLUSIONS

In this Chapter we have investigated various applications of Bose-Einstein condensates as sources for experiments in atom optics. The evolution of Bose-Einstein condensates falling under gravity and bouncing off a mirror formed by a far-detuned sheet of light has been studied. After reflection, the atomic density profile develops splitting and interference structures which depend on the drop height, on the strength of the light sheet, as well as on the initial mean field energy and size of the condensate. We compare experimental results with simulations of the Gross-Pitaevskii equation. A comparison with the behaviour of bouncing thermal clouds allows to identify quantum features specific for condensates.

Bose-Einstein condensates of ^{87}Rb have been loaded to a linear waveguide for atomic deBroglie waves. The waveguide is created by the optical dipole force of a far off-resonant, blue detuned Laguerre-Gaussian laser beam of high order. The atomic cloud can be transported inside this waveguide over long distances.

We have discussed the creation of vortices using a phase imprinting method. This method was further used to create dark soliton states in Bose-Einstein condensates.

ACKNOWLEDGMENTS

Part of the work presented here has been done in fruitful and stimulating cooperation with G. Birkl, S. Dettmer, L. Dobrek, M. Gajda, M. Kovacev, M. Lewenstein, K. Rzążewski, A. Sanpera, and G.V. Shlyapnikov.

This work is supported by SFB 407 of the Deutsche Forschungsgemeinschaft.

REFERENCES

1. Anderson M.J., Ensher J.R., Matthews M.R., Wieman C.E., and Cornell E.A., 1995, *Science*, **269**: 198.
2. Davis K.B., Mewes M.-O., Andrews M.R., van Druten N.J., Durfee D.S., Kurn D.M., and Ketterle W., 1995, *Phys. Rev. Lett.*, **75**: 3969.

3. Bradley C.C., Sackett C.A., Tollett J.J., and Hulet R.G., 1995, *Phys. Rev. Lett.*,**75**: 1687; Bradley C.C. et al., *Phys. Rev. Lett.*, 1997, **78**: 985.
4. Fried D.G., Killian T.C., Willmann L., Landhuis D., Moss S.C., Kleppner D., and Greytak T.J., 1998, *Phys. Rev. Lett.*, **81**: 3811.
5. *For an overview see*: Dalfovo F., Giorgini S., Pitaevskii L.P. and Stringari S., 1999, *Rev. Mod. Phys.*, **71**: 463.
6. *For an overview see*: articles by Ketterle W., et al. and by Cornell E., et al., July 1998, In *Proceedings of the International School of Physics* "Enrico Fermi" (M. Inguscio, S. Stringari, and C. Wieman, eds.), Varenna.
7. Mewes M.-O., Andrews M.R., Kurn D.M., Durfee D.S., Townsend C.G., and Ketterle W., 1997, *Phys. Rev. Lett.*, **78**: 582.
8. Anderson B.P., and Kasevich M.A., 1998, *Science*, **282**: 1686.
9. Hagley E. W., Deng L., Kozuma M., Wen J., Helmerson K., Rolston S. L., and Phillips W. D., 1999, *Science*, **283**: 1706.
10. Bloch I., Hänsch T.W., and Esslinger T., 1999, *Phys. Rev. Lett.*, **82**: 3008.
11. Deng L., Hagley E.W., En J.W., Trippenbach M., Band Y., Julienne P. S., Simsarian J. E., Helmerson K., Rolston S.L., and Phillips W. D., 1999, *Nature*, **398**: 218.
12. Kasevich M.A., Weiss D., and Chu S., 1990, *Opt. Lett.* **15**: 607; Aminoff C.G., Steane A. M., Bouyer P., Desbiolles P., Dalibard J., and Cohen-Tannoudji C., 1993, *Phys.Rev. Lett.* **71**: 3083; Roach T.M., Abele H., Boshier M.G., Grossman H.L., Zetie K.P., and Hinds E.A., 1995, *Phys. Rev. Lett.* **75**: 629; Szriftgiser P., Guèry-Odelin D., Arndt M., and Dalibard J., 1996, *Phys. Rev. Lett.* **77**: 4; Landragin A., Courtois J.-Y., Labeyrie G., Vansteenkiste N., Westbrook C. I., and Aspect A., 1996, *Phys. Rev. Lett.* **77**: 1464; Sidorov A., et al., 1996, *Quant. Semiclass. Opt.* **8**: 713.
13. Bongs K., Burger S., Birkl G., Sengstock K., Ertmer W., Rzążewski K., Sanpera A., and Lewenstein M., 1999, *Phys. Rev. Lett.*, **83** (18): 3577.
14. *For an overview see*: Dowling J.P. and Gea-Banacloche J., 1997, *Adv. At. Mol. Opt. Phys.*, **37**:1.
15. Dobrek L., Gajda M., Lewenstein M., Sengstock K., Birkl G., and Ertmer W., 1999, *Phys. Rev. A.*, **60** (5): R 3381.
16. *For an overview see:* Adams C.S., Sigel M., and Mlynek J., 1994, *Phys. Rep.*, **240**: 143.
17. Bongs K., Fühner C., Burger S., Birkl G., Sengstock K., and Ertmer W., to be published.
18. Mewes M.-O., Andrews M.R., van Druten N.J., Kurn D.M., Durfee D.S., and Ketterle W., 1996, *Phys. Rev. Lett.*, 77 (3): 416.
19. Renn M.J., Montgomery D., Vdovin O., Anderson D.Z., Wieman C.E., and Cornell E.A., 1995, *Phys. Rev. Lett.*, **75**: 3253.
20. Renn M.J. Donley E.A., Cornell E.A., Wieman C.E., and Anderson D.Z., 1996, *Phys.Rev. A.*, **53**: R648.
21. Ito H., Nakata T., Sakaki K., Ohtsu M., Lee K. I., and Jhe W., *Phys. Rev. Lett.*, 1996, **76**: 4500.
22. Wokurka G., Keupp J., Sengstock K., and Ertmer W., 1998, *Procs. EQEC,* QFG 5,235.
23. Schiffer M., Rauner M., Kuppens S., Zinner M., Sengstock K., and Ertmer W., 1998, *Appl. Phys. B*, **67**: 705.
24. Kuppens S., Rauner M., Schiffer M., Sengstock K., and Ertmer W., 1998, *Phys. Rev.A.*, **58** (4): 3068.
25. Kuppens S., Rauner M., Schiffer M., Wokurka G., Slawinski T., Zinner M., Sengstock K., and Ertmer W., 1997. In Burnett K., editor, OSA *TOPS*(7), pp. 102-107.
26. Jackson B., McCann J.F., and Adams C.S., 1998, *Phys. Rev. Lett.*, **80**: 3903.
27. Caradoc-Davies B.M., Ballagh R.J., and Burnett K., cond-mat/9902092.

28. Marzlin K.-P., and Zhang W., 1998, *Phys. Rev. A.*, **57**: 3801; *ibid.* 4761.
29. Dum R., Cirac J.I., Lewenstein M., and Zoller P., *Phys. Rev. Lett.*, 1998, **80**: 2972.
30. Marzlin K.-P., Zhang W., and Wright E.M., 1997, *Phys. Rev. Lett.*, **79**: 4728.
31. Rokhsar D.S., 1997, *Phys. Rev. Lett.* **79**: 2164 and cond-mat/9709212; Svidzinsky A.A., and Fetter A.L., 1998, *Phys. Rev.* A **58**: 3168; Pu H., Law C.K., Eberly J.H., and Bigelow N.P., 1999, *Phys. Rev.* A **59**: 1533; Isoschima T. and Machida K., 1999, *Phys. Rev.* A **59**: 2203; Fedichev P.O. and Shlyapnikov G. V., cond-mat/9902204.
32. Marshall R.J., New G.H.C., Burnett K., and Choi S., 1995, *Phys. Rev.* A., **59**: 2085.
33. Drummond P.D. and Cornay J.F., cond-mat/9806315.
34. Dalfovo F. and Stringari S., 1996, *Phys. Rev.* A., **53**: 2477.
35. Butts D.A. and Rokhsar D.S., 1999, *Nature*, **397**: 327.
36. Feder D.L., Clark Ch.W., and Schneider B.I., 1999, *Phys. Rev. Lett.*, **82**: 4956.
37. Matthews M.R., Anderson B.P., Haljan P.C., Hall D.S., Wieman C.E. and Cornell E.A., 1999, *Phys. Rev. Lett.*, **83** (13): 2498.
38. Raman C., Köhl M., Onofrio R., Durfee D.S., Kuklewicz C.E., Hadzibabic Z., and Ketterle W., 1999, *Phys. Rev. Lett.*, **83** (13): 2502.
39. Bolda E.L. and Walls D.F., 1998, *Phys. Rev. Lett.*, **81**: 5477.
40. Pines D. and Nozières Ph., 1966, *Theory of quantum liquids*. Benjamin.
41. Staliunas K., Weiss C.O., and Slekys G., *Horizons of World Physics*, **228**: ed. Vasnetsov M. and Staliunas K., (Nova Science Publishers, 1999); Basistiy I.V., *et al.*, 1993, Opt. Comm. **103**: 422; Kreminskaya L.V., Soskin M.S., Khizhnyak A.I., 1998, Opt. Comm. **145**: 377; Nye J.F. and Berry M.V.,1974, Proc. R. Soc. Lond. A **336**: 165.
42. Tempere J. and Devreese J. T., 1998, *Solid State Comm.*, **108** (12): 993.
43. Andrews M.R., Townsend C.G., Miesner H.-J., Durfee D.S., Kurn D.M., and Ketterle W., 1997, *Science*, **275**: 637.
44. Kozuma M., Deng L., Hagley E.W., Wen J., Lutwak R., Helmerson K., Rolston S.L., and Phillips W.D., 1999, *Phys. Rev. Lett.*, **82**: 871.
45. Stenger J., Inouye S., Chikkatur A.P., Stamper-Kurn D.M., Pritchard D.E., and Ketterle W., 1999, *Phys. Rev. Lett.*, **82** (23): 4569.
46. Fedichev P.O., Muryshev A.E. and Shlyapnikov G.V., 1999, *Phys. Rev.* A., **60** (4): 3220.
47. Muryshev A.E., van Linden van den Heuvell H.B., and Shlyapnikov G.V., 1999, *Phys. Rev.* A., **60**: R2665.
48. Hong T., Wang Y.Z., and Huo Y.S., 1998, *Phys. Rev.* A., **58** (4): 3128.
49. Busch Th., and Anglin J., cond-mat/9809408.
50. Burger S., Bongs K., Dettmer S., Ertmer W., Sengstock K., Sanpera A., Shlyapnikov G.V., and Lewenstein M.,1999, *Phys. Rev. Lett.*, **83**: 5198.
51. Kivshar Y.S. and Luther-Davies B.,1998, *Physics Reports*, **298**: 81-197.
52. Madison K.W., Chevy W., Wohlleben W., and Dalibard J., 2000, *Phys. Rev. Lett.*, **84** (5): 806.

Generating And Manipulating Atom Laser Beams

T. ESSLINGER, I. BLOCH, M. GREINER, AND T. W. HÄNSCH
Sektion Physik, Ludwig-Maximilians-Universität, Schellingstraße 4/III, D-80799 Munich, and Max-Planck-Institut für Quantenoptik, D-85748 Garching, Germany

1. INTRODUCTION

Four decades ago the first optical lasers[1] were demonstrated[2,3], marking a scientific breakthrough: coherent optical radiation had been produced, resulting in the ultimate control over frequency, intensity and direction of optical waves. Since then, lasers have found innumerable applications, both for scientific and general use. With the realisation of Bose-Einstein condensates[4-6] and atom lasers[7-10], tremendous progress has been made in achieving similar control over matter waves.

The quest for highly coherent matter wave beams has been a central goal in atomic physics over the last decade. Efficient techniques to cool atoms with laser light have been developed[11-14] and several schemes have been suggested[15-19] and experimentally tested[20-23] as to how an atom laser could be realised with these techniques. So far, however, all atom laser generate their output from Bose-Einstein condensates. The condensates[24,25] are prepared in magnetic traps by evaporative cooling, which can be regarded as the pump mechanism of the atom laser. In evaporative cooling, atoms with higher than average energy are selectively removed from the trap, while those which remain are rethermalized through elastic collisions. Since the average energy is reduced in this process, the new state of equilibrium corresponds to a lower temperature. As the critical temperature for Bose-Einstein condensation is reached, a macroscopic atomic population in the ground state builds up and the scattering rate into the ground state is enhanced by bosonic stimulation[26].

Bose-Einstein Condensates and Atom Lasers
Edited by Martellucci *et al.*, Kluwer Academic/Plenum Publishers, 2000

The output of an atom laser is created by coherently coupling the bound quantum state of the condensate to the continuum of propagating output modes. Various methods have recently been demonstrated to obtain pulsed[7,8]or quasi-continuous[9] output from Bose-Einstein condensates. In a recent experiment with rubidium 87 atoms we have succeeded in demonstrating an atom laser with a continuous output coupler[10], a mechanism which allows for a monoenergetic and highly collimated output beam. In this article we will first describe the basic idea of the continuous output coupling mechanism and present experimental data showing the collimation of the output beam. Output coupling with two frequencies and a versatile method for manipulating atom laser beams are presented.

2. ATOM LASER WITH A CONTINUOUS OUTPUT COUPLER

A continuous output coupler can be accomplished with a monochromatic rf-field that resonantly transfers condensed atoms from the magnetically trapped into the untrapped state. The mechanism requires that the level of fluctuations in the magnetic trapping field is much less than the change of the magnetic trapping field over the spatial size of the condensate. Let us consider the geometry of the output coupling process in more detail. The magnetically trapped Bose-Einstein condensate is subjected to a radio wave of frequency ω. The geometry of the magnetic field $B(\mathbf{r})$ of our trap[27] gives rise to a harmonic potential which confines the condensate in the shape of a cigar, with its long axis oriented perpendicular to the gravitational force. The radial diameter of the condensate is 10 μm. Due to gravity, both the minimum of the trapping potential and the condensate are displaced by 13 μm relative to the minimum of the magnetic field. Atoms are predominantly transferred into the untrapped state at the intersection of the condensate with the shell that is determined by the electron spin resonance condition $1/2\,\mu_B\,|B(\mathbf{r})| = \hbar\omega$, with μ_B being the Bohr magneton[28]. Over the size of the trapped atomic cloud the height at which the resonance condition is fulfilled varies only by a small amount for a given radio frequency ω. The experimental situation is therefore well described by a one-dimensional model, which is illustrated in Fig. 1. In the region of the condensate the output waves can be approximated by Airy-functions Ai(ζ), which are the eigenfunctions of massive particles in the gravitational potential[29]. These functions are characterised by an energy $E = mgz_0$, where z_0 is the apex of the corresponding classical trajectory, m the atomic mass and g the gravitational acceleration. The scaled parameter ζ is given by $\zeta = (z_0 - E/mg)$

$/\ell$, with a natural length scale $\ell^3 = \hbar^2/2m^2g$. Outside the condensate region the asymptotic behaviour of the outgoing matter wave ϕ_{out} is given by[29]:

$$\phi_{out}(\zeta) \propto |\zeta|^{-1/4} \exp\left(i\frac{2}{3}|\zeta|^{3/2}\right) \qquad (1)$$

An output wave of sharply defined energy is produced if the coupling into the untrapped state is weak and sustained for a long enough period[30]. Such output coupling can be described by a rate proportional to the square of the overlap integral between the output state and the trapped state wavefunction, where the maximum contribution to the integral is localised in the region determined by the resonance condition. For our trapping parameters the mean-field interaction energy of the condensate causes only a slight distortion in the gravitational potential so that Airy functions are a good approximation for the output waves.

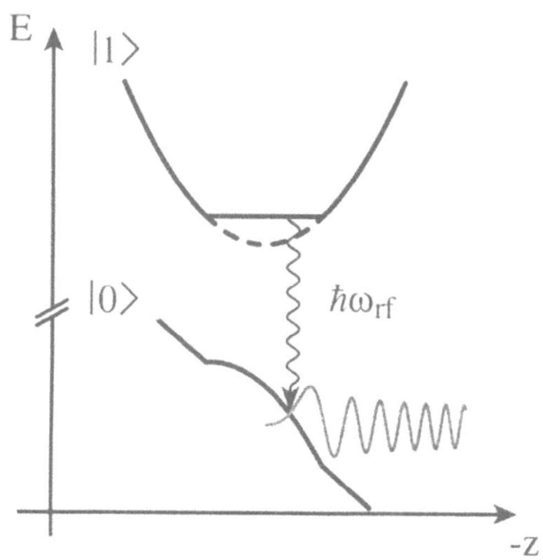

Figure 1. Principle of the atom laser. A radio frequency field transfers condensed atoms from the trapped state |1> into the untrapped state |0>. For a weak coupling rate the energy of the output wave (grey line) is sharply defined. The potential curves for the states |0> and |1> (thick lines) are modified by the mean-field of the condensate.

To produce Bose-Einstein condensates we use the same experimental set-up as described in our previous work. In brief, rubidium 87 atoms are trapped and cooled in a magneto-optical trap. Then the atoms are optically pumped into the hyperfine state with total angular momentum F=1 and the magnetic quantum number $m_F = -1$. In this state the atoms are magnetically trapped and cooled by rf-induced evaporation. The Bose-Einstein condensates typically contain 10^6 atoms at a temperature of 50 nK.

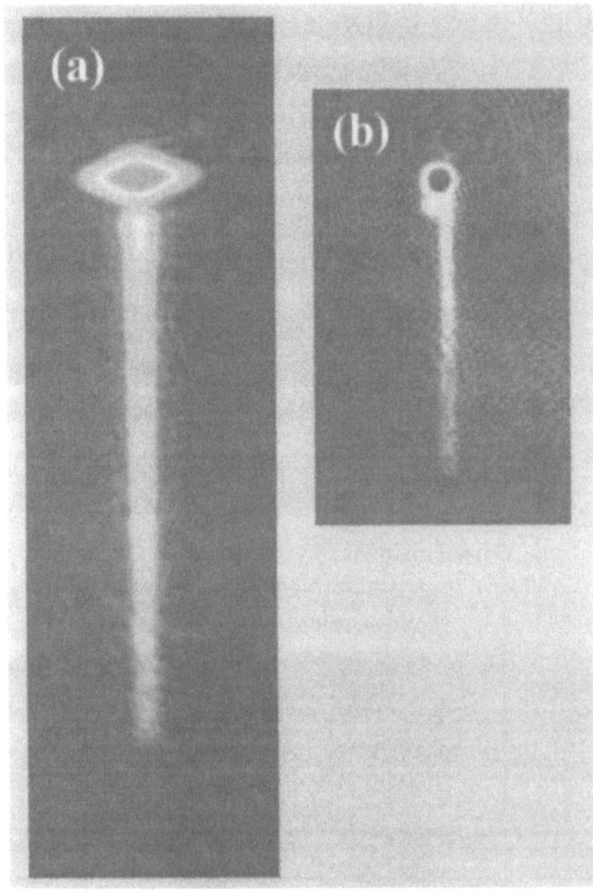

Figure 2. Absorption images of the atom laser output. Image (a) shows the density distribution after 13 ms of output coupling. The weakly confining axis of the magnetic trap is oriented horizontally. Image (b) shows the density distribution after 8 ms of output coupling. The weakly confining axis is perpendicular to the image plane. The magnetic trapping field was switched off before the absorption images were taken. During this process the condensate is shifted to the left due to a residual magnetic field gradient, image (b). The image sizes are 2.2 mm x 0.7 mm (a) and 1.1 mm x 0.7 mm (b).

For magnetic trapping a very compact magnetic trap is employed which combines the quadrupole with the Ioffe geometry[27]. It is placed inside a magnetic shield enclosure so that field fluctuations due to the environment are reduced to a level below 0.1 mG. The radial and axial trapping frequencies are $\omega_{\perp} = 2\pi \times 140$ Hz and $\omega_{ax} = 2\pi \times 13$ Hz, respectively. The rf-field for the output coupler is produced by a synthesiser and is radiated from the same coil as used for evaporative cooling. The magnetic field vector of the rf is oriented in the horizontal plane, perpendicular to the magnetic bias field of the trap.

The following procedure is used to obtain the atom laser output in a typical experiment. We cool the trapped rubidium gas to a temperature of around 50 nK and then we switch off the rf-field used for evaporative cooling. After a delay of 50 ms the radio frequency of the output coupler is switched on for 13 ms, at a frequency of 1.62 MHz and with an amplitude of B_{rf}=0.4 mG. Over this period atoms are extracted from the trapped gas and accelerated by gravity. Subsequently, the magnetic trapping field is switched off and 3 ms later the atomic distribution is measured by absorption imaging. The density distribution of the atom laser output is shown in Fig. 2. (Both for the radial and the axial direction.)

3. TWO-MODE ATOM LASER

The mechanism for continuous output coupling has the unique feature that the atoms are extracted from the condensate in a spatially localised region. By applying a radio wave field composed of two frequencies ω and ω', transitions between trapped and untrapped states are induced in two spatially separated regions. The resulting output from the trap then consists of two matter wave beams with different energies. Due to gravity the beams are collimated and propagate downwards. In this geometry, almost complete overlap between both beams can be achieved, leading to a high contrast interference pattern in the density distribution of the output beam, as shown in Fig. 3. The two outgoing beams have a difference in energy of $\Delta E = \hbar(\omega-\omega')$. This results in a spatial separation $\Delta z = \Delta E/mg$ between the classical turning points of the corresponding Airy functions. Using the asymptotic approximation[29] the atomic density distribution n_{out} of the output waves is given by:

$$n_{out}(z) = |\phi_{out}(\zeta) + \phi_{out}(\zeta')|^2 \propto \frac{1}{\sqrt{|z|}}\left\{2 + 2\cos\left(q\sqrt{|z|} + (\omega-\omega')t\right)\right. \quad (2)$$

where the variable q is given by $q = m\Delta z\sqrt{2g}/\hbar$. The envelope of the density distribution decreases as $|z|^{-1/2}$. The interference term in Eq. 2 is also obtained when two point sources for atomic deBroglie waves are considered which are positioned in the gravitational potential with a distance Δz between them.

In the experiment (Fig. 3) two matter wave beams were extracted from the condensate over a period of 13 ms. The two components of the radio wave field had the same amplitudes (0.4 mG ± 0.1 mG) and a frequency difference of 1 kHz. This corresponds to a spatial separation of the output coupling regions of 465 nm. After a time delay of 2 ms the magnetic trapping field was switched off and the absorption image was taken another 3 ms later. Eq. 2 was used to fit the data. To quantify the contrast of the interference pattern, the second term on the right hand side of Eq. 2 has been multiplied by a visibility V. From the fit we obtained a visibility of V=0.95. In a recent experiment we determined the spatial correlation function of the trapped Bose gas by measuring the visibility V as a function of Δz for temperatures above and below the transition temperature[31].

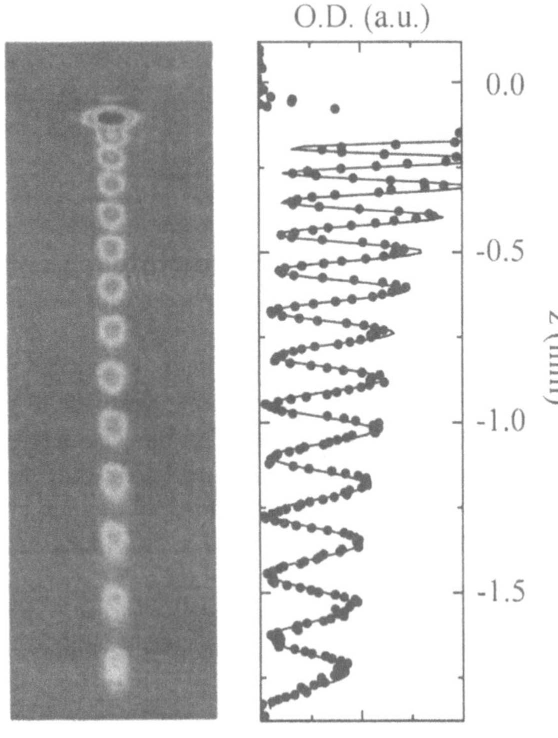

Figure 3. Interference pattern of matter wave beams emitted from two spatially separated regions of a trapped Bose gas. The plot next to the image shows a 21 μm wide vertical cut through the absorption image, with the optical density being plotted horizontally.

4. SPIN-FLIP MIRROR

Highly monoenergetic and coherent atom laser beams will open up a new regime in atom optics[32] and challenge the level of precision achievable with current atom optic elements. Atom optic mirrors are a key element for the manipulation of atom laser beams. Two classes of such mirrors have so far been demonstrated. One class of mirrors uses the dipole force which arises when the induced electric dipole moment of an atom interacts with the light field of a laser beam[33,34]. The laser beam, tuned to a frequency higher than the atomic transition frequency, is totally internally reflected in a prism, producing an evanescent field just above the glass surface. Atoms approaching the surface experience a force towards low light intensity and can be reflected by the evanescent wave field. Similarly, a sheet of light, produced by an intense laser source, has recently been used to reflect a Bose-Einstein condensate released from the magnetic trap[35].

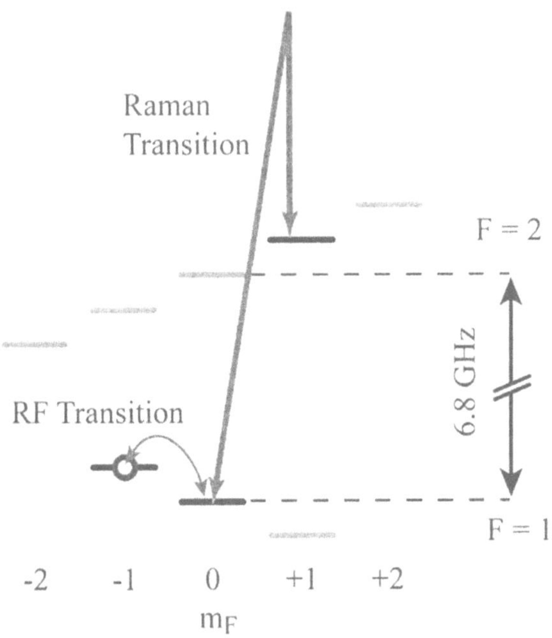

Figure 4. Hyperfine Raman transitions in rubidium 87 in the presence of the magnetic trapping field. Two optical photons (thick arrows) are used to induce transitions between the magnetically untrapped state |F=1, m_F=0> and the magnetically trapped state |F=2, m_F=1>. The atom laser beam is generated by inducing rf transitions (thin line with arrows) between the trapped state |F=1, m_F=-1> and the untrapped state |F=1, m_F=0>.

The second class of mirrors utilises the force that an atom with a nonzero magnetic moment experiences in an inhomogeneous magnetic field, as, for example, in the famous Stern-Gerlach experiment, where silver atoms were deflected. More recently, the reflection of atoms near a surface has been demonstrated using periodically magnetised data storage media[36] or, alternatively, suitable patterns of current carrying wires[37].

Here we demonstrate a novel method for manipulating and reflecting neutral atoms in an inhomogeneous magnetic field. Two-photon Raman transitions between different hyperfine ground states of rubidium 87 are used to selectively transfer the magnetically untrapped state |F=1, m_F=0> into the magnetically trapped state |F=2, m_F=1>, as illustrated in Fig. 4 (F: total angular momentum, m_F: magnetic quantum number). Due to the inhomogeneous magnetic trapping field the resonance condition for the Raman transition is satisfied only in a spatially localised region. Accurate control over the frequency difference between the two Raman photons (see below) combined with a stable magnetic field allows us to precisely control the interaction of the atoms with the magnetic trapping field.

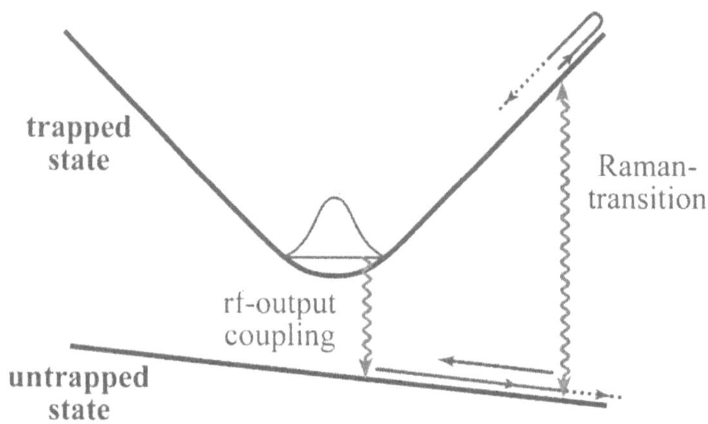

Figure 5. Spin-flip mirror.

The experimental implementation of this manipulation scheme is sketched in Fig.5. The continuous output coupling mechanism transfers magnetically trapped condensate atoms from the trapped state |F=1, m_F=-1> into the untrapped state |F=1, m_F=0>. The untrapped atoms are accelerated by gravity and propagate downwards. In the region where the resonance condition for the two-photon Raman transition is satisfied the atoms can make a transition into the |F=2, m_F=1> state. The probability for this process depends on how strongly the Raman transition is driven. In the |F=2, m_F=1> state the atoms experience the confining potential of the magnetic trap and are decelerated until their initial kinetic energy is transferred into potential energy. Then the atoms reverse direction and are accelerated upwards. Passing through the Raman transition region for the second time, the atoms either remain in the |F=2, m_F=1> state, or are transferred back into the |F=1, m_F=0> state. In the first case the atoms are recaptured and their motion continues to be determined by the potential of the magnetic trap. The latter case corresponds to a reflection of the atoms since they propagate further upwards on a ballistic trajectory which is unperturbed by the trapping potential. We have experimentally explored both cases: recapturing and reflection of an atom laser beam.

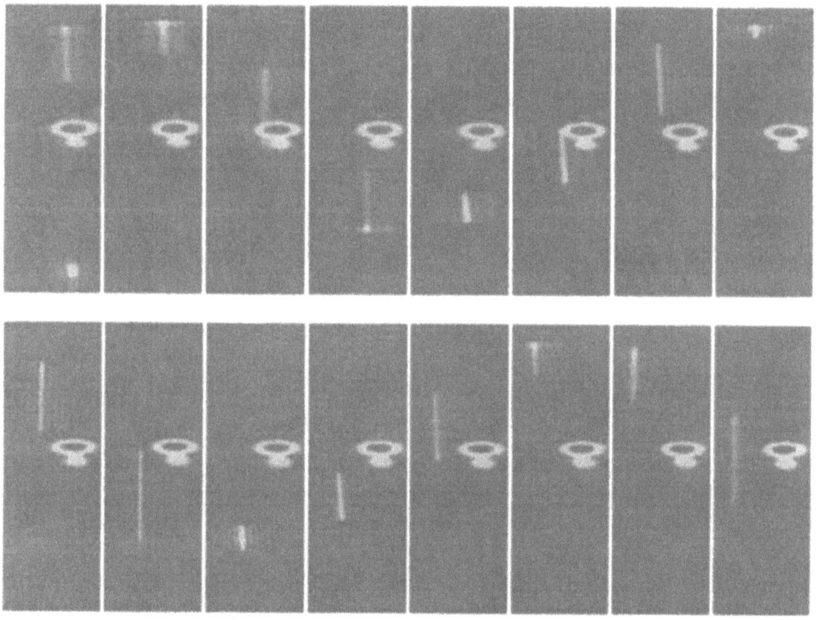

Figure 6. Recapturing of an atom laser beam in the magnetic trap. The elliptical cloud in the centre of each absorption image is the condensate. The stripe of varying width and length is the atom laser beam, which oscillates in the trap. Part of the atom laser beam which has not been recaptured can be seen in the first image of the upper row. Each of the images has a size of 2.2 mm x 0.8 mm.

The two-photon Raman transitions are induced by two copropagating and overlapping laser beams with mutually perpendicular linear polarisations. Both beams are derived from grating stabilised diode laser[38] which are tuned to frequencies approximately 50 GHz to the red of the D_1-line of rubidium 87. The frequency difference between the two laser beams is phase-locked[38,39] to a radio-frequency which can be tuned around 6.8 GHz. The waists of the two overlapping beams are 200 μm in the horizontal direction and are adjustable in the vertical direction. The beams are typically located 400 μm below the condensate with the propagation direction parallel to the long axis of the condensate.

To recapture the atom laser beam we apply the Raman beams only for a short time during the 13 ms period of output coupling. The atoms which make a transition into the trapped state during the propagation downwards will therefore not interact with the Raman beams for a second time when they move upwards. This results in efficient recapturing of the atom laser beam, as illustrated in the sequence of images shown in Fig. 6. The data was obtained from identical repetitions of the experiment. Only the time delay at which the absorption images were taken had been increased in 2 ms steps. For these images the direction of view was perpendicular to the long axis of the condensate. The influence of the trapping potential in the horizontal plane results in the atom laser beam being focused, as can be seen in the second image from the left in the second row of Fig. 6. This process has not the same periodicity as the vertical oscillation of the atoms. We observed focusing of the atom laser beam to the resolution limit of our imaging system, which is 8 μm.

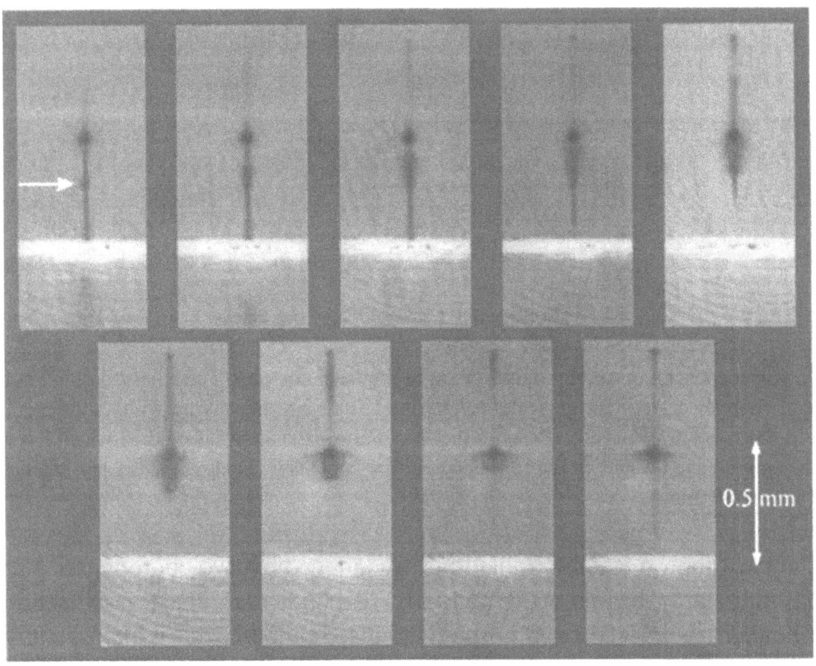

Figure 7. Reflection of an atom laser beam. The round dark cloud in the centre of each absorption image is the magnetically trapped condensate. The white, horizontally oriented, stripe is caused by the laser beams which induce the Raman transition. The white arrow points to the reflected atoms in the |F=1, m_F=0> state. The long thin stripe which is oriented vertically shows atoms in the |F=2, m_F=1> state which have been recaptured.

The atom laser beam is reflected when the Raman beams are applied long enough for the atoms to change their magnetic moment during the propagation both downwards and upwards. A typical sequence showing the reflection of an atom laser beam is shown in Fig. 7. The reflected atoms are marked with an arrow showing the motion to the apex of their trajectory. In these images the direction of view was parallel to the long axis of the condensate.

5. CONCLUSIONS AND OUTLOOK

New levels of control over the atomic motion have been attained through the development of the atom laser. The unique properties of these coherent atomic sources open up a wide range of new possibilities in the fields of atom optics and precision measurement. In this article we have discussed the high contrast interference pattern obtained from an atom laser with two output modes of different energies. It should be possible to create a highly complex interference pattern in the output beam of an atom laser by applying a correspondingly tailored radio wave field for output coupling.

A novel atom optical manipulation method allowed us to reflect, recapture and focus an atom laser beam. Due to its simplicity and precision the technique is ideally suited for creating an interferometer for the atom laser beam. It seems feasible to split the atom laser beam into two components and recombine it after delaying one of the two components. The resulting interference pattern would reveal the coherence properties of the atom laser beam.

ACKNOWLEDGMENTS

We would like to thank Olaf Mandel for experimental assistance with the phase-locked laser diodes.

REFERENCES

1. Schawlow, A.L., and Townes, C.H., 1958, *Phys. Rev.* **112**: 1940.
2. Maiman, T.H., 1960, *Nature* **187**: 493.
3. Javan, A., Bennett Jr., W.B., and Herriott,D.R., 1961, *Phys. Rev. Lett.* **6**: 106.
4. Anderson, M.H., Ensher, J.R., Matthews, M.R., Wieman, C.E., and Cornell, E.A.,1995, *Science* **269**: 198.
5. Davis, K.B., Mewes, M.-O., Andrews, M.R., van Druten, N.J., Durfee, D.S., Kurn, D.M., and Ketterle, W.,1995, *Phys. Rev. Lett.* **75**: 3969.
6. Bradley, C.C., Sackett, C.A., Tollet, J.J., and Hulet, R.G., 1995, *Phys. Rev. Lett.* **75**: 1687; and Bradley, C.C., Sackett, C.A., and Hulet, R.G., 1997, *Phys. Rev. Lett.* **78**. 985.
7. Mewes, M.-O., Andrews, M.R., Kurn, D.M., Durfee, D.S., Townsend, C.G., and Ketterle, W., 1997, *Phys. Rev. Lett.* **78**: 582.

8. Anderson, B.P., and Kasevich, M.A., 1998, *Science* **282**: 1686.
9. Hagley, E., Deng, L., Kozuma, M., Wen, J., Helmerson, K., Rolston, S.L., and Phillips, W.D., 1999, *Science* **283**: 1706.
10. Bloch, I., Hänsch, T.W., and Esslinger, T., 1999, *Phys. Rev. Lett.* **82**: 3008.
11. Adams, C.S., and Riis, E., 1997, *Progr. Quantum Electron.* **21**: 1.
12. Chu, S., 1998, *Rev. Mod. Phys.* **70**: 685.
13. Cohen-Tannoudji, C., 1998, *Rev. Mod. Phys.* **70**: 707.
14. Phillips, W.D., 1998, *Rev. Mod. Phys.* **70**: 721.
15. Wiseman, H., and M. Collett, Physics Letters A **202**: 246 (1995).
16. Olshannii, M., Castin, Y., and Dalibard, J., 1996, *Proceedings of the 12th International Conference on Laser Spectroscopy, World Scientific*, 7.
17. Spreeuw, R., Pfau, T., Janicke, U., and Wilkens, M., 1996 *Europhys. Lett.* 32, 469.
18. Wiseman, H., 1997, *Phys. Rev.* A **56**: 2068.
19. Castin, Y., Cirac, J., and Lewenstein, M., 1998, *Phys. Rev. Lett.* **80**: 5305.
20. Lee, H., Adams, C., Kasevich, M., and Chu, S., 1996, *Phys. Rev. Lett.* **76**: 2658.
21. Boiron, D., Michaud, A., Fournier, J.M., Simard, L., Sprenger, M., Grynberg, G., and Salomon, C., 1998, *Phys. Rev.* A **57**: R4106.
22. Vuletic, V., Chin, C., Kerman, A., and Chu, S., 1998, *Phys. Rev. Lett.* **81**: 5768.
23. Katori, H., Ido, T., Isoya ,Y., and Kuwata-Gonokami, M., 1999, *Phys. Rev. Lett.* **82**: 1116.
24. Ketterle, W., Durfee, D.S., and Stamper-Kurn, D.M., *Proceedings of the International School of Physics "Enrico Fermi", Course CXL*, Inguscio, M., Stringari, S., and Wieman, C., (Eds.) IOS Press, Amsterdam 1999, and references therein.
25. Dalfovo, F., Giorgini, S., Pitaevskii, L. P., Stringari, S., 1999 *Rev. Mod. Phys.* **71**: 463, and references therein.
26. Miesner, H.-J., Stamper-Kurn, D., Andrews, M., Durfee, D., Inouye, S., and Ketterle, W., 1998 *Science* **279**: 1005.
27. Esslinger, T., Bloch, I., and Hänsch, T.W., 1998, *Phys. Rev. A* **58**: R2664-R2667.
28. Bloch, I., Hänsch, T.W., Esslinger, T., 1999 *Phys. Rev. Lett.* **82**: 3008-3011.
29. Flügge, S., 1974, *Practical Quantum Mechanics.* Springer-Verlag, 101-107.
30. Band, Y.B., Julienne, P.S., Trippenbach, M., 1999, *Phys. Rev. A.* **59**: 3823-3831.
31. Bloch, I., Hänsch, T. W., and Esslinger, T., 2000, *Nature* **403**: 166-170.
32. Adams, C.S., Sigel, M., Mlynek, J., 1994, *Physics Reports* **240**. 143.
33. Cook R.J., and Hill, R.K., 1982, *Opt. Commun.* **43**: 258.
34. Balykin, V.I., Letokhov, V.I., Ovchinnikov, Y.B., and Sidorov, A., 1987, *JETP Lett.* **45**: 353.
35. Bongs, K., Burger, S., Birkl, G., Sengstock, K., Ertmer, W., Rzazewski, K., Sanpera, A., and Lewenstein, M., 1999 *Phys. Rev. Lett.* **83**: 3577.
36. Hinds, E.A., and Hughes, I.G., 1999, *J. Phys. D: Appl. Phys.* **32**: R119, and references therein.
37. Johnson, K.S., Drndic, M., Thywissen, J.H., Zabow, G., Westervelt, R.M., and Prentiss, M., 1998, *Phys. Rev. Lett.* **81**: 1137.
38. Ricci, L., Weidemüller, M., Esslinger, T., Hemmerich, A., Zimmermann, C., Vuletic, V., König, W., and Hänsch, T.W., 1995, *Optics Commun.* **117**: 541.
39. Santarelli, G., Clairon, A., Lea, S.N., Tino, G.M., 1994, *Optics Commun.* **104**: 339.

Multiple ^{87}Rb Condensates And Atom Lasers By RF Coupling

F. MINARDI, C. FORT, P. MADDALONI, AND M. INGUSCIO
INFM - European Laboratory for Non Linear Spectroscopy (L.E.N.S.) – Dipartimento di Fisica dell'Università di Firenze L.go E. Fermi 2, I-50125 Firenze, Italy

1. INTRODUCTION

One of the major goals in the study of Bose-Einstein condensation (BEC) in dilute atomic gases has been the realisation and development of atom lasers. An atom laser may be understood as a source of coherent matter waves. One can extract coherent matter waves from a magnetically trapped Bose condensate. Schemes to couple the atomic beam out of the magnetic trap have been demonstrated[1-4]. In the experiments of Refs. [1,2] the output coupling is performed by the application of a radio-frequency (RF) field that induces atomic transitions to untrapped Zeeman states. The atom laser described in Ref. [3] is based on the Josephson tunnelling of an optically trapped condensate and in Ref. [4] a two-photon Raman process is described that allows directional output coupling from a trapped condensate.

Characterizing the output coupler is necessary to understand the atom laser itself. The literature dealing with theoretical descriptions of output couplers for Bose-Einstein condensates has focused both on the use of RF transitions[5-8] and Raman processes[9].

RF output coupling is based on single– or multi– step transitions between trapped and untrapped atomic states. As a consequence, a rich phenomenology arises that include the observation of "multiple" condensates corresponding to atoms in different atomic states. These may display varied dynamical behaviour while in the trap. Also, both pulsed and continuous output-coupled coherent matter beams have been observed. The phenomenology is made even more varied by the possibility of out-coupling solely under gravity and also of magnetically pushed out beams. The apparatus operated by the Florence group[10] offers the possibility to investigate various aspects of output-coupling achieved by RF transitions of atoms in a magnetically trapped ^{87}Rb.

2. EXPERIMENTAL PRODUCTION OF THE CONDENSATE

We bring a ^{87}Rb sample to condensation using the now standard technique of combining laser cooling and trapping in a double magneto-optical trap (MOT) and evaporative cooling in a magneto-static trap. Our apparatus had been originally designed for potassium, as presented by C. Fort in this book[11].

Our double MOT set-up consists of two cells connected in the horizontal plane by a 40 cm long transfer tube with an inner diameter of 1.1 cm. We maintain a differential pressure between the two cells in order to optimise conditions in the first cell for rapid loading of the MOT (10^{-9} Torr) while the pressure in the second cell is sufficiently low (10^{-11} Torr) to allow for the long trapping times in the magnetic trap necessary for efficient evaporative cooling.

Laser light for the MOTs is provided by a cw Ti:sapphire laser (Coherent model 899-21) pumped with 8 W of light coming from an Ar^+ laser. The total optical power of the Ti:sapphire laser on the Rb D_2 transition at 780 nm is 500 mW. The laser frequency is locked to the saturated absorption signal obtained in a rubidium vapour cell. The laser beam is then split into four parts each of which is frequency and intensity controlled by means of double pass through an AOM: two beams are red detuned respect to the F=2 -> F'=3 atomic resonance and provide the cooling light for the two MOTs. Another beam, resonant on the F=2 -> F'=3 cycling transition, is used both for the transfer of cold atoms from the first to the second MOT and for resonant absorption imaging in the second cell. Finally a beam, resonant with the F=2 -> F'=2 transition, optically pumps the atoms in the low field seeking F=2, $m_F=2$ state immediately before switching on the magneto–static trap. 5 mW of repumping light for the two MOTs resonant on the F=1 -> F'=2 transition are provided by a diode laser (SDL-5401-G1) mounted in external cavity configuration.

In the first MOT, with 150 mW of cooling light split into three retroreflected beams (2 cm diameter), we can load 10^9 atoms within a few seconds. However, every 300 ms we switch off the trapping fields of the first MOT and we flash on the "push" beam (1 ms duration, few mW) in order to accelerate a fraction of atoms through the transfer tube into the second cell.

Permanent magnets placed around the tube generate an hexapole magnetic field that guides the atoms during the transfer. In the second cell the atoms are recaptured by the second MOT which is operated with six independent beams (diameter=1 cm) each with 10 mW of power. The overall transfer efficiency between the two MOTs is ~30 %, and after 50 shots we have typically loaded 1.2 10^9 atoms in the second MOT. The final part of laser cooling in the second MOT is devoted to maximising the density and minimizing the temperature just before loading the magnetic trap. Firstly the atomic density is increased with 30 ms of Compressed-MOT[13] and this is followed by 8 ms of optical molasses to reduce the temperature. Soon after, we optically pump the atoms into the low-field seeking F=2, m_F=2 state by shining the σ^+-polarised F=2 ->F'=2 beam for 200 μs, together with the repumping light. At this point we switch on the magneto-static trap in the second cell where we perform evaporative cooling of the atoms.

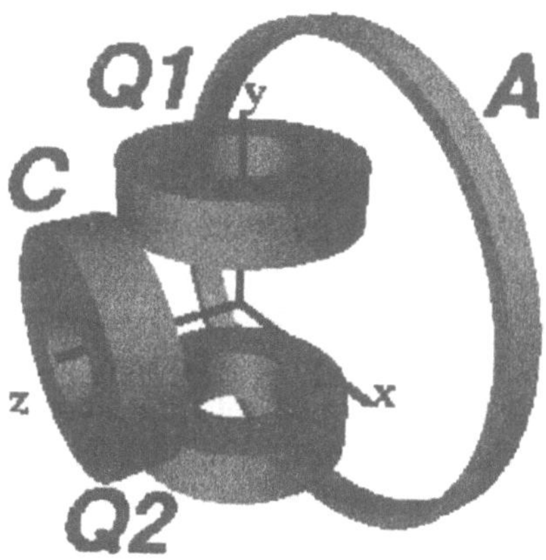

Figure 1. 4–coils magnetic trap: Q1 and Q2 coils produce the quadrupole field, the curvature coil (C) provides the axial confinement and the antibias coil (A) reduces the bias field thus increasing the radial confinement.

The magneto-static trap is created by passing DC current through 4-coils (see Fig. 1), which gives rise to a cigar-shaped harmonic magnetic potential elongated along the *z* symmetry axis (Ioffe-Pritchard type). Our magnetic trap is inspired by the scheme first introduced in Ref. [12], but is operated with a higher current. The coils are made from 1/8-inch, water cooled, copper tube. The three identical coils consist of 15 windings with diameters ranging from 3 cm to 6 cm. The fourth coil consists of 6 windings with a diameter of 12 cm. The coils Q1 and Q2 (Fig. 1) generate a quadrupole field symmetric around the vertical *y*-axis, and in this direction the measured field gradient is 10 mT/(A·m). These two coils operated together at low current (~10 A) also provide the quadrupole field for operation of the MOT. The *curvature* (C) and *antibias* (A) coils produce opposing fields in the z direction. The modulus of the magnetic field during magnetic trapping has a minimum displaced by 5 mm from the centre of the quadrupole field (toward the curvature coil) and the axial field curvature is 0.46 $T/(A\cdot m^2)$. The coils are connected in series and fed by a Hewlett Packard 6681A power supply. By means of MosFET switches we can, however, disconnect coils A and C. The maximum current is 240 A, corresponding to an axial frequency of ν_z=13 Hz for atoms trapped in the F=2, m_F=2 state. The radial frequency ν_r can be adjusted by tuning B_b=min(|**B**|), the bias field at the center of the trap: $\nu_r=(2.18/(B_b[T])^{1/2})$ Hz. With typical operating values of B_b from 0.14 mT to 0.18 mT, ν_r ranges from 160 Hz to 180 Hz. In addition, a set of three orthogonal pairs of Helmholtz coils provide compensation for stray magnetic fields.

The transfer of atoms from the MOT to the magneto-static trap is complicated by the fact that the MOT (centred at the minimum of the quadrupole field) and the minimum of the harmonic magnetic trap are 5 mm apart. The transfer of atoms from the MOT to the magnetic trap consists of a few steps. We first load the atoms in a purely quadrupole field with a gradient of 0.7 T/m (I=70 A), roughly corresponding to the "mode-matching" condition (magnetic potential energy equals the kinetic energy) which ensures minimum losses in the phase-space density. Then we adiabatically increase the gradient to 2.4 T/m by ramping the current to its maximum value I=240 A in 400 ms. Finally, the quadrupole potential is adiabatically (750 ms) transformed into the harmonic one by passing the current also through the *curvature* (C) and *antibias* (A) coils, hence moving the atoms 5 mm in the z direction. At the end of this procedure, 30% of atoms have been transferred from the MOT into the harmonic magnetic trap and we start RF forced evaporative cooling with $4\cdot 10^8$ atoms at 500 μK. We estimate the elastic collision rate to be $\gamma \sim 30\ s^{-1}$ and this, combined with the measured lifetime in the magnetic trap of 60 s, gives a ratio of "good" to "bad" collisions of~1800. This is sufficiently high to perform the evaporative cooling and reach BEC.

The RF field driving the evaporative cooling is generated by means of a 10 turn coil of diameter 1-inch placed 3 cm from the center of the trap in the x direction and fed by a synthesiser (Stanford Research DS345). The RF field is first ramped for 20 s with a exponential-like law from 20 MHz to a value which is only 100 kHz above the frequency that empties the trap, $\nu^0_{rf}=\mu_0 B_b/2h$. Then a 5 s linear ramp takes the RF closer to ν^0_{rf}: the BEC transition takes place roughly 5 kHz above ν^0_{rf}.

We analyse the atomic cloud using resonant absorption imaging (Fig. 2). The atomic sample is released by switching off the current through the trapping coils in 1ms. The cloud then falls freely under gravity and after a delay of up to 25 ms, we flash a probe beam, resonant with the F=2 -> F'=3 transition, for 150 μs and at one tenth of the saturation intensity. The shadow cast by the cloud is imaged onto a CCD array (pixel size=24 μm) with two lenses, giving a magnification of 6. However our resolution is 7 μm, due to the diffraction limit of the first lens (f=60 mm, N.A.=0.28). We process three images to obtain the two dimensional column density $\tilde{n}(y,z)=\int n(x,y,z)\,dx$. The column density is then fitted assuming that n(x,y,z) is the sum of a gaussian distribution corresponding to the uncondensed fraction and an inverted parabola, which is solution of the Gross-Pitaevskii equation in the Thomas-Fermi approximation (condensed fraction). The effect of free expansion, which is trivial for the Gaussian part, is taken into account also for the condensate as a rescaling of the cloud radii, according to Ref. [14]. The temperature is obtained from the Gaussian widths of the thermal cloud.

We observe the BEC transition at a temperature T_c=200 nK with $2\cdot10^5$ atoms, the peak density n_c being $7\cdot10^{19}$ m^{-3}. The number of condensed atoms shows fluctuations of 20% from shot to shot. We may attribute this to thermal fluctuations of the magnetic trap coils giving rise to fluctuation of the magnetic field.

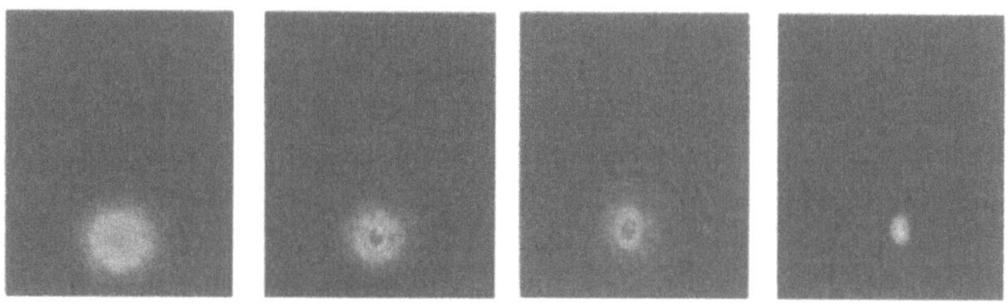

Figure 2. Absorption imaging picture of the atomic cloud after a time of flight of 20 ms. From the left to the right the RF ramp was stopped respectively at 0.94 MHz, 0.92 MHz, 0.90 MHz and 0.88 MHz. The symmetric expansion in the first image is typical of a thermal cloud. By further evaporation the high density peak characteristic of the condensate emerges, finally giving rise to a pure condensate.

3. RADIO-FREQUENCY OUTPUT COUPLING

After producing the condensate in the F=2, m_F=2 state, the same RF field used for evaporation is also employed to coherently transfer the condensate into different m_F states of the F=2 level. Multistep transitions take place at low magnetic field where the Zeeman effect is approximately linear. This means that the RF field couples all the Zeeman sublevels of F=2. Two of these, m_F=2 and m_F=1, are low-field seeking states and stay trapped. m_F=0 is untrapped and falls freely under gravity, while m_F=-1 and m_F=-2 are high-field seeking states and are repelled from the trap. Different regimes may be investigated by changing the duration and amplitude of the RF field. The absorption imaging with a resonant beam tuned on the F=2 -> F'=3 transition allows us to detect at the same time all Zeeman sublevels of the F=2 state.

It is worth noting that the spatial extent of the condensate results in a broadening of the RF resonance. Due to their delocalization density atoms experience a magnetic field that is non-uniform over their spatial extent. Our condensate is typically 40 μm in the axial direction and 4 μm in the radial one. The corresponding resonance broadening is of the order of 1 kHz. This means that RF pulses shorter than 0.2 ms interact with all the atomic cloud, while for longer pulses, and sufficiently small amplitudes, only a slice of the condensate will be in resonance with the RF field.

4. PULSED REGIME

We investigate the regime of "pulsed" coupling characterised by RF pulses shorter than 0.3 ms. In particular, Fig. 3 shows the effect of a pulse of 10 cycles at ~1.2 MHz (B_b=0.17 mT) with an amplitude B_{rf}=7 μT. After the RF pulse, we leave the magnetic trap on for a time Δt and then switch off the trap, thus allowing the atoms to expand and fall under gravity for 15 ms. Pictures from the left to the right correspond to trap times after the RF pulse of Δt=2,3,4,5 and 6 ms.

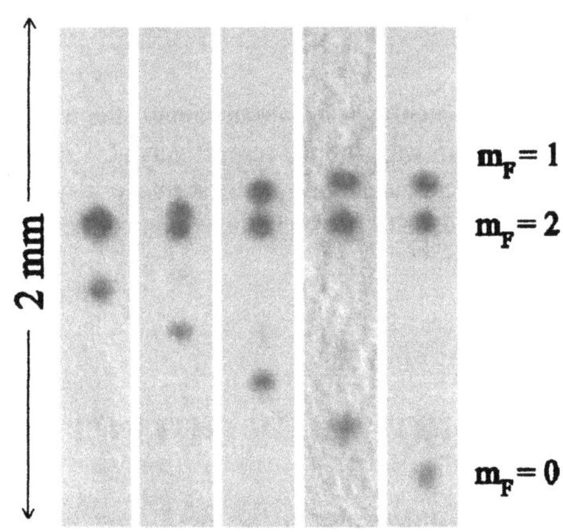

Figure 3. Time of flight images of multiple condensates produced by a RF pulse. Taken after 15 ms of free fall, the pictures refer to different evolution times Δt in the magnetic trap, respectively 2,3,4,5 and 6 ms. The m_F=2 condensate is below the centre of oscillation of the m_F=1, opposite to what one expects considering the gravitational sagging. This can be explained by the presence of a gradient during the switching off of the magnetic field.

Three distinct condensates are visible (Fig. 3): we observe that one is simply falling freely in the gravitational field and hence we attribute to the condensate atoms being in the $m_F=0$ state. The other two condensates initially overlap and then separate. However, we point out that the pictures are always taken after an expansion in the gravitational field. The initial position of the condensate in the trap may be found by applying the equation of motion for free-fall under gravity.

Leaving the magnetic field on for longer times after the RF pulse allows us to identify the condensates in different m_F state by their different center of mass oscillation frequency in the trap. Considering that the images are taken after a free fall expansion of $t_{exp}=15$ ms, one can deduce the oscillation amplitude in the trap, a, from the observed oscillation amplitude, A, by using the relation

$$A = a\sqrt{1+\omega_i^2 t_{\exp}^2} \tag{1}$$

where ω_i is the oscillation frequency for atoms in the m_F=i level.

From Fig. 3 we note that the position of the centre of mass of the condensate in $m_F=2$ is fixed (at the level of resolution) while the condensate in $m_F=1$ oscillates at the radial frequency of the corresponding trap potential, with a measured amplitude of $a=8.7\pm0.4$ μm. This can be explained by considering the different trapping potentials experienced by the two condensates. The total potential results from the sum of the magnetic and the gravitational potentials, so that the minima for the two states in the vertical direction are displaced by $\Delta y=g/\omega_r^2$ ("sagging"). With the experimental parameter of Fig. 3 ($\omega_r=2\pi(171\pm4\text{Hz})$ for atoms in F=2, $m_F=2$ state) Δy equals 8.5 ± 0.4 μm. This is in very good agreement with the measured centre of mass oscillation amplitude. The $m_F=1$ condensate is produced at rest in the equilibrium position of the $m_F=2$ condensate and begins to oscillate around its own potential minimum with an amplitude equal to Δy.

Fig. 3 shows the situation where the RF pulse is adjusted to equally populate the two trapped states, $m_F=1$ and $m_F=2$. In general, the relative population in different Zeeman sublevels can be determined by varying the duration of the RF pulse. This is clearly illustrated in Fig. 4, where the relative population of the $m_F=2,1$ and 0 condensates are shown as a function of the pulse duration.

The theoretical curves, calculated for a Rabi frequency of 26 kHz corresponding to the amplitude of our oscillating RF magnetic field $B_{rf}=3.6$ μT, are shown together with the experimental data.

The population of each Zeeman state is calculated by solving the set of the Bloch equation in the presence of an external RF coupling field. These results clearly show that we can control in a reproducible way the relative populations of the multiple condensates.It is worth noting that the use of a static magnetic trap allows a straightforward explanation of the phenomenon; similar investigations recently reported for a time-dependent TOP trap [15] show that the theory is more complicated in presence of a time varying magnetic field.

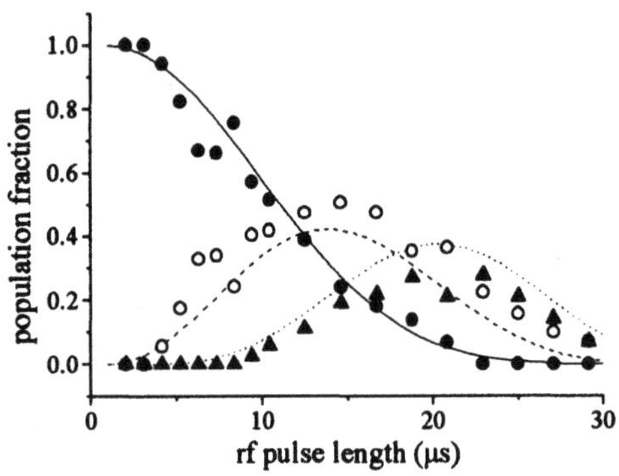

Figure 4. Measured population fraction of m_F=2 (solid circle), m_F=1 (open circle) and m_F=0 (triangle) states as a function of RF pulse duration. The curves correspond to the calculated Rabi oscillations of the relative population in m_F=2 (solid line), m_F=1 (dashed line) and m_F=0 (dotted line).

The m_F=-1,-2 sublevels are also populated by RF induced multistep transitions. However, these condensates are quickly expelled by the magnetic potential and the effect can be observed for shorter times after the RF pulse. This is evident in the image in Fig. 5 which is taken under the same conditions of Fig. 3 but with a shorter time (Δt=1.5 ms) in the magnetic trap.

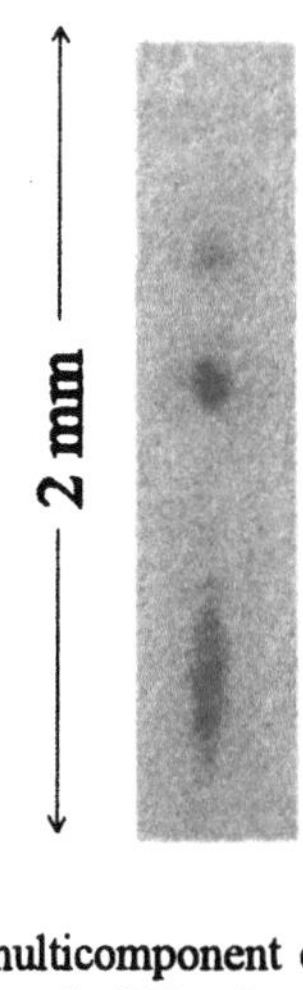

Figure 5. Absorption imaging of the multicomponent condensate produced by a RF pulse. The magnetic trap was on for 1.5 ms after the RF pulse and the picture was taken after 20 ms of free expansion.

As expected, in addition to the free-falling m_F=0 condensate atoms coupled-out simply by gravity, an elongated cloud appears, corresponding to atoms in the high-field seeking states that are repelled from the trap.

5. CW ATOM LASER

Continuously coupling atoms out of a Bose condensate with resonant RF radiation was first proposed by W. Ketterle *et al.*[1]. In their paper on the RF output coupler they discuss this scheme and point out the necessity to have a very stable magnetic field. I. Bloch *et al.*[2] realized a cw atom laser based on RF output coupling using an apparatus with a very well controlled magnetic field. They placed a μ-metal shield around the cell where the condensate forms, achieving residual fluctuations below 10^{-8} T.

Figure 6. Absorption imaging of the atom laser obtained applying a RF field for 10 ms and illuminating the atoms 5 ms after switching off the trap.

We explored the regime of continuous coupling by leaving the RF field on for at least 10 ms. In this case we observed a stream of atoms escaping from the trap (Fig. 6). The experimental configuration is similar to the one described in Ref. [2], except for the fact that our apparatus is not optimized to minimize magnetic field fluctuations, that are at the level of 10^{-6} T. Nevertheless, our observation demonstrate that these fluctuations do not prevent the operation of a cw atom laser.

Fig. 6 shows an absorption image taken after an RF pulse 10 ms long with an amplitude B_{rf}=0.36 μT. The first picture corresponds to the temperature of the rubidium atoms being above the critical temperature (T_c) for condensation. In this case a very weak tail of atoms escaping from the magnetically trapped cloud is observed. Decreasing the temperature below T_c (second picture of Fig. 6) the beam of atoms leaving the trap becomes sharper and more collimated.

We have demonstrated a cw atom laser. However, the fluctuations in our magnetic field strongly influences not only the reproducibility but also the quality of the extracted beam. The high magnetic field stability achieved by the Münich group allowed the measurement of the spatial coherence of a trapped Bose gas[16,17] by observing the interference pattern of two matter waves out-coupled from the condensate using a RF field composed of two frequencies.

6. CONCLUSIONS

We have illustrated the rich phenomenology arising from the interaction of an RF field with a ^{87}Rb condensate originally in the F=2, m_F=2 state. We have shown that condensates can be produced in each of the five Zeeman sublevels and that the relative populations can be controlled by varying the duration and amplitude of the RF pulse. We investigated the behaviour of both trapped and untrapped condensates as a function of the time in the magnetic trap. At short times we recorded the different behaviour between the atoms output-coupled under gravity only (m_F=0), and those with an additional impulse due to the magnetic field (m_F=-1, -2). We have also produced a cw atom laser by simply increasing the time duration of the RF pulse. In our apparatus no particular care is devoted to the shielding of unwanted magnetic field. The stability and homogeneity requirements seem to be less stringent than those predicted in the pioneering work of Ref. [1] and of those of the magnetic field implemented by the original cw atom laser apparatus[2]. This could make the cw atom laser based on RF out–coupling more generally accessible. Most of the observed phenomena may be understood using a simple theoretical model; a more detailed and complete description of the multicomponent condensate should take into account also the mean field potential and interaction between different condensates.

Future applications of the experimental set-up we are currently operating can be foreseen, for instance for the study of collective excitations induced by the sudden change in the atom number. The interaction between condensates in different internal states may possibly be investigated as well as time-domain matter-wave interferometers using a sequence of RF pulses.

ACKNOWLEDGMENTS

We would like to thank M. Prevedelli for his contribution in setting up the experiment. This work was supported by the INFM "Progetto di Ricerca Avanzata" and by the CNR "Progetto Integrato". We would like to thank also D. Lau for careful reading of the manuscript.

REFERENCES

1. Mewes, M. -O., Andrews, M. R., Kurn, D. M., Durfee, D. S., Towsend, C. G., and Ketterle, W., 1997, An output coupler for Bose condensed atoms, *Phys. Rev. Lett.* **78**: 582.
2. Bloch, I., Hänsch, T. W. and Esslinger, T., 1999, An Atom Laser with a cw Output Coupler, *Phys. Rev. Lett.* **82**: 3008.
3. Anderson, B. P., and Kasevich, M. A., 1998, Macroscopic Quantum Interference from Atomic Tunnel Arrays, *Science* **282**: 1686.
4. Hagley, E. W., Dung, L., Kozuma, M., Wen, J., Helmerson, K., Rolston, S. L., and Phillips, W. D., 1999, A well Collimated Quasi-Continous Atom Laser, *Science* **283**: 1706.
5. Naraschewski, M., Schenzle, A., and Wallis, H., 1997, Phase diffusion and the output properties of a cw atom-laser, *Phys. Rev. A* **56**: 603.
6. Ballagh, R. J., Burnett, K., and Scott, T. F., 1997, Theory of an Output Coupler for Bose-Einstein Condensed Atoms, *Phys. Rev. Lett.* **78**: 1607.
7. Steck, H., Naraschewski, M., and Wallis, H., 1998, Output of a pulsed Atom Laser, *Phys. Rev. Lett.* **80**: 1.
8. Band, Y. B., Julienne, P. S., and Trippenbach, M., 1999, Radio-frequency output coupling of the Bose-Einstein condensate for atom lasers, *Phys. Rev. A* **59**: 3823.
9. Edwards, M., Griggs, D. A., Holman, P.L., Clark, C. W., Rolston, S. L., and Phillips, W. D.,1999, Properties of a Raman atom-laser output coupler, *J. Phys. B* **32**: 2935.
10. Fort, C., Prevedelli, M., Minardi, F., Cataliotti, F. S., Ricci, L., Tino, G. M., and Inguscio, M.,2000, Collective excitations of a ^{87}Rb Bose condensate in the Thomas Fermi regime, *Eur. Phys. Lett.* **49**: 8.
11. Fort, C., Experiments with potassium isotopes, in this Volume.
12. Esslinger, T., Bloch, I., and Hänsch, T. W., 1998, Bose-Einstein condensation in a quadrupole-Ioffe-configuration trap, *Phys. Rev. A* **58**: R2664.
13. Petrich, W., Anderson, M. H., Ensher, J. R., and Cornell, E. A., 1994, Behavior of atoms in a compressed magneto-optical trap, *J. Opt. Soc. Am. B* **11**: 1332.
14. Castin, Y., and Dum, R., 1996, Bose-Einstein Condensates in Time-Dependent Traps, *Phys. Rev. Lett.* **77**: 5315.
15. Martin, J. L., McKenzie, C. R., Thomas, N. R., Warrington, D. M., and Wilson, A. C., Production of two simultaneously trapped Bose-Einstein condensates by RF coupling in a TOP trap, cond-mat/9912045.
16. Esslinger, T., Bloch, I., Greiner, M., and Hänsch, T. W., Generating and manipulating Atom Lasers Beams, in this Volume.
17. Bloch, I., Hänsch, T. W., and Esslinger, T., 2000, Measurement of the spatial coherence of a trapped Bose gas at the phase transition, *Nature* **403**: 166.

Theory Of A Pulsed RF Atom Laser

J. SCHNEIDER, AND A. SCHENZLE
Max-Planck-Institut für Quantenoptik, Hans-Kopfermann-Straße 1, 85748 Garching: Sektion Physik, Theresienstraße 37, 80333 München, Germany

1. INTRODUCTION

Atom laser, *i.e.* sources of coherent beams of atoms, are a quite new field of research (see Ref. [l] and references therein). Most of the models for such devices are based on a coherent outcoupling mechanism that transforms a coherent trapped Bose-Einstein condensate of atoms into a coherent atomic beam.

The prototype of such a setup was realized some time ago in the MIT-group[2]. It used strong radio frequency pulses to flip the spins of magnetically trapped sodium atoms that were then leaking out of the trap. More recently, two other groups were able to produce continuous[3] or at least quasi-continuous[4] atomic beams with a much weaker outcoupling rate based either on radio-frequency[3] or Raman outcoupling[4].

Recent theoretical work on atom lasers has focused along two major lines: the work in[5-16] starts with the coupled Gross-Pitaevskii equations (GPE) and analyzes their solutions either numerically or analytically. On the other hand, the studies in[17-22] concentrate on the characterization of atom lasers using master equations analogous to the work on optical lasers and calculate properties like the linewidth of atom lasers.

Both theory and experiment have concentrated so far on two types of outcoupling: either a radio-frequency (RF) is used to flip the spins of magnetically trapped atoms to an untrapped state[2-5] or the transition to an untrapped state is accomplished by a Raman transition[4,23]. Whereas the

Bose-Einstein Condensates and Atom Lasers
Edited by Martellucci *et al.*, Kluwer Academic/Plenum Publishers, 2000

atoms are always falling downwards in RF-outcoupling they might get a momentum kick in any direction during the Raman -outcoupling process.

Here, we want to describe our results on a slightly generalized RF-type outcoupling scheme that was recently realized experimentally[24]: instead of using one radio frequency one can use two different ones. This leads to two different energies for the atoms in the coherently outcoupled beam, which - as a consequence - exhibits pulses.

In the following we analyze the output of such a two-mode atom laser in the limit of weak RF coupling. Further details can be found in a new article[25].

2. COUPLED GROSS-PITAEVSKII EQUATIONS

To describe our setup we use coupled Gross-Pitaevskii equations[5]. These have proven to be very useful both for the description of Bose condensates at zero temperature[26] and for analyzing atom lasers[5,8,9,13,14,16].

We try to model the situation in Ref. [24] where Rb-87 was trapped in its F=1 hyperfine-manifold. Though the experiment uses naturally a three-dimensional trap elongated in one horizontal direction (y-axis), we use a 1D version of the coupled GPE where the axis of interest is the vertical trap axis (z-axis). This should describe the main physics of the device due to the strong influence of gravity on the outcoupling process. The other two axes are taken care of via an effective mean field coupling g_{1D}[16] that is chosen to result in the same chemical potential as in 3D.

The 1D GPE couple the macroscopic wavefunctions ψ_m of the three Zeemansublevels with $m \in \{-1, 0, 1\}$ via the time dependent coupling term

$$\tilde{W}_{m,m'}(t) = 2\hbar\Omega\cos(\delta_{\rm rf}t/2)\left(\delta_{m,m'+1} + \delta_{m,m'-1}\right) \tag{1}$$

The Rabi frequency Ω is determined by the magnetic field strength of the radiofrequency field $\Omega = g_F \mu_{\rm Bohr} B_{\rm rf} / \left(\sqrt{2\hbar}\right)$ (where $g_F = 1/2$). $\delta_{\rm rf}$ denotes the difference between the two radio frequencies $\omega_{\rm rf,1}$, $\omega_{\rm rf,2}$ that are used for outcoupling. This form of the coupling term is obtained after a transformation $\psi_m(t) \rightarrow e^{-im\bar{\omega}_{\rm rf}t}\psi_m(t)$ (where $\bar{\omega}_{\rm rf} = \left(\omega_{\rm rf,1} + \omega_{\rm rf,2}\right)/2$) and after applying the rotating wave approximation.

The total set of equations then reads

$$i\hbar\frac{\partial}{\partial t}\psi_m(z,t)=\left(-\frac{\hbar^2}{2M}\frac{\partial^2}{\partial z^2}+V_{m,\text{eff}}(z,t)\right)\psi_m(z,t)+ +\sum_{m'}\widetilde{W}_{m,m'}(t)\psi_{m'}(z,t) \tag{2}$$

with

$$V_{\text{m,eff}}(z,t)=-mM\omega_z^2z^2/2+m\hbar\Delta-Mgz+g_{1D}\|\psi(z,t)\|^2 \tag{3}$$

being the total effective potential seen by the atoms including the mean field term and gravitation (see Fig. 1). Here, we have assumed that all Zeeman-sublevels interact with the same s-wave scattering length (which enters g_{1D}); $\|\psi(z,t)\|^2=\Sigma_m|\psi_m(z,t)|^2$ denotes the total density in the trap divided by the number N of particles. $\Delta=\hbar\overline{\omega}_{\text{rf}}-V_{\text{off}}$ denotes the detuning from the transitions at the trap center (V_{off} = - $g_F\mu_{\text{Bohr}}\text{B}_{\text{off}}$).

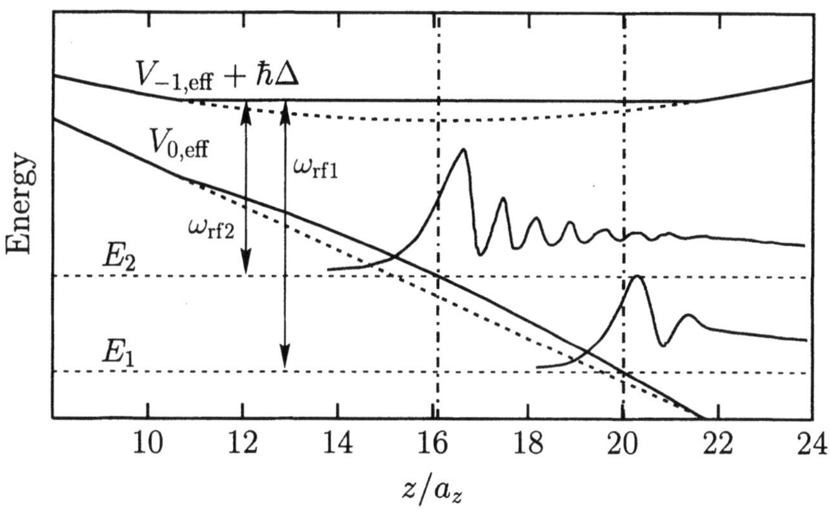

Figure 1. Potentials $V_{m,\text{eff}}$ - $m\hbar\Delta$ (where m = *-1*, 0) and energies $E_{1,2}$ of the outcoupled atoms. The two wavefunctions indicate the situation for stationary outcoupling with either one of the two radio frequencies. The two vertical lines denote the resonance points $z_{\text{rf},1,2}=\sqrt{2(\Delta\pm\delta_{\text{rf}}/2)/(\hbar\omega_z)}a_z$, (where $a_z=\sqrt{\hbar/(M\omega_z)}$).

In the weak coupling region $\Omega<\omega_z$ we are considering here, the state m=1, which is also entrapped, is hardly populated at all. For this reason we will not discuss it further; nevertheless, it is included in all numerical calculations.

3. WAVEFUNCTIONS FOR FALLING ATOMS

3.1 Airy Functions

Atoms with $m = 0$ leaving the trap region are in principle influenced by two forces: gravitation and the mean field force due to their own s-wave interactions. But as the density of the outcoupled atoms is very low, we only have to consider gravity which is described by a linear potential. We need a wavefunction for falling particles that has to be complex for this reason. A solution of the Schrödinger equation with energy E then reads[27]

$$\begin{aligned}\psi_E(z,t) &= \{\mathrm{Ai}(-\xi_E(z)) - i\mathrm{Bi}(-\xi_E(z))\}e^{-iEt/\hbar} \\ &= \mathcal{M}(\xi_E(z))e^{-i[\Theta(\xi_E(z))+Et/\hbar]}\end{aligned} \tag{4}$$

The argument $\xi_E(z,t)$ simply shifts and rescales the z-axis

$$\xi_E(z) = \frac{1}{l}\left(z + \frac{E}{Mg}\right) \qquad l = \left(\frac{\hbar^2}{2M^2 g}\right)^{1/3} \tag{5}$$

where l is a length scale. This solutions corresponds to freely falling atoms flowing steadily to $z \to \infty$.

To describe an atom laser with two radio frequencies, we need two of the wavefunctions in Eq. (4) with different energies as indicated in Fig. 1. With the two detunings $\Delta_{1,2} = \Delta \pm \delta_{rf}$ one gets

$$E = \mu - \frac{1}{2}\frac{Mg^2}{\omega_z^2} - \hbar\Delta_{1,2} \tag{6}$$

for the energies in the atomic beam (where μ is the chemical potential[26]).

If we now assume that the outcoupling is coherent and consider only the direct outcoupling process (no higher orders in the coupling strength $\hbar\Omega$), then the total wavefunction for the $m = 0$-state is given by

$$\psi_0(z,t) = \mathcal{N}\left(\psi_{E1}(z,t) + s\,\psi_{E2}(z,t)\right), \tag{7}$$

where $\mathcal{N}$ is an overall normalization constant and s accounts for the relative strength of the two contributing waves.

The most relevant function to compare with experimental data is the density distribution $|\psi_0(z,t)|^2$ at fixed time. To compute it, we assume that we can Taylor-expand around the mean energy $\bar{E} = (E_1+E_2)/2$ and we use an asymptotic form of the modulus $\mathcal{M}(\xi)=\pi^{-1/2}\xi^{-1/4}$ of Airy functions[28]. This leads to

$$|\psi_0(z,t)|^2 = \tilde{\mathcal{N}}^2 \mathcal{M}^2(\xi_{\bar{E}}(z)) \times \{2 + 2P\cos(\phi(z,t) - \alpha) \tag{8}$$

with the phase

$$\phi(z,t) = \delta_{\mathrm{rf}}t - \frac{\hbar\delta_{\mathrm{rf}}}{Mgl}\sqrt{\xi_{\bar{E}}(z)}, \tag{9}$$

the modified norm

$$\tilde{\mathcal{N}}^2 = \mathcal{N}^2\frac{1+|s|^2}{2}, \tag{10}$$

and

$$P = \frac{2|s|}{1+|s|^2}. \tag{11}$$

Eq. (8) has the form of an interference pattern (apart from the $\mathcal{M}$ factor) and P is the visibility. The falling of the pulses in the outcoupled beam is described by the maxima of this function[25]. The pulse frequency is, as one might expect, $\nu_{\mathrm{pulse}} = \delta_{\mathrm{rf}}/(2\pi)$. We use the phase α appearing in Eq. (8) only as a way to shift the interference pattern in time to compare with the numerical results to be presented in Sect. 4.

3.2 Outcoupling Rates

The normalization factor $\mathcal{N}$ in Eq. (7) determines the outcoupling rate of the atom laser. We calculate it by noting that the current density j_z in 1D is a rate and therefore must be equal to the outcoupling rate at least outside the trap. j_z is given by

$$j_z(z,t) = |\psi_0(z,t)|^2 v_z(z) \tag{12}$$

where $v_z(z)$ is the mean velocity of the atoms. If again we consider only the asymptotic behavior at large z and average over time, we get

$$j_z = \frac{2\hbar}{\pi M l} \tilde{\mathcal{N}}^2 \tag{13}$$

The outcoupling rate of a 3D, isotropic atom laser without gravity was calculated in Ref. [9]; a similar treatment is possible also in 1D. For one resonance point $z_{\rm res}$ (with detuning Δ) this yields

$$\Gamma^{\rm 1d}_{z_{\rm res}} = 2\pi \left(\frac{\Omega}{\omega_z}\right)^2 \frac{|\psi_{-1}(z_{\rm res})|^2}{\sqrt{2\Delta/\omega_z}} a_z \omega_z \tag{14}$$

If we now assume that the rates at the two resonance points $z_{\rm rf,1,2}$ add up to the total outcoupling rate and thus must be equal to j_z, we finally arrive at an expression for the normalization

$$\tilde{\mathcal{N}}^2 = \frac{\pi M l}{2\hbar}\left(\Gamma^{\rm 1d}_{z_{\rm rf,1}} + \Gamma^{\rm 1d}_{z_{\rm rf,2}}\right) \tag{15}$$

where $\Gamma^{\rm 1d}_{z_{\rm rf,1,2}}$ is obtained from the detunings $\Delta \pm \delta_{\rm rf}/2$.

Inspection of Eq. (7) shows that $|s|$ should reflect the relative outcoupling rates at the two resonance points. We define it via

$$|s|^2 = \frac{\Gamma^{\rm 1d}_{z_{\rm rf,1}}}{\Gamma^{\rm 1d}_{z_{\rm rf,2}}}, \tag{16}$$

which leads to

$$P = 2\frac{\sqrt{\Gamma^{\rm 1d}_{z_{\rm rf,1}}\Gamma^{\rm 1d}_{z_{\rm rf,2}}}}{\Gamma^{\rm 1d}_{z_{\rm rf,1}} + \Gamma^{\rm 1d}_{z_{\rm rf,2}}} \tag{17}$$

for the visibility. The numerical results presented in the next section coincide with this analytic formula in a very nice way.

For small radio frequency difference $\delta_{\rm rf}$ we can write Eq. (17) in an appealing way:

$$P = 2\frac{\left|\psi_{-1}\left(z_{\rm rf,1}\right)\right|\left|\psi_{-1}\left(z_{\rm rf,2}\right)\right|}{\left|\psi_{-1}\left(z_{\rm rf,1}\right)\right| + \left|\psi_{-1}\left(z_{\rm rf,2}\right)\right|} \tag{18}$$

i. e., P is related to the correlation function of the trapped Bose gas[24].

4. NUMERICAL RESULTS

The GPE in Eq. (1) can be solved numerically in an efficient way by using a split-operator technique on a spatial grid[25]. The time propagation start always with the ground state of the m = -1 atoms in the trap. The trap parameters are taken from the Munich experiment[3], the trapping frequency for the m = 1 state is $\omega_z = 2\pi\times$ 127 Hz. We often use harmonic oscillator units, the length unit is a_z = 0.95 μm, time is measured in $1/\omega_z$ = 1.3 ms.

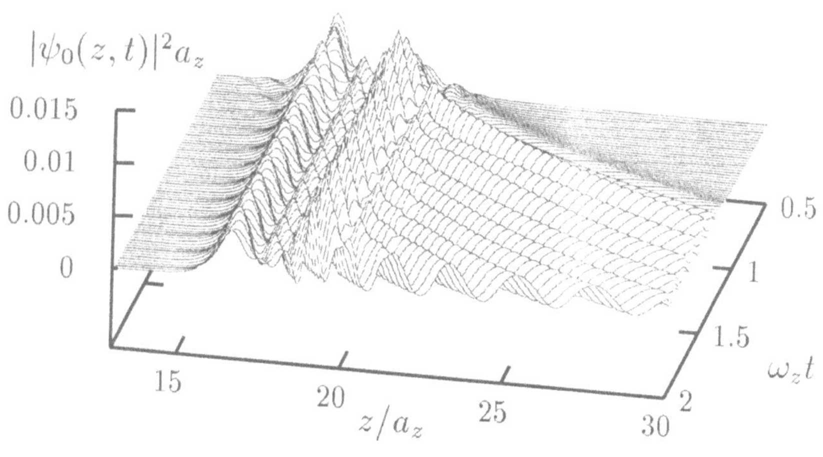

Figure 2. Plot of the outcoupled density. The atoms in the free m = 0 state emerge primarily at the two resonance points $z_{\rm rf,1,2}$ and then fall downwards (to the right in the plot). The radio frequency detunings in this plot are $\Delta_{1,2} = 2\pi \times (16.5 \pm 3.0)$ kHz.

In Fig. 2 we show a plot of the outcoupled density $|\psi_0(z,t)|^2$ for some typical experimental parameters. All results shown here were obtained for N=5 × 10^5 atoms in the condensate. We choose the detunings in such a way that the resonance point $z_{\rm rf,2}$ is always located at the middle of the condensate

($\Delta_l = 2\pi \times 16.5$ Hz) and Δ_l is fixed by the difference frequency δ_{rf}. The figure clearly demonstrates the pulsing nature of the output and shows how the atomic pulses are falling downwards due to the constant acceleration by gravity. The coupling strength represented by the Rabi frequency Ω is kept small in our calculations which enables us to use the weak coupling result of the Eq. (14) for the outcoupling rate. This ensures also that the output is continuous without having a pump mechanism to keep the number of condensate atoms constant. Nevertheless, the output rate is still decreasing slowly in time. We usually take a magnetic field strength of $B_{rf} = 0.1$ mG leading to $\Omega = 2\pi \times 70$ Hz. These values are also used in the experiments[24].

We now turn to the comparison of the analytical predictions of the preceding section and the numerical results. The interference pattern predicted by Eq. (8) for fixed time is shown in Fig. 3 together with numerical data. The analytical curve was obtained by calculating the normalization $\tilde{\mathcal{N}}^2$ and the visibility P with the help of Eq. (14). One may take the values for $|\psi_{-1}(z_{rf,1,2})|$ either from the numerics or calculate them in the Thomas-Fermi (TF) approximation [26]. In both cases $\tilde{\mathcal{N}}^2$ and P differ so slightly that the difference can not be seen in the plot of the graph in Fig. 3.

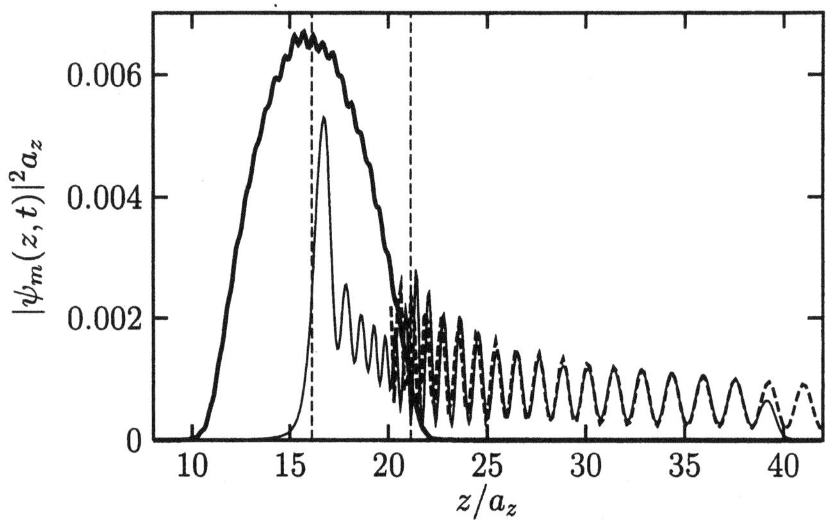

Figure 3. Plot of the densities of the condensate and of the outcoupled ($m = 0$) atoms. This plot is made after 5 ms of operation with $\delta_{rf} = 2\pi \times 12000$ Hz. The resonance point $z_{rf,1}$ is in the wing of the condensate (cf. vertical dashed lines), the visibility $P = 0.61$ is therefore small. The dashed line is a plot of the function in Eq. (8) with the normalization of Eq. (15). The density of the untrapped state goes down to zero around $z = 40\ a_z$ due to absorbing boundary conditions used in the calculations.

The next step is to check whether the dependence of the visibility as obtained from the numerics is described by Eq. (17). For this purpose we have propagated the initial state for a fixed time but with different RF differences δ_{rf}. Then we made fits of the analytic function for $|\psi_0(z,t)|^2$ in Eq. (8) to the numerical density distributions of the atomic beam (like in Fig. 3) to get values for the visibility P. The circles in Fig. 4 denote the points obtained in this way. One can also determine $P(\delta_{rf})$ by using either the TF-approximation (full line in the plot) or the numerical values of $|\psi_{-1}(z_{rf,1,2})|$ (triangles). The data obtained by fitting follows very well these (semi-) analytic values.

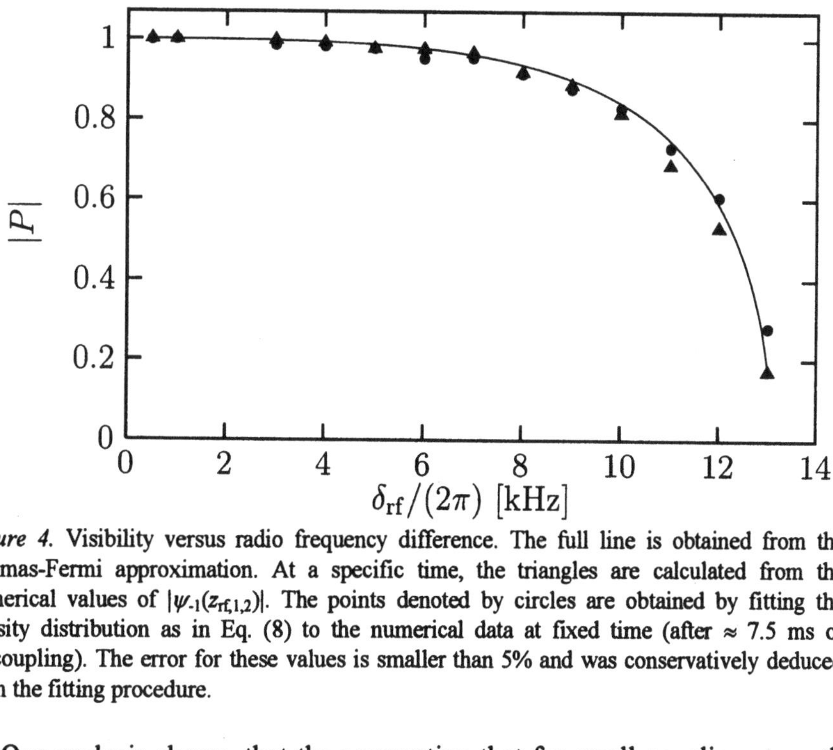

Figure 4. Visibility versus radio frequency difference. The full line is obtained from the Thomas-Fermi approximation. At a specific time, the triangles are calculated from the numerical values of $|\psi_{-1}(z_{rf,1,2})|$. The points denoted by circles are obtained by fitting the density distribution as in Eq. (8) to the numerical data at fixed time (after ≈ 7.5 ms of outcoupling). The error for these values is smaller than 5% and was conservatively deduced from the fitting procedure.

Our analysis shows, that the assumption that for small coupling strength the atoms are coupled out independently at the two resonance points is indeed correct. Still there are small, particle-like oscillations of the trapped condensate that manifest themselves as small wiggles on top of the condensate density distribution (see Fig. 3).

5. CONCLUSIONS

An atom laser based on RF-outcoupling can be operated in a coherently pulsed way by using two radio frequencies. Such a two-mode atom laser can be described by coupled Gross-Pitaevskii equations which are valid for very low temperatures (T ≈ 0). We have derived analytic expressions for the output rate and the visibility of the pulsed atomic beam in the weak coupling regime. A comparison between the analytical and numerical results shows a good agreement. This proves our assumption that the outcoupling process may be seen as the coherent outcoupling of two independent atomic beams.

Though our model is only one-dimensional, we expect also a good agreement with the full 3D situation of experiment; the estimate for the value of the output rate might be only in the right order of magnitude[16].

Summarizing, the manipulation of trapped Bose condensates with electromagnetic waves opens a wide new area for the field of coherent atom optics.

ACKNOWLEDGMENTS

It is a pleasure to thank Immanuel Bloch and Tilman Esslinger for a lot of nice and fruitful discussions. We acknowledge financial support by DFG under Grant Nr. SCHE 128/7-1.

REFERENCES

1. Parkins, A. S., and Walls, D. F., 1998, *Phys. Rep.* **303**:1.
2. Mewes, M.-O., *et al.*, 1997, *Phys. Rev. Lett.* **78**: 582.
3. Bloch, I, Hänsch, T. W., and Esslinger, T., 1999, *Phys. Rev. Lett.* **82**: 3008.
4. Hagley, W. W., *et al.*, 1999, *Science* **283**: 1706.
5. Ballagh, R. J., Burnett, K., and Scott, T. F., 1997, *Phys. Rev. Lett.* **78**: 1607.
6. Zhang, W, and Walls, D. F., 1998, *Phys. Rev. A* **57**: 1248.
7. Zhang, W, and Walls, D. F., 1998, *Phys. Rev. A* **58**: 4248.
8. Naraschewski, M., Schenzle A., and Wallis, H., 1997, *Phys. Rev. A* **56**: 603.
9. Steck, H., Naraschewski, M., and Wallis, H., 1998, *Phys. Rev. Lett.* **80**: 1.
10. Kneer, B., 1998, *et al., Phys. Rev. A* **58**: 4841.
11. Hutchinson, D. A. W., 1999, *Phys. Rev. Lett.* **82**: 6.
12. Japha, Y, Choi, S., Burnett, K., and Band, Y. B., 1999, *Phys. Rev. Lett.* **82**: 1079.
13. Graham R., and Walls, D. F., 1999, *Phys. Rev. A* **60**: 1429.
14. Band, Y B., Julienne, P. S., and Trippenbach, M., 1999, *Phys. Rev. A* **59**: 3823.
15. Edwards, M., 1999, *et al., J. Phys. B*: At. Mol. Opt. Phys. **32**: 2935.
16. Schneider. J., and Schenzle, A., 1999, *Appl. Phys. B* **69**: 353.
17. Hope, J.J., 1997, *Phys. Rev. A* **55**: 82531.
18. Moy, G.M., Hope, J.J., and Savage, C.M., 1999, *Phys. Rev. A* **59**: 667.
19. Jack, M.W., Naraschewski, M., Collett, M.J., and Walls, D.F., 1999, *Phys. Rev. A* **59**: 2962.
20. Hope, J., Moy, G.M., Collett, M.J., and Savage, C.M., 1999, condmat/9901073.
21. Hope, J., Moy, G.M., Collett, M.J., and Savage, C.M., 1999, condmat/9907023.
22. Breuer, H.P., Faller, D., Kappler, B., and Petruccione, F., 1999, *Phys. Rev. A* **60**: 3188.
23. Moy, G.M., Hope, J.J., and Savage, C.M., 1997, *Phys. Rev. A* **55**: 3631.

24. Bloch, I., Hànsch, T.W., and Esslinger, T., 1999, *Nature*, accepted.
25. Schneider, J., and Schenzle, A., 1999, cond-mat/9910442.
26. Dalfovo, F., Giorgini, S., Pitaevskii, L.P., and Stringari, S.,1999, *Rev. Mod. Phys.* **71**: 463.
27. Flügge, S., 1974, *Practical Quantum Mechanics*. Springer, Berlin.
28. *Handbook of Mathematical Functions*, 1964, (M. Abramowitz, and I. A. Stegun, eds.), National Bureau of Standards, Washington, chapter 10.4.

The Atomic Fabry-Perot Interferometer

[1,3]I. CARUSOTTO AND [2,3]G. C. LA ROCCA
[1]*Scuola Normale Superiore, P.za dei Cavalieri 7, I-56126 Pisa, Italy;* [2]*Dipartimento di Fisica, Università di Salerno, 1-84081 Baronissi (Sa), Italy;* [3]*INFM, Scuola Normale Superiore, P.za dei Cavalieri 7, I-56126 Pisa, Italy.*

1. INTRODUCTION

Since the realisation in 1995 of an atomic Bose-Einstein condensate (BEC), atom optics has began to take advantage of the remarkable properties of such a coherent matter wave[1]: indeed, the large number of atoms sharing the same quantum state can be described as a classical C-number matter wave[2]. The relation of the atomic field of a BEC to a non-degenerate thermal cloud is analogous to that of a laser field to the light of a lamp[3].

In the last few years, a great effort has been focussed on the extraction of coherent *atom laser* beams from trapped BECs[4,5] and, very recently, on the study of nonlinear atom optical effects[6], for which the classical wave character of the atom laser pulses plays a fundamental role. The atomic analog of the nonlinear susceptibility of optical media is given by atom-atom interactions[7,8]: the two-body elastic collisions are in fact responsible for a nonlinear cubic term in the equations of motion for the atomic field which has the same form as a Kerr-like nonlinear refractive index[9].

A few different nonlinear atom optical effects have already been studied by several authors, e.g. atomic four-wave mixing[6,10] and gap-solitons[7]. Our present contribution is a proposal of a new concept of Fabry-Perot interferometer for atomic matter waves to be used for cavity atom optics. The physical importance of such a device is obvious: indeed, not only it can effectively filter the atoms in terms of their velocity, but also it can coherently localize matter waves and hence enhance nonlinear interactions, in analogy to what done with photons in semiconductor microcavities[11].

The main requirement is that the discrete resonances in the linear transmission spectra due to resonant tunneling on bound states be sufficiently separated in energy as well as tightly localized in space. Our arrangement should be able to satisfy both conditions, since the mode spacing can be as large as one tenth of the Bragg frequency of the lattice and the spatial confinement as tight as a few tens of lattice periods.

The effect of nonlinearity in our proposed arrangement can be modelled as a blue-shift of the bound states induced by the (repulsive) atom-atom interactions; at a fixed incident frequency ω_{inc}, this can lead to a feed-back on the transmittivity which can be either positive of negative, depending on the relative position of ω_{inc}, and the empty cavity resonance ω_c, In the former case, *atom optical bistability* can occur; in the latter, *atom optical limiting*[12].

2. THE SIMPLE OPTICAL LATTICE: LOCALIZED MODES AND TUNNELING RESONANCES

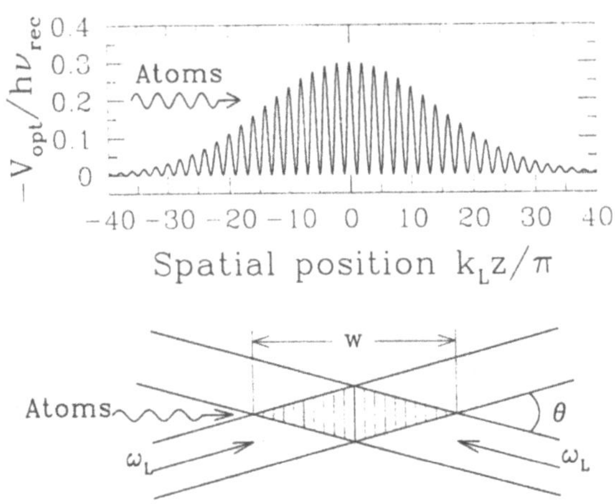

Figure 1. The simple optical lattice; lower panel: sketch of the proposed arrangement. Upper panel: qualitative behaviour of the optical potential experienced by the atoms

Consider a linearly polarized gaussian laser beam of frequency ω_L, beam-center Rabi frequency Ω_L and beam waist ω_o, back reflected by a system of mirrors along a nearly counterpropagating direction with the same linear polarization as in the lower panel of Fig. 1; let θ be the (small) angle between the two beams. In the region of space where they overlap, a *lin* || *lin* standing wave pattern is formed, with a profile similar to the one reproduced in the upper panel of Fig.1: a periodic potential of wavevector $k_l = k_L \cos\theta/2$ with a broad gaussian envelope of length[14] $w = w_o/\sin(\theta/2)$.

A monochromatic and collimated atom beam is sent on the optical lattice along its axis; assuming the laser frequency ω_L to be red-detuned and far off-resonance with respect to the atomic optical transition frequency ω_{at}, the atoms experience an optical potential $V^{opt}(z)$ proportional to the effective local light intensity

$$V^{opt}(z) = \hbar\Omega_o e^{-z^2/2w^2} \cos^2 k_l z \tag{1}$$

with $\Omega_o = \Omega_L^2/(\omega_L - \omega_{at})$. The atomic dynamics in the transverse plane has been neglected assuming a transversally wide optical lattice, so that the atomic beam can be considered as a plane wave and the calculations can be performed in a one-dimensional longitudinal approximation. A truly one-dimensional geometry can however be obtained by using an atomic waveguide to confine the transverse motion[15].

At linear regime, a simple numerical integration of the Schrödinger equation in the potential $V_i^{opt}(z)$ leads to the transmission spectra reproduced in the left panel of Fig. 2. Given the slowly modulated periodic shape of the optical potential, the main features of the spectra can be explained using concepts mutuated from solid state physics[16]. The atomic dispersion in a periodic potential is characterized by allowed bands and forbidden gaps, the first of which opens up close to the Bragg frequency $\omega_{Br} = \hbar k_l^2/2m$ of the lattice[17], the effect of the small modulation can be treated in the so-called *envelope function approach*, in which the atomic wave function is approximated by the product of the fast oscillating band edge Bloch wavefunction and a slow envelope. In this way, the motion in a complicated potential can be reduced to a Schrödinger-like equation for the envelope function, in which the potential is given by the slowly varying band edge energy and the mass is the effective mass $m^* = \hbar(d^2\omega/dk^2)^{-1}$.

In our case of a lattice potential very weak with respect to its Bragg frequency ω_{Br}, a *nearly free atom* approximation can be performed for the calculation of the first two atomic bands, leading to the values

$$\hbar\omega_{c,v}(z) = \hbar\omega_{Br} - \hbar\Omega_o\left(\frac{1}{2} \mp \frac{1}{4}\right) e^{-z^2/2w^2} \tag{2}$$

for the band edges and to the values

$$\frac{1}{m^*_{c,v}} = \frac{1}{m}\left(1 \pm \frac{8\omega_{Br}}{\Omega_o}\right) \tag{3}$$

for the effective masses at the band extrema. As usual, the mass of conduction band particles is positive while the mass of valence ones is negative; in modulus, they are both much smaller than the free-space one; for the parameters of Fig. 2, the reduction amounts to a factor 80. Quantitatively, the Bragg frequency ω_{Br} is of the same order as the recoil frequency $\omega_{rec} = \hbar\omega_{at}^2/2mc^2$; this means that the Raman output coupling scheme described in Ref. [4] can be used to create the incident atomic beam.

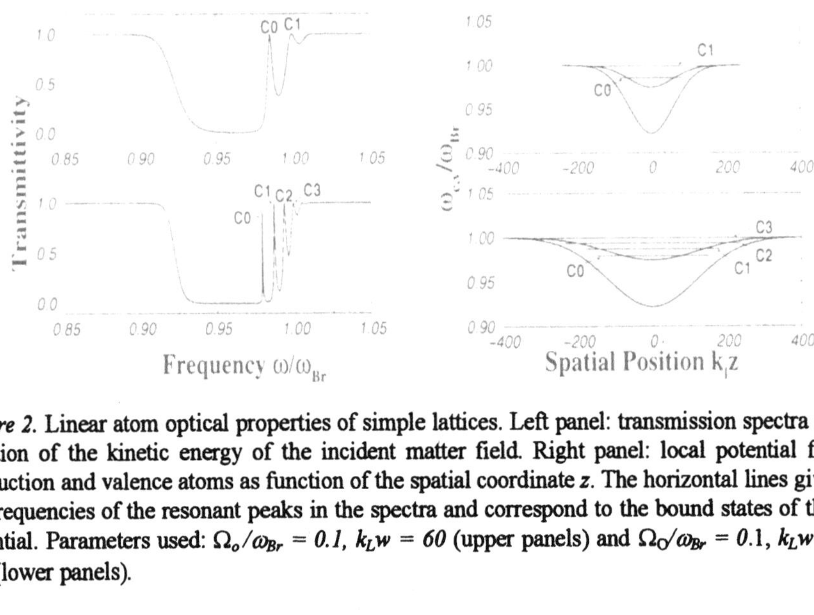

Figure 2. Linear atom optical properties of simple lattices. Left panel: transmission spectra as function of the kinetic energy of the incident matter field. Right panel: local potential for conduction and valence atoms as function of the spatial coordinate z. The horizontal lines give the frequencies of the resonant peaks in the spectra and correspond to the bound states of the potential. Parameters used: $\Omega_o/\omega_{Br} = 0.1$, $k_L w = 60$ (upper panels) and $\Omega_o/\omega_{Br} = 0.1$, $k_L w = 120$ (lower panels).

In the right panel of Fig. 2, we have plotted the spatial profiles of the band edge frequencies, which give the potential to be used in an eventual envelope function calculation; the Gaussian envelope of the lattice reflects in the inverted Gaussian shape of the band edges: outside the lattice, they both tend to ω_{Br}, while they show a minimum at the center of the lattice. Given their negative effective mass, valence band atoms are repelled by such a potential minimum, while conduction band atoms, whose mass is instead positive, are attracted and can eventually be trapped in it. Since the effective mass in the lattice is much smaller than the free-space one, the energy spacings predicted for the bound states of our arrangement result much larger than in the previous models of atomic Fabry-Perot interferometers[18].

The main features of the spectra reproduced in Fig. 2., can be simply explained by this model: for frequencies located below the minimum of the valence band edge frequency (i.e. its value at $z = 0$), transmission is complete: incident atoms can in fact couple to valence band states and freely propagate through the lattice. For energies above the Bragg frequency $\omega > \omega_{Br}$, transmission is again complete, since conduction band states are available for the propagation through the lattice. For frequencies in the gap between the two bands at $z = 0$, there is complete reflection, since the propagation in valence band can occur only up to the point where the valence band edge becomes equal to w. Afterwards, propagation is forbidden and atoms have to be reflected back.

Starting from analogous considerations we would expect a complete reflection also for incident frequencies comprised between the minimum of the conduction band edge frequency ω_c ($z = 0$) and the Bragg frequency $\omega_c (z = 0) < \omega < \omega_{Br}$; if the envelope function approximation was exact, this would be the case, since the discrete state in the conduction band potential would not be coupled to the incident states. But the fact that the lattice is not uniform gives a small but finite amplitude to non-adiabatic *interband* transitions which can couple the valence band incident atoms to the conduction band bound states of the potential well. Resonant tunneling processes can thus occur, giving the resonant peaks in the transmission spectra that can be observed in Fig. 2. The resonant enhancement compensates for the small amplitude of the interband jumps, giving transmission peaks which, in our case of a symmetric potential, grow from a nearly vanishing transmission up to a nearly complete one; since the coupling to the continuum of incident and transmitted states is very weak, the peak linewidth is however very narrow.

In Fig. 3, we have plotted the density profiles of the atomic wavefunctions for incident frequencies tuned at exact resonance with the quantized modes; the behaviour is the typical one of a quantum-mechanical particle in a one dimensional potential well; the mode wavefunction is localised in the well: the larger the order of the mode, the wider its spatial extension, as well as the number of its nodes. The enhanced particle density in the potential well compared to the external one is a clear signature of resonant behaviour: the peak density is in fact strictly related to the Q-factor of the localized mode, i.e. the inverse of the linewidth; in our specific case, this results a decreasing function of the mode order.

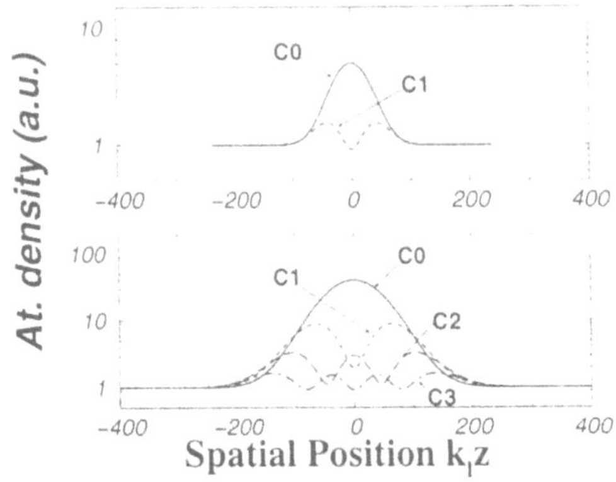

Figure 3. Density profile of the localized mode wavefunctions. The parameters are the same as in Fig. 2.

The difference between upper and lower panels of Fig. 2, consists only in a different value of the lattice width w, which has two main consequences: the increased spatial size of the potential well implies a reduced spacing of quantized modes and an increased total number of modes; at the same time, the slower spatial modulation of the lattice reduces the amplitude of non-adiabatic jumps, meaning a weaker coupling to the localized mode.

Using a bichromatic optical lattices, the performances of the device can be improved: a tighter confinement of the discrete mode wavefunction can be achieved, as well as a more efficient coupling to the incident and transmitted beams. The more complicate shape of the potential will allow for localized states also in valence band, for which the coupling to propagating modes is based on resonant tunneling across a potential barrier, without the need of interband transitions.

3. THE BICHROMATIC OPTICAL LATTICE

Consider a *bichromatic* optical lattice[12], formed by a pair of laser beams containing two distinct frequencies $\omega_{1,2} = \omega_L \pm \delta\omega_L/2$ ($\delta\omega_L << \omega_L$), both far off-resonance from the atomic transition at ω_{at}; provided the energy separation $\hbar\delta\omega_L$ of the two components of the light field is much larger than both the hyperfine splitting of the electronic ground state and the typical kinetic energies of the atoms (which are of the order of ω_{Br}), the optical potential can be written as an incoherent sum of the potentials due to the two frequency components

$$V^{\rm opt}(z)=\left\{\hbar\Omega_1^o\cos^2\left(k_L+\frac{\delta k_L}{2}\right)+\right.$$

$$\left.+\hbar\Omega_2^o\cos^2\left[\left(k_L-\frac{\delta k_L}{2}\right)z+\phi_o\right]\right\}e^{-z^2/2w^2} \tag{4}$$

where the effective Rabi frequencies $\Omega_{1,2}$ depend on the single beam Rabi frequencies and detunings according to $\Omega_{1,2}^o=(\Omega_{1,2}^L)^2/(\omega_{1,2}-\omega_{at})$. Interference terms due to processes in which the atom absorbs a photon at ω_1 and reemits a photon at ω_2 can be safely neglected on the base of energy conservation arguments.

The superposition of standing waves patterns given by Eq. (4) gives a periodically modulated optical lattice, in which both the amplitude and the lattice constant are periodic functions of the spatial coordinate z with a period $l_{mod}=\pi/\delta k_L$ and a phase given by the relative phase ϕ_o, of the two standing wave patterns, which can be controlled acting by the position of the backreflecting mirror. As previously, the Gaussian envelope of the laser beams imposes an overall Gaussian profile to the lattice, setting the total length to w and limiting the number of oscillations actually present. By choosing the appropriate values for the lattice parameters $\left(\phi_o=\pi/2,\Omega_1^o\geq\Omega_2^o\right)$ the symmetric configuration of Fig. 4 can be achieved, in which the lattice amplitude has a single minimum at its center.

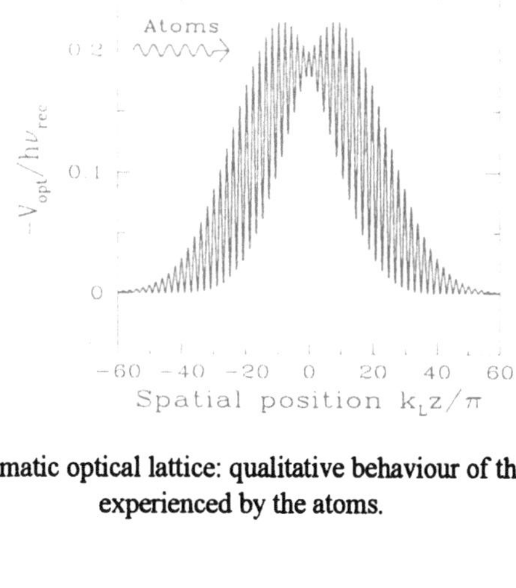

Figure 4. The bichromatic optical lattice: qualitative behaviour of the optical potential experienced by the atoms.

As previously, the transmission spectra through the bichromatic lattice can be obtained by numerically integrating the Schrödinger equation with the appropriate optical potential Eq. (4); an example of spectrum is shown in the left panel of Fig. 5. Again, the results can be explained by the envelope function model, together with the band edge profile reproduced in the right panel: given the opposite signs of the effective masses, both conduction and valence atoms can now be trapped, leading to two series of quantized bound states on which resonant tunneling processes can occur. As for the simple optical lattice, the coupling of conduction band bound states to the propagating ones is due to interband jumps, while for valence band states it is given by tunneling processes across a potential barrier.

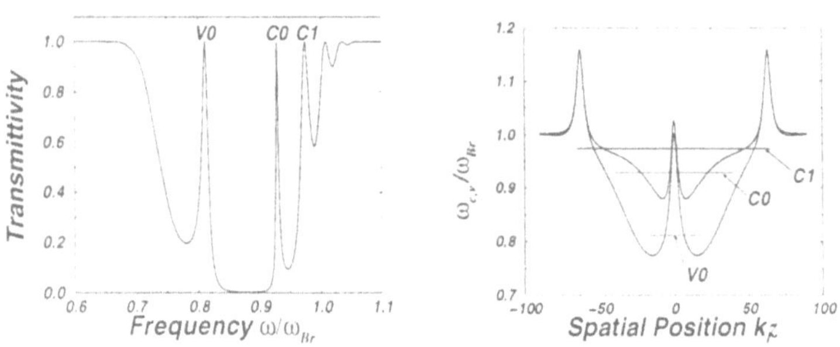

Figure 5. Linear atom optical properties of bichromatic optical lattices. Left panel: transmission spectra as function of the kinetic energy of the incident matter field. Right panel: local potential for conduction and valence atoms as function of the spatial coordinate z. The horizontal lines give the frequencies of the resonant peaks in the spectra and correspond to the bound states of the potential. Since the nearly degeneracy of valence and conduction bands for $|k_L z| \geq 60$, the potential hills on the edges of the lattice have no physical relevance. Used parameters $\Omega_1^o/\omega_{Br} = 0.2$, $\Omega_2^o/\omega_{Br} = 0.15$ and $\phi_o = \pi/2$; $\delta\omega_L/\omega_L = 0.05$ and $k_L w = 30$.

For frequencies above the maximum value of the conduction band edge frequency $\omega > \omega_c^{(max)}$ and below the minimum value of the valence band edge frequency $\omega < \omega_v^{(min)}$ transmission is again complete, since atoms find propagating states at such a frequency for every z; for frequencies comprised between these two values, transmission is instead negligible but for the resonance peaks corresponding to the discrete bound states; as previously, the weaker the coupling to the propagating modes, the narrower the linewidth and the smaller the integrated peak intensity. The unity value of the transmittivity at exact resonance is a consequence of the symmetry of the lattice: the smaller the peak intensity, the larger the sensitivity to asymmetries[20].

4. CONCLUSIONS AND PERSPECTIVES: NONLINEAR AND QUANTUM ATOM OPTICS

If we assume the frequency spacing of the bound states to be much larger than their nonlinear shift due to (repulsive) atom-atom interactions, our system can be described in the *single mode* approach by the Hamiltonian

$$\mathcal{H} = \hbar\omega_c \hat{a}^+\hat{a} + \frac{\hbar\omega_{nl}}{2}\hat{a}^+\hat{a}^+\hat{a}\hat{a} + \hbar k_{inc} e^{-i\omega_{inc}t}\hat{a}^+ + h.c. \tag{5}$$

the first term defines the empty cavity oscillation frequency ω_c the second one the nonlinear shift per particle ω_{nl}, the last ones describe the driving of the cavity by the incident field at ω_{inc}. The transmission losses have to be inserted in the master equation according to the general theory of cavity damping[19].

If we perform the so-called *mean field* approximation, all the field operators $\hat{a}$ can be substituted by their mean values $a = \langle \hat{a} \rangle$ making the operator products to factorize[2]. This leads to the classical results for the nonlinear oscillator[13]: depending on the sign of $\delta = \omega_{inc} - \omega_{cav}$, *atom optical bistability* ($\delta > 0$) or *atom optical limiting* ($\delta \leq 0$) has been obtained[12].

Such a mean-field approach is only valid if the nonlinear coupling per particle ω_{nl} is much smaller than the cavity linewidth γ; otherwise the theory has to keep track of the discrete nature of the atoms and strongly nonclassical features can arise in transmitted beam. As a preliminary result, in the limiting case $\omega_{nl} >> \gamma$, if the driving is resonant with the empty cavity $\omega_{inc} \approx \omega_c$, expect the statistical properties of the transmitted atoms to be the same as the ones of the scattered light by a single two-level atom. Hence, the beam of transmitted atoms results strongly antibunched and, even for a strictly monochromatic driving, the transmitted spectrum can show a triplet of peaks, the so-called Mollow triplet[3]. The fact that atom-atom interactions are much stronger than the corresponding photon-photon ones in a nonlinear dielectric material suggests that the observation of quantum optical effects should be more easily performed in the atom optical systems[21] than in the classical photonic ones.

ACKNOWLEDGMENTS

It is a pleasure to thank A. Minguzzi, M.L. Chiofalo, C. Henkel, F. Minardi and F.S. Cataliotti for useful discussions. I. Carusotto acknowledges partial financial support from M. Inguscio.

REFERENCES

1. Ketterle, W. and Miesner, H.-J., 1997, Coherence properties of Bose-Einstein condensates and atom lasers, *Phys.Rev.A* **56**: 3291; Wiseman, H.M., 1997, Defining the (atom) laser, *Phys.Rev. A* **56**: 2068.
2. Dalfovo, F., Giorgini, S., Pitaevskii L.P., and Stringari, S., 1999, Theory of trapped Bose-condensed gases, *Rev.Mod.Phys.* **71**: 463.
3. Loudon, R., 1973, The quantum theory of light, *Clarendon Press*, Oxford.
4. Hagley, E.W., Deng, L., Kozuma, M., Wen, J., Helmerson, K., Rolston, S.L., and Phillips, W.D., 1999, A well-collimated quasi-continuous atom laser, *Science* **283**: 1706.
5. Bloch, L, Hànsch T.W., and Esslinger, T., 1999, An atom laser with a cw output coupler, *Phys.Rev.Lett.* **82**: 3008 ;Mewes, M.O, Andrews, M.R., Kurn, D.M., Durfee, D.S., Townsend C.G., and Ketterle, W., 1997, Output coupler for Bose-Einstein condensed atoms, *Phys.Rev.Lett.* **78**: 582.
6. Deng, L., Hagley, E.W., Wen, J., Trippenbach, M., Band, Y., Julienne, P.S., Simsarian, J.E., Helmerson, K., Rolston S.L., and Phillips, W.D., 1999, Fourwave mixing with matter waves, *Nature* **398**: 218.
7. Lenz, G., Meystre P., and Wright, E.M., 1993, Nonlinear atom optics, *Phys.Rev.Lett.* **71**: 3271; Lenz, G., Meystre P., and Wright, E.M., 1994, Nonlinear atom optics: general formalism and atomic solitons, *Phys.Rev. A* **50**: 1681; Zobay, O., Pótting, S., Meystre, P., and Wright, E.M., 1999, Creation of gap solitons in Bose-Eistein condensates, *Phys.Rev.A* **59**: 643; Schernthanner, K.J., Lenz G., and Meystre, P., 1994, Influence of spontaneous emission on atomic solitons, *Phys.Rev.A* **50**: 4170.
8. Zhang, W.,and Walls, D.F., 1994, Quantum field theory of interaction of ultracold atoms with a light wave: Bragg scattering in nonlinear atom optics, *Phys.Rev.A* **49**: 3799; Zhang W., and Walls, D.F., 1995, Bosonic-degeneracy-induced quantum correlation in a nonlinear atomic beam splitter, *Phys.Rev.A* **52**: 4696.
9. Boyd, R.W., 1992, *Nonlinear optics.* Academic Press, San Diego.
10. Trippenbach, M., Band YB., and Julienne, P., 1998, Four-wave mixing in the scattering of Bose-Einstein condensates, *Opt.Express* **3**: 530; Goldstein E.V., and Meystre, P.,1999, Phase conjugation of trapped Bose-Einstein condensates, *Phys.Rev.A* **59**:1509.
11. Carusotto, I., and La Rocca, G.C., 1997, Optical response of linear and nonlinear photonic superlattices near the first photonic band gap, *Phys.Stat.Sol.***164**: 377; Carusotto, I., and La Rocca, G.C., 1998, Nonlinear optics of coupled semiconductor microcavities, *Phys.Lett.A* **243**: 236.
12. Carusotto, I., and La Rocca, G.C., 2000, The modulated optical lattice as an atomic Fabry-Perot interferometer, *Phys.Rev.Lett* **84**: 399.
13. Landau, L.D., and Lifshitz, E.M., 1960,*Mechanics.* Pergamon Press, Oxford.
14. Friedman, N., Ozeri R., and Davidson, N., 1998, Quantum reflection of atoms from a periodic potential, *J.Opt.Soc.Am. B* **15**: 1749.
15. Fortagh, J., Grossmann, A., Zimmermann C., and Hänsch, T.W., 1998, Miniaturized wire trap for neutral atoms, *Phys.Rev.Lett.* **81**: 5310 (1998); Renn, M.J., Montgomery, D., Vdovin, O., Anderson, D.Z., Wieman C.E., and Cornell, E.A., 1995, Laser-guided atoms in hollow-core optical fibers, *Phys.Rev.Lett.* **75**: 3253.
16. Yu, P., and Cardona, M., 1996, ***Fundamentals of semiconductors: physics and material properities***. Springer, Berlin; Ashcroft N.W., and Mermin, N.D., 1976, ***Solid state physics.*** Saunders College Publishing, Orlando; Bassani F., and Pastori Parravicini, G., 1975, ***Electronic states and optical transitions in solids.*** Pergamon Press, Oxford.

17. Wilkens, M., Schumacher E., and Meystre, P., 1991, Band theory of a common model in atom optics, *Phys.Rev.A* **44**: 3130; Oberthaler, M.K., Abfalterer, R., Bernet, S., Schmiedmayer J., and Zeilinger, A., 1996, Atom waves in crystals of light, *Phys.Rev.Lett.* **77**: 4980.
18. Wilkens, M., Goldstein, E., Taylor B., and Meystre, P., 1993, Fabry-Perot interferometer for atoms, *Phys.Rev.A* **47**: 2366.
19. Cohen-Tannoudji, C., Dupont-Roc J., and Grynberg, G., 1988, *Processus d'interaction entre photons et atones.* InterEditions/Editions du CNRS, Paris.
20. Born, M., and Wolf, E., 1964, *Principles of Optics.* Pergamon Press, London.
21. Moore, M.G., and Meystre, P., 1999, Optical control and entanglement of atomic Schrödinger fields, *Phys.Rev.A* **59**: 81754; Loudon, R., 1998, Fermion and boson beam-splitter statistics, *Phys.Rev. A* **56**: 4904; Goldstein, E.V., and Meystre, P., 1999 Quantum theory of atomic four-wave mixing in Bose condensates, *Phys.Rev.A* **59**: 3896.

RF-Induced Evaporative Cooling And BEC In A High Magnetic Field

[1]P. BOUYER, [1]V. BOYER, [1]S.G. MURDOCH, [1]G. DELANNOY, [1]Y. LE COQ, [1]A. ASPECT AND [2]M. LÉCRIVAIN
[1]Groupe d'Optique Atomique Laboratoire Charles Fabry de l'Institut d'Optique UMRA 8501 du CNRS Orsay, France. [2]L.E.Si.R. URA 1375 du CNRS - ENS Cachan, France

1. INTRODUCTION

Bose-Einstein condensates[1-3] are very promising for atom optics[4-6], where they are expected to play a role as important as lasers in photon optics, since they are coherent sources of atoms, with a very large luminosity. In view of applications, it is crucial to develop apparatuses that produce BEC faster - the average production rate of a condensate is 0.01 Hz - and with more versatile designs, by reducing, for example, the power consumption of the electromagnets.

For this purpose, we have developed a magnetic trap for atoms based on an iron core electromagnet, in order to avoid the large currents, electric powers, and high pressure water cooling, required in schemes using simple coils. The latest developments allow us to achieve a very high confinement that will permit to achieve much higher production rates.

In this Chapter, we will first present the design of the iron core electromagnet and how to solve the specific experimental problems raised by this technique. After presenting the experimental set-up, we will address the interruption of runaway evaporative cooling when the Zeeman effect is not negligible compared to the hyperfine structure. We will then present two ways to circumvent this problem: use of multiple RF frequencies and sympathetic cooling. Another method, hyperfine evaporation, was used in

Ref. [2]. In conclusion, we will present some applications of this high magnetic fields.

2. IRON-CORE ELECTROMAGNET TRAP FOR ATOMS

Our iron-core electromagnet is shown in Fig. 1. It follows the scheme of Tollett *et al.*[7]. Instead of using permanent magnets, we use pure iron pole pieces excited by coils, which allows us to vary the trap configuration. The use of ferromagnetic materials was reported in Ref. [8]. The role of the pole pieces is to guide the magnetic field created by the excitation coils far away from the center of the trap towards the tips of the poles. To understand this effect, let us consider the magnetic circuit represented in Fig. 2.

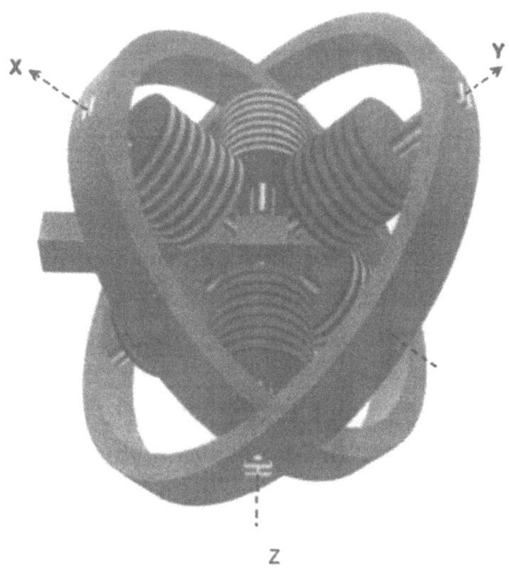

Figure 1. Diagram showing the position of the pole pieces of the electromagnet. The tip-to-tip spacing is 3 cm and the section of the cell is 1×2 cm^2.

The two tips are separated by a gap e of a few centimeters. The ferromagnetic structure has a total length l and a section S. The whole structure is excited by a coil of $2N$ loops driven by a current I, leading to an excitation $2NI$. From the Ampere's theorem[9], we can introduce the reluctance R_{iron}:

$$R_{\text{iron}} = \oint_{\Gamma_i} \frac{dl}{\mu_r \mu_0} \cong \frac{l}{\mu_r \mu_0} \tag{1}$$

inside the iron core and R_{gap}:

$$R_{\text{gap}} = \oint_{\Gamma_g} \frac{dl}{\mu_0} \cong \frac{e}{\mu_0} \tag{2}$$

in the gap between the tips. A simple relation between the excitation $2NI$ and the magnetic flux BS can be written:

$$BS = \frac{2NI}{R_{\text{iron}} + R_{\text{gap}}} \tag{3}$$

In our case, the gap e and the size of the ferromagnets l are comparable. Since μ_r is very important ($\mu_r > 10^4$) for ferromagnetic materials, only the gap contribution is important. The case of very small gap where $R_{\text{iron}} > R_{\text{gap}}$ was studied in Ref. [10]. In this case, the ferromagnetic materials amplify the magnetic field in the gap. A more complete calculation shows that the field created in the gap is similar to that created with two coils of excitation NI placed close to the tips as represented[10] in Fig. 2. Thus, guiding of the magnetic field created by arbitrary large coils far away from the rather small trapping volume is achieved. All this demonstration is only valid if a yoke links a north pole to a south pole. If not, no guiding occurs and the field in the gap is significantly reduced.

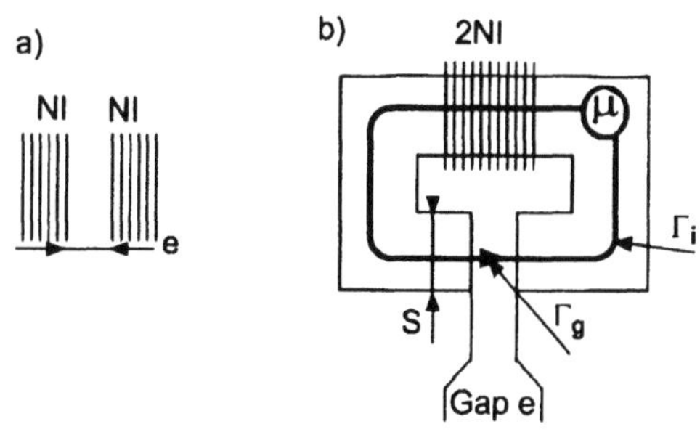

Figure 2. a) System of two anti-Helmholtz coils of excitation NI separated by e. b) Equivalent magnetic circuit. l: length of the ensemble poles + yoke. e: gap between the tips.

We will focus now on our Ioffe-type trap for Rubidium 87 (Fig. 1), which consists of a superposition of a linear quadrupole field and dipole field. The linear quadrupole field of gradient G is produced by two pairs of coils in a anti-Helmholtz configuration, along the x- and y-axis, and the dipole field of curvature C is produced by two coils along the z-axis in a Helmholtz configuration[8]. The magnitude of the total magnetic field can be approximated by:

$$|\mathbf{B}| = B_0 + \left(\frac{G^2}{2B_0} - \frac{C}{2}\right)\left(x^2 + y^2\right) + Cz^2 \tag{4}$$

leading to an anisotropic axial harmonic potential for trapping states, in the linear Zeeman effect regime.

The use of ferromagnetic materials raises several problems:

Geometry. As mentioned previously, a ferromagnetic yoke has to link a north pole to a south pole. A bad coupling between two poles can result in reduced performances of the electromagnet. Our solution optimizes the optical access around the vacuum cell while keeping the required coupling efficient.

In addition, the geometry of the magnetic field relies on the shape of the pole pieces rather than the geometry of the exciting coils. This results in the fact that the bias field B_0 cannot be easily decreased without also canceling the dipole curvature C. In the systems using coils, an additional compensation coil is used the reduce the bias field. In our case, this additional external excitation would couple into the ferromagnetic structure, decreasing both the bias field and the curvature of the field. This implies a high bias field, for which the Zeeman shift is no longer linear in the magnetic field, due to a contamination between hyperfine levels. In fact, a more complicated design of the poles along the z-axis allows for canceling the bias field while keeping an important dipole curvature. This new design will be discussed in the last section of this Chapter.

Hysteresis. Hysteresis prevents from returning to zero magnetic field after having switched ON and OFF the electromagnet. A remanent field of a few Gauss remains, as shown in Fig. 3.

This remanent magnetization needs to be cancelled in order, for example, to release the atoms and make a velocity (temperature) time-of-flight (TOF) measurement. Extra coils around the magnetic poles (see Fig. 4), carrying a DC current will shift the hysteresis cycle so that it crosses zero again. The current is adjusted to provide the coercive excitation which cancel exactly the remanent magnetic field when the large coils are switched off. This compensation is valid as long as we remain on the same excitation cycle. This stability is achieved thanks to a computer control of the experiment.

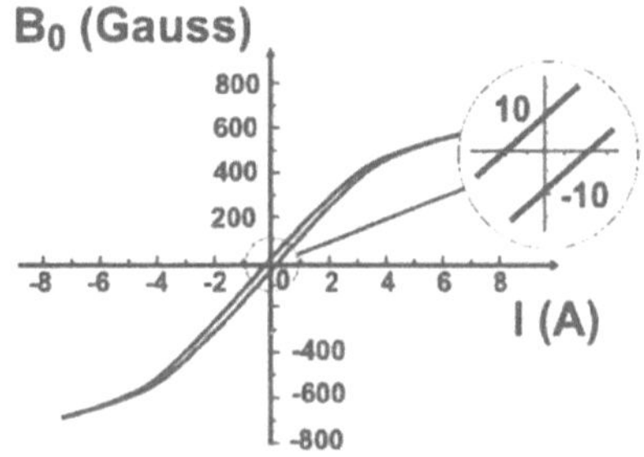

Figure 3. Hysteresis cycle of a ferromagnetic structure. The insert shows the typical remanent field value before compensation.

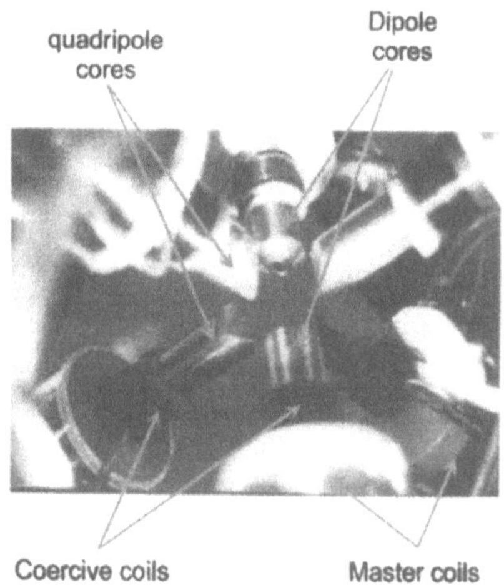

Figure 4. Photo of the poles, the excitation coils and the compensation coils.

Dynamic properties. The use of big coils (lots of loops) results in a big inductance ($\cong$ 100 mH), leading to a switching time $\tau = L/R \cong 100$ ms, too long to allow a good transfer of atoms into the magnetic trap. By assisting the switching with a capacitor, we are able to reduce τ to less than a millisecond.

Eddy currents are expected to seriously slow down the switching, and indeed a field decay constant of more than 10 ms was found in our first electromagnet[8]. The use of laminated material (stacked 100 μm/1 mm thick layers of ferromagnetic materials isolated by epoxy) solves this problem and allows to switch ON or OFF the field within 100 μs.

3. EXPERIMENTAL SETUP

The experimental setup is shown in Fig. 5. The electromagnet is placed around a glass cell of inner section of 1 × 2 cm^2, pumped with two ion pumps and a titanium sublimation pump. The background pressure is of the order of 10^{-11} mbar. The tip to tip spacing is 3 cm for the poles of the dipole, and 2 cm for the poles of both quadrupoles. The power consumption is 25 W per coil for a gradient of 900 Gauss/cm, and the maximum gradient at saturation is 1400 Gauss/cm.

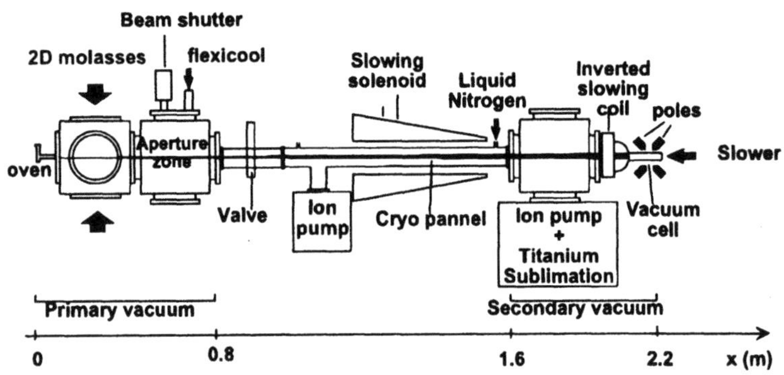

Figure 5. Experimental apparatus.

Our source of atoms is a Zeeman slowed atomic beam of ^{87}Rb. The beam is collimated with a transverse molasses and is decelerated in a partially reversed solenoid. It allows us to load a MOT with 1.5×10^9 atoms in 5 s. In order to increase the density, we then switch to a forced dark MOT by suppressing the repumper in the center and adding a depumper tuned to the $F = 2 \rightarrow F' = 2$ transition[12,13]. We obtain 8×10^8 atoms at a density of 1.5×10^{11} cm^{-3}. After additional molasses cooling, we optically pump the atoms into either the $F = 2$ or the $F = 1$ state. We then switch on the electromagnet in a configuration adapted to the phase space density of the atomic cloud. The bias field B_0 is fixed to ~ 140 Gauss for $F = 2$, or to ~ 207 Gauss for $F = 1$. The corresponding oscillation frequency is $\Omega/2\pi = 21$ Hz for $F = 2$ and $\Omega/2\pi = 18$ Hz for $F = 1$. We end up with $N = 4 \times 10^8$ trapped atoms at a temperature of 120 μK, with a peak density of 5×10^{10} cm^{-3}. All this information is obtained by conventional absorption imaging on a CCD camera.

4. INTERRUPTED EVAPORATIVE COOLING IN A HIGH MAGNETIC FIELD

To achieve Bose-Einstein Condensation (BEC), we use RF-induced evaporative cooling of the ^{87}Rb atoms confined in the magnetic trap.

In the approximation of the magnetic moment of the atom adiabatically following the direction of the field during the atomic motion, the magnetic potential is a function of the modulus of the field and the projection $\hbar m$ of the total angular momentum on the field axis. Depending on the sign of m, the Zeeman sublevel will be confined towards (trapping state) or expelled from (non-trapping state) a local minimum of the field modulus. RF-induced evaporative cooling consists of coupling the trapping state to a non-trapping state with a radio-frequency field (RF knife), in order to remove the most energetic atoms from the trap.

Efficient evaporative cooling[14-16] relies on fast thermal relaxation, and thus on the ability of increasing the collision rate by adiabatic compression of the atomic cloud. The most widely used mean to increase the curvature of the trapping potential is to partially cancel the bias field B_0 with two additional coils in Helmholtz configuration along the z-axis. As seen in Eq. (4), this increases the radial curvature without changing the axial curvature. Typical values of the compensated bias field in previous experiments are 1 to 10 Gauss. One can also radially compress the atomic cloud by increasing the gradient G without modifying the bias field. This is the approach for our trap. However, the quadratic Zeeman effect is not negligible anymore.

Defining

$$\varepsilon = \frac{\mu_B B_0}{2\hbar\omega_{\mathrm{hf}}} = \frac{\omega_{\mathrm{L}}}{\omega_{\mathrm{hf}}} \tag{5}$$

as the linear Zeeman effect ω_{L} divided by the hyperfine splitting of the ground state ω_{hf}, we may write the Zeeman effect for ^{87}Rb to the second order in ε as

$$E_{F,mf}(\varepsilon) - E_F = (-1)^F \left\{\hbar\omega_{\mathrm{hf}} m_F \varepsilon + \hbar\omega_{\mathrm{hf}} \left(4 - m_F^2\right)\varepsilon^2\right. \tag{6}$$

with

$$E_F = \frac{\hbar\omega_{\mathrm{hf}}}{2}\left(1 + (-1)^F\right) \tag{7}$$

the hyperfine structure. It gives a quadratic Zeeman effect of about 2 MHz at 100 Gauss (a typical bias field for our trap). An immediate consequence is that the ($F = 2$, $m_F = 0$) state is a trapping state. For a small magnetic field, as in most experiments, the second order in Eq. (6) is negligible. The RF coupling between adjacent Zeeman sublevels ($|\Delta_m| = 1$) results in an adiabatic multiphoton transition to the non-trapping state, leading to efficient RF induced evaporative cooling. In the case of a high bias field, the RF couplings are not resonant at the same location because of the quadratic Zeeman effect. Depending on the hyperfine level, this effect leads to different scenarios[17], that we have experimentally identified[18] thanks to our magnetic trap allowing strong confinement with a high bias field.

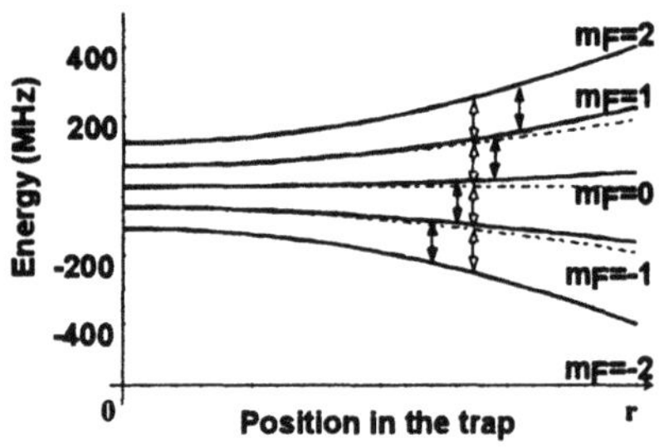

Figure 6. Forced evaporative cooling, in the presence of the quadratic Zeeman shift. We have plotted the magnetic energy of the ($F = 2, m_F = 0$) sublevels as a function of a trap coordinate r (the magnetic field $B(r) = B_0 + \beta r^2$). For a given RF frequency, the various transitions do not happen at the same position. We have indicated with black arrows the resonant transitions and with white arrows what would happen in the absence of quadratic corrections.

For atoms in the ($F = 2$, $m_F = 2$) state, forced evaporative cooling will be subject to unwanted effects as the RF knife gets close to the bottom of the potential well. Indeed, we can only cool down the sample to about 50 μK until the atoms cannot be transferred to a non-trapping state and cooling stops. In addition, a careful analysis of the evaporation shows that it can only be optimized to rather poor efficiency. In order to give an insight of the efficiency of the forced evaporation in such a situation, let us consider an atom initially in the $m_F = 2$ trapping state, and following the path represented in Fig. 7 to connect to the $m_F = -1$ non-trapping state.

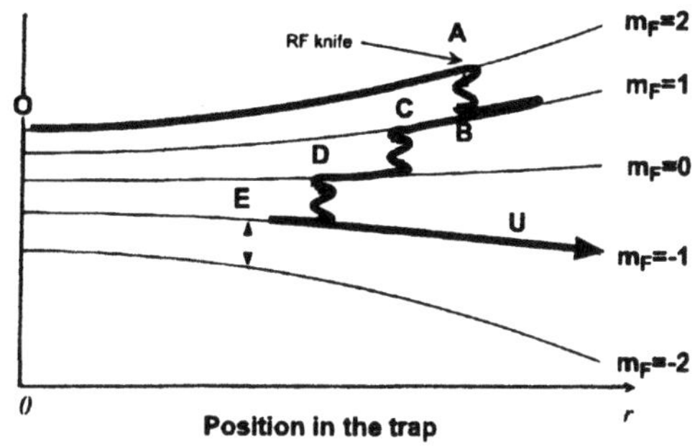

Figure 7. One short path connecting the trapping state ($F = 2$, $m_F = 2$) to the non-trapping state ($F = 2$, $m_F = -1$). The atom crosses 5 times the RF knife without making a transition at point **B** and **E**.

The atom, travelling from the center of the trap, reaches the RF knife at A, and makes a transition to the $m_F = 1$ state at **B** with a transition probability P. From there, it continues to move away from the center. When it comes back towards the center of the trap, the atom passing on **B** must not make a transition in order to reach the RF knife on **C**. The probability to reach **BC** from **OA** is $P(1 - P)$. Assuming the same probability P for all the RF transitions, the probability that the atom follows the path shown in Fig. 7 and leaves the trap on **EU** is $P^3(1 - P)^2$. There are 4 analogous paths involving 5 crossings of the RF knife. Consequently, neglecting interference effects, the total probability associated to these 4 short evaporation paths is $4P^3 (1 - P)^2$. This probability has a maximum value of about 10% for a transition probability P of $\sqrt{3/5}$, and is associated to a precise value of the atomic velocity. When considering all possible velocities, the probability of leaving the trap on $m_F = -1$ averages to less than 10%, much less than for the standard situation where the adiabatic passage has 100% efficiency for almost all velocities[17,19]. The experimental observation in Fig. 8 supports this simple analysis: when we increase the RF power, the efficiency of the evaporation reaches a maximum and then decreases. Of course, with sufficient RF power (P > 100W), we would eventually reach a situation where all the various transitions merge, and a direct adiabatic transition to a non-trapping state with 100% efficiency would be obtained.

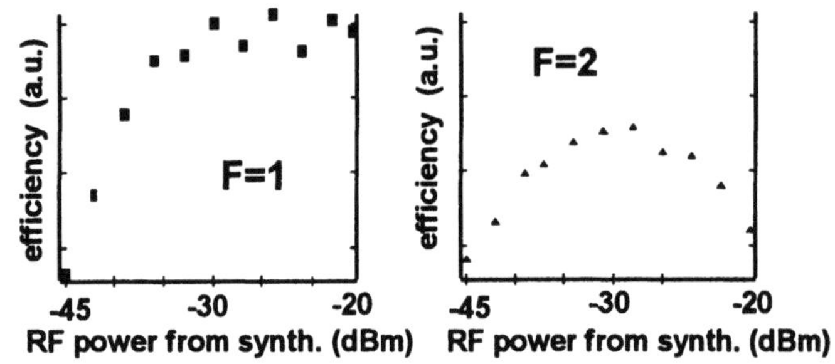

Figure 8. Efficiency of the evaporation for $F = 1$ (left) and $F = 2$ (right) as observed via the optical thickness of the atomic cloud. A clear optimum can be observed in the $F = 2$ case (the relative height between $F = 1$ and $F = 2$ is arbitrary).

In addition, all paths longer than the 5 crossings path as in Fig. 7 contribute to build up a macroscopic population in the intermediate levels $m_F = 1$ and $m_F = 0$, as soon as evaporation starts. This results in the presence of the atoms intermediate sublevels during the evaporation and the observation of a heating of 5 μK/s of the $m_F = 2$ cloud, when we remove the RF knife at the end of the evaporation[19].

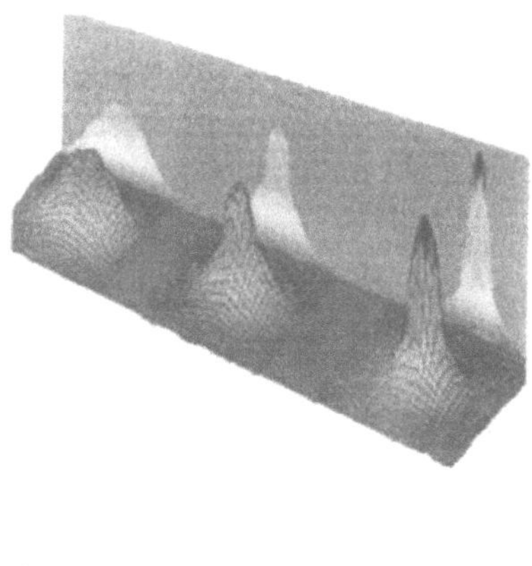

Figure 9. Bose Einstein condensation in $F = 1$, $m_F = -1$

Forced evaporative cooling for atoms trapped in ($F = 1$, $m_F = -1$) is not adversely affected by the quadratic Zeeman effect at a bias field B_0 of 207 Gauss, since the $m_F = 0$ state is non-trapping because of the sign of the quadratic term. The RF power has to be large enough to ensure an adiabatic transfer to $m_F = 0$ with an efficiency close to 1. For atoms in the $F = 1$ state, after adiabatic compression, the oscillation frequencies are $\Omega_{\|}/2\pi = 18$ Hz along the dipole axis, and $\Omega_{\perp}/2\pi = 55$ Hz along both quadrupole axis. We could successfully cool down the sample, and we obtained a condensate of a few 10^6 atoms as shown in Fig. 9.

5. REACHING BEC IN F=2 IN HIGH MAGNETIC FIELD

Several strategies can circumvent the adverse consequences of the quadratic Zeeman effect, and achieve efficient forced evaporative cooling of ^{87}Rb in $F = 2$.

5.1 Evaporation with 3 RF Knives

When evaporating the ($F = 2$ and $m_F = 2$) state of ^{87}Rb in a high bias field trap like ours, the RF couplings between the adjacent magnetic sublevels are not resonant at the location in the trap and thus the transfer of atoms from trapping to non-trapping states is inefficient (or even non-existent). This problem can be overcome if we evaporate with three distinct RF frequencies chosen so that a direct transition to a non trapping state is always possible, as shown in Fig. 10.

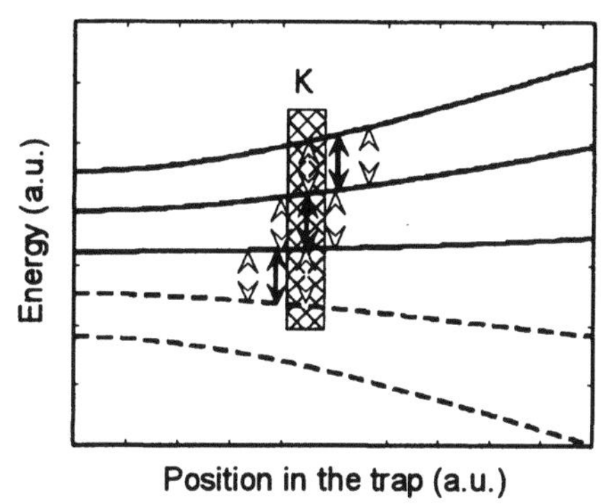

Figure 10. Implementation of 3 RF knives to evaporate in a high magnetic field. All possible transitions are represented. Frequency matching is only achieved at the RF knife *k*.

However the requirement of three independent frequency sources is not very practical. Apart from the technological complexity, the mixing of 3 different frequencies can generate sidebands that will induce stray RF knives, reducing the evaporation efficiency. Rather a simpler solution involving the mixing of two frequencies will compensate the quadratic term of the Zeeman shift. As calculated by Eq. (6), the difference between successive RF transitions is the same. Thus, compensation is achievable by mixing one independent frequency source (the carrier) with another RF frequency source of frequency $\delta\nu$ to obtain a pair of sidebands of fixed detuning and approximately the same power as the carrier. Additional sidebands of much lower power are also generated. Consequently, this forced us to reduce the RF power of the 3 knives to avoid unwanted evaporation effects. This detuning by is chosen as to align the three knives perfectly only at the end of the evaporation ramp.

This approach will be limited to magnetic fields where the higher order Zeeman terms are not significant. Of course, even then the evaporation efficiency at the start of the ramp will not be optimal as the frequency detuning of sidebands has been optimised for magnetic field at the end of the ramp. Indeed, one has to compare the quadratic correction in Eq. (6) with the well-known Breit-Rabi formula:

$$E_{F,m_F}(B)-E_F=-m_F\,g_I\,\mu_n B+(-1)^F\frac{\hbar\omega_{\mathrm{hf}}}{2}\left(\sqrt{1+m_F\xi+\xi^2}-1\right) \quad (8)$$

with

$$\xi=(g_s\mu_B-g_I\mu_n)\frac{B}{\hbar\omega_{\mathrm{hf}}}\cong 4\varepsilon \quad (9)$$

where g_s and g_I are the electronic and nuclear g-factors, and μ_n the nuclear magneton. The RF frequencies between the sublevels calculated from Eq. (8) are shown in Table 1 We list only the transitions required to transfer the atoms to the first non-trapping state ($F = 2$, $m_F = -1$).

Table 1. Zeeman effect without approximations for different bias fields

B	$\|2,2\rangle\rightarrow\|2,1\rangle$	$\|2,1\rangle\rightarrow\|2,0\rangle$	$\|2,0\rangle\rightarrow\|2,-1\rangle$
56 (Gauss)	39.08-0.44 (MHz)	39.08 (MHz)	39.08+0.45 (MHz)
111	77.12-1.66	77.12	77.12-1.78
207	142.18-5.43	142.18	142.18+6.12

From Table 1 we can immediately see that for a bias field of 207 Gauss this approach will not work, as it is impossible to choose a sideband detuning $\delta\nu$ for which either the ($F = 2$, $m_F = 2$) $\rightarrow$ ($F = 2$, $m_F = 1$) or the ($F = 2$, $m_F = 0$) $\rightarrow$ ($F = 2$, $m_F = -1$) transition will not be detuned from resonance by at least 500 kHz.

This is much larger than the available RF power broadening estimated to be of about 10 kHz. Indeed, experimentally when evaporating ($F = 2$, $m_F =$

2) with the 3 RF knifes in this bias field we are unable to cool the atoms below 15 μK. However this is an order of magnitude lower than the lowest temperature we can obtain when evaporating ($F = 2$, $m_F = 2$) with only one RF knife (100 μK). In a bias field of 111 Gauss the situation is already better with an optimum detuning of the two sidebands from their respective resonance of 50 kHz. Here we can cool the ($F = 2$, $m_F = 2$) cloud down to 500 nK, and obtain a phase space density of 0.1. We believe that with just a little more RF power or better initial conditions for the evaporation the condensation of ($F = 2$, $m_F = 2$) should be possible for this technique for this bias field. When we again lower the bias field by a factor of two to 56 Gauss the effect of the nonlinear terms of the Zeeman shift higher than the quadratic correction becomes negligible compared to the RF power broadening. Here we were able to cool atoms below 100 nK and could attain BEC in ($F = 2$, $m_F = 2$) as desired. It should be noted that the effect of the quadratic correction to the Zeeman shift is significant here, since with one RF knife we are unable to cool the cloud below 10 μK. Fig. 11 shows a graph of the measured number of atoms in a condensate of ($F = 2$, $m_F = 2$) as a function of the sideband detuning $\delta\nu$. The optimal detuning of the RF sidebands from the central carrier is measured to be 0.45 MHz in good agreement with the prediction of Table 1. The width of the curve in Fig. 11 is in good agreement with the estimated Rabi frequency and with the residuals calculated with Eq. (8). From this, we can conclude that the average Rabi frequency of our RF knives is indeed of the order of 10 kHz.

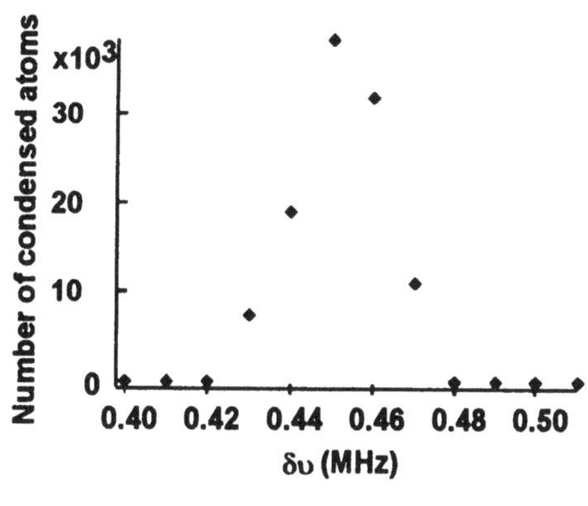

Figure 11. Bose Einstein condensation with 3 RF knives: number of atoms in the condensate versus $\delta\nu$.

5.2 Sympathetic Cooling

Another possible path to condensation in the ($F = 2$, $m_F = 2$) state of ^{87}Rb in a high magnetic field is to use sympathetic cooling[20]. In sympathetic cooling, one evaporatively cools one species of atom, a second species being cooled simply by thermal contact with the former. In our case, we evaporate ($F = 1$, $m_F = -1$) which we know may be efficiently cooled by the standard method, even in a high magnetic field, and use them to cool atoms in ($F = 2$, $m_F = 2$). This cooling method is nearly lossless for the ($F = 2$, $m_F = 2$) atoms as they are evaporated in a potential twice as strong as the ($F = 1$, $m_F = -1$) atoms.

The efficiency of sympathetic cooling can be estimated with a simple model assuming that the two species are always at thermal equilibrium. The total energy of the system can be written

$$E = 3N_1 k_B T + 3N_2 k_B T\,. \tag{10}$$

The energy taken away by dN_1 atoms evaporated at height[14-16] η is

$$dE = 3dN_1(\eta - 1)k_B T \tag{11}$$

and the energy of the atoms remaining trapped after evaporation of these dN_1 atoms is

$$E - dE = 3(N_1 - dN_1)k_B(T - dT) + 3N_2 k_B(T - dT) \tag{12}$$

since the number N_2 of atoms in ($F = 2$, $m_F = 2$) is nearly constant during evaporation. If we assume that evaporation is performed at fixed height η, one can simply integrate Eq. (12) by replacing E and dE by their expression in Eqs. (10) and (11). This results in the equation

$$\frac{T^{\mathrm{f}}}{T^{\mathrm{i}}} = \left(\frac{N_1^{\mathrm{f}} + N_2}{N_1^{\mathrm{i}} + N_2}\right)^{\frac{(\eta-2)}{3}} \tag{13}$$

relating the ratio between initial temperature T^{i} and final temperature T^{f} with the loss of atoms in ($F = 1$, $m_F = -1$).

For example, if we choose η to be 5 (a typical value for experiments) and if we suppose that $N_2 \ll N_1^{\mathrm{i}}$ one can immediately see that the minimum achievable temperature T_{min} scales as the initial ratio N_2 / N_1^{i}. We can now estimate if the initial conditions are sufficient to achieve BEC. For that, we need to compare $T_{\min}$ to the critical temperature

$$T_c(F=1)=\sqrt[3]{\frac{N_1^{\mathrm{f}}}{1.202}}\left(\hbar\sqrt{\frac{\mu_B G^2}{2B_0 M}}\right)=K\sqrt[3]{N_1^{\mathrm{f}}} \tag{14a}$$

$$T_{\mathrm{c}}(F=2)=\sqrt{2}\,K\sqrt[3]{N_2} \tag{14b}$$

for each of the 2 species. We can easily see that the initial ratio N_2 / N_1^{i} can be chosen to either condense ($F = 1$, $m_{\mathrm{F}} = -1$) before ($F = 2$, $m_{\mathrm{F}} = 2$), or ($F = 2$, $m_{\mathrm{F}} = 2$) before ($F = 1$, $m_{\mathrm{F}} = -1$). If N_2 is too large, no condensation is possible and if N_2 is too small, only the ($F = 1$, $m_{\mathrm{F}} = -1$) atoms can be condensed. This happens for a critical number of atoms N_2^c

$$N_2^{\mathrm{c}} \cong \left(\frac{2\sqrt{2}K^3\left(N_1^{\mathrm{i}}\right)^{\eta-2}}{T_{\mathrm{i}}^3}\right)^{1/(\eta-3)} \tag{15}$$

In order to keep evaporative cooling efficient all the way towards BEC, one has to insure that the atoms remain in good thermal contact. Because of gravity the ($F = 1$, $m_{\mathrm{F}} = -1$). cloud is centered below the ($F = 2$, $m_{\mathrm{F}} = 2$). cloud as it is more weakly trapped. This displacement between the two clouds is given by

$$D=\frac{Mg}{\mu_B}\frac{B_0}{G^2} \tag{16}$$

For a fixed gradient G = 900 Gauss/cm this gives a variation with B_0, D = 0.16 μm/Gauss. The RMS width of a thermal cloud decreases with the square root of the cloud temperature during the cooling, so assuming the two clouds must be within one RMS width of each other for good thermal contact we can obtain an estimate for the minimum temperature to which ($F = 2$, $m_{\mathrm{F}} = 2$) can be sympathetically cooled by atoms in ($F = 1$, $m_{\mathrm{F}} = -1$), namely,

$$T_{\mathrm{limit}}=\frac{M^2 g^2}{24 k_B \mu_B G^2} B_0 \tag{17}$$

proportional to the bias field B_0. For a gradient G = 900 Gauss/cm (Eq. (17)) may be evaluated to give a variation with B_0 of $T_{\mathrm{limit}} \cong 1.4$ nK/Gauss. The physical interpretation of this is simple; for our trap, the higher the value of the bias field, the weaker the confinement of the quadrupole, the larger the displacement between the two species and hence the higher the minimum possible temperature.

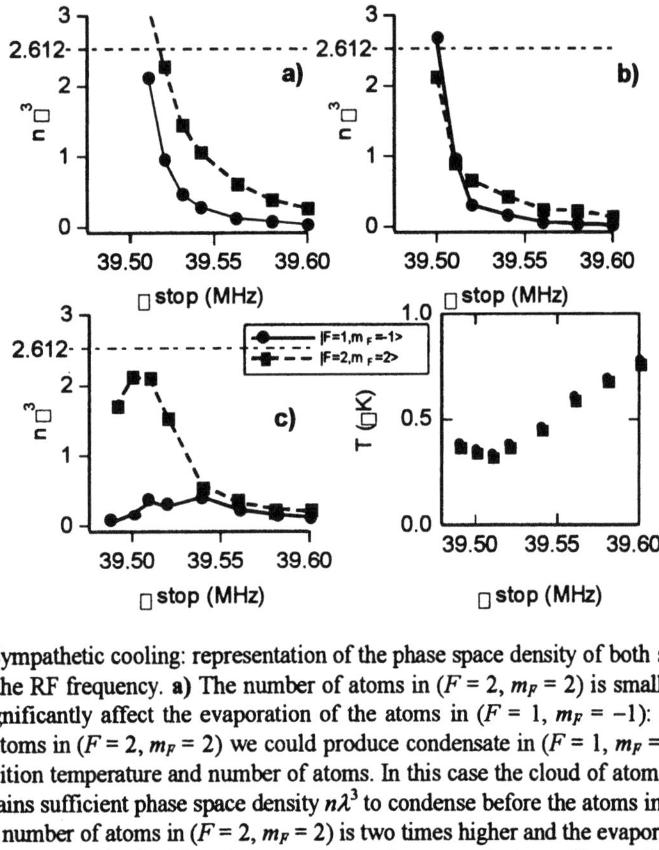

Figure 12. Sympathetic cooling: representation of the phase space density of both species as a function of the RF frequency. **a)** The number of atoms in ($F = 2$, $m_F = 2$) is small enough so as not to significantly affect the evaporation of the atoms in ($F = 1$, $m_F = -1$): in the total absence of atoms in ($F = 2$, $m_F = 2$) we could produce condensate in ($F = 1$, $m_F = -1$) with a similar transition temperature and number of atoms. In this case the cloud of atoms in ($F = 1$, $m_F = -1$) attains sufficient phase space density $n\lambda^3$ to condense before the atoms in ($F = 2$, $m_F = 2$). **b)** The number of atoms in ($F = 2$, $m_F = 2$) is two times higher and the evaporation of the ($F = 1$, $m_F = -1$) cloud is significantly hampered by the sympathetic cooling of ($F = 2$, mF = 2) atoms. Here the atoms in ($F = 2$, $m_F = 2$) attain sufficient phase space density to condense before the ($F = 1$, $m_F = -1$), as even though there are less of them they are more tightly confined in the magnetic trap. **c)** The number of atoms in ($F = 2$, $m_F = 2$) is to large too be cooled sympathetically to the temperature required for condensation. **d)** Diagram showing the 2 clouds remain in good thermal equilibrium during all the evaporation.

Experimentally this simple theory was in good agreement with our observations. By a careful adjustment in the optical pumping cycle during the transfer to the magnetic trap we could start the evaporation with a small but controllable fraction of the atoms in the ($F = 2$, $m_F = 2$) state and the rest in the ($F = 1$, $m_F = -1$) state. For a bias field of 207 Gauss we found we were able to cool the atoms in ($F = 2$, $m_F = 2$) down to a temperature of 400 nK, in rough agreement with the simple estimate of Eq. (17) of 290 nK. The phase space density for the ($F = 2$, $m_F = 2$) cloud at this point was 0.05. Further cooling the ($F = 1$, $m_F = -1$) cloud did not reduce the temperature of the atoms in ($F = 2$, $m_F = 2$).

When we repeated this experiment for a bias field of 56 Gauss, we were able to condense ($F = 2$, $m_F = 2$) sympathetically in the presence of ($F = 1$, $m_F = -1$) for a sufficiently small initial number of atoms in ($F = 2$, $m_F = 2$). When the proportion of atoms in the ($F = 2$, $m_F = 2$) state is too large their rethermalization heats the cooling atoms in ($F = 1$, $m_F = -1$) too much for an efficient evaporation. Fig. 12 shows the phase space density in each state as a function of the final frequency of the evaporation ramp, for three different initial numbers of atoms in the ($F = 2$, $m_F = 2$) state.

6. AN APPLICATION OF HIGH BIAS FIELD: COUPLING BETWEEN 2 POTENTIAL WELLS

The quadratic Zeeman effect can be an asset rather than a nuisance once condensation is reached. For instance, one can make a selective transfer of part of the condensate from the ($F = 1$, $m_F = -1$) state to the ($F = 2$, $m_F = 0$) state by using a 6.8 GHz pulse. Thanks to the quadratic Zeeman effect, the ($F = 2$, $m_F = 0$) state is a very shallow trapping state (for a bias of 56 Gauss, the oscillation frequencies are $\Omega_\perp^0 \leq 40\text{Hz}$ along the quadrupole and $\Omega_z^0 \leq 10\text{Hz}$ along the dipole), some features of a trapped Bose gas can eventually be observed more easily. We studied the weak coupling between those two states by turning on a weak 6.8 GHz RF knife (Rabi frequency of the order of 100 Hz).

The two-coupled potential wells are represented in Fig. 13. Because of gravity, the centers of these two harmonic traps are displaced by typically 300 μm.

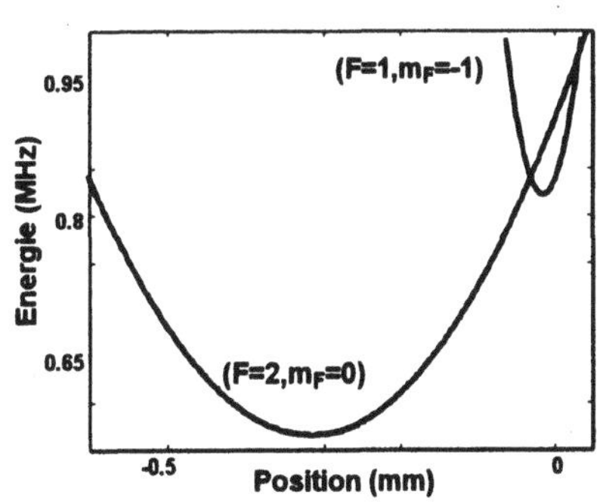

Figure 13. Energy diagram of the ($F = 1$, $m_F = -1$) and ($F = 2$, $m_F = 0$) states.

We start with a condensate in the (F = 1, m_F = -1) state. We will restrict ourselves to the vertical dimension, where the two traps are offset. In the Thomas-Fermi approximation, this condensate is described by the wave function

$$|\Phi(x)|^2 = \frac{\mu - U_1(x)}{\tilde{U}} \qquad (18)$$

where μ is the chemical potential, a the scattering length and $U_1(x)$ the trapping potential for the atoms in F = 1. We define

$$\tilde{U} = \frac{4\pi\hbar^2 a}{M}. \qquad (19)$$

This state has a size of typically $\sigma_x \cong 10\mu$m. The origin of coordinates is taken at the center of this trap. We will neglect the interactions in the (F = 2, m_F = 0) potential well. Consequently, the weak RF knives will couple the wave function $\Phi(x)$ to the eigenstates $\Psi_n(x - D)$ of the F = 2 trapping potential. In good approximation, this potential is that of a harmonic oscillator of oscillation frequency $\Omega_\perp^0$ offset down by D from the Bose-Einstein condensate. Thus, we can write

$$\Psi_n(x - D) = \frac{1}{2^n n!}\left(\frac{\beta^2}{\pi}\right)^{1/4} e^{-\frac{\beta^2 x^2}{2}} H^n(\beta x) \qquad (20)$$

where H^n (x) is a Hermite Polynomial of order n and

$$\beta = \sqrt{\frac{m\Omega_\perp^0}{\hbar}} \qquad (21)$$

a scaling parameter. The size of these eigenstates is given by

$$\Delta x_n = \frac{\sqrt{(n + 1/2)}}{\beta}. \qquad (22)$$

The coupling efficiency is proportional to the overlap integral between $\varphi(x)$ and $\Psi_n(x)$. Roughly, only the eigenstates n such as $D-\sigma_x \leq \Delta x \leq D+\sigma_x$ will be efficiently coupled. Since the condensate is highly coherent, the resulting wavefunction will be the coherent sum of those coupled eigenstates. Experimental studies[21] showed that the trapped condensate is coherent over its full length. This allows us to evaluate the atomic density created in the ($F = 2$, $m_F = 0$) state

$$\Pi(x) \cong \left| \sum_{n \backslash \Delta x_n \cong D \pm \sigma_x} \Psi_n(x) \right|^2 \tag{23}$$

as shown in Fig. 14. One clearly sees beatnotes between the different atomic modes. On the contrary, in the case of a thermal cloud of $F = 1$ atoms with approximately the same size, the resulting density distribution in the $F = 2$ trap will be the sum of the single eigenstates density profiles, since the coupled eigenstates will incoherently add-up.

$$\Pi(x) \cong \sum_{n \backslash \Delta x_n \cong D \pm \sigma_x} \left| \Psi_n(x) \right|^2 \tag{24}$$

this results in the disappearance of the periodic structure.

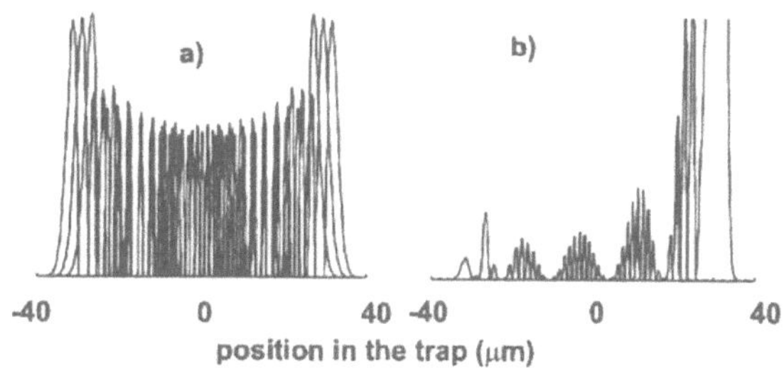

Figure 14. Simple picture of the effect of a weak coupling between the ($F = 1$, $m_F = -1$) state and the ($F = 2$, $m_F = 0$) state. Left, incoherent superposition of different eigenstates of the ($F = 2$, $m_F = 0$) trapping potential. Right, atomic density profile $\Pi(x)$ of the coherent sum of the coupled eigenstates.

A more complete analysis can be done by computing a numerical solution of the coupled Gross-Pitaevskii equations.

$$i\hbar \frac{d}{dt}\varphi_1 = -\frac{\hbar^2}{2M}\nabla^2\varphi_1 + U_1\varphi_1 + \tilde{U}\left(|\varphi_1|^2 + |\varphi_2|^2\right)\varphi_1 + \frac{\hbar\Omega_{RF}}{2}\varphi_2 \tag{25a}$$

$$i\hbar\frac{d}{dt}\varphi_2 = -\frac{\hbar^2}{2M}\nabla^2\varphi_2 + U_2\varphi_2 + \tilde{U}\left(|\varphi_1|^2 + |\varphi_2|^2\right)\varphi_2 + \frac{\hbar\Omega_{RF}}{2}\varphi_1 \quad (25b)$$

For simplicity, we supposed that the scattering length is the same for any binary elastic collision. A comparison of the numerical calculation and of preliminary experimental results is shown in Fig. 15.

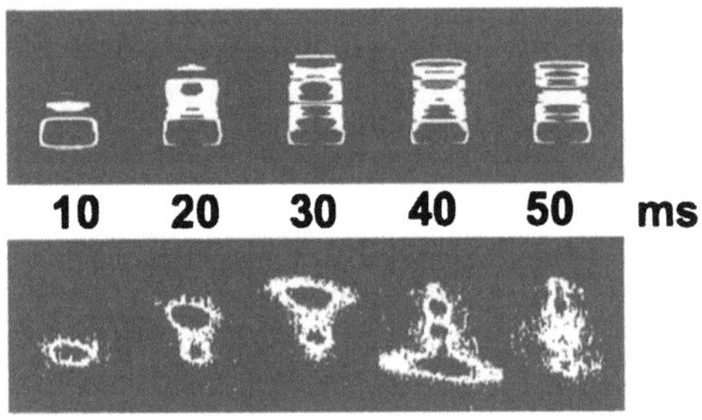

Figure 15. Effect of a weak coupling between the (F = 1, m_F = -1) state and the (F = 2, m_F = 0) state. Top, result of a numerical integration of the coupled Gross-Pitaevskii equation. The image takes into account corrections from the resolution of the imaging system and the camera. Bottom, experimental result of the weak coupling versus the time after the coupling was switched ON. One can clearly see the apparition of beatnotes between the atomic mode. With a cold but thermal cloud, no similar feature could be observed.

7. IMPROVED IRON CORE ELECTROMAGNET TRAPS

A new design of the pole pieces allow for a compensated bias field B_0 - on the order of 1 Gauss - while keeping a significant value for the curvature C - on the order of 100 Gauss/cm^2. This, combined with an improved quadrupole gradient to 2400 Gauss/cm allows for a very high compression ratio. Depending on the initial number of atoms, this would allow to reach BEC in a few seconds.

The parameters of this new trap will also allow for studying new properties of BEC. Given a bias field of 80 mG, this trap has a transverse field curvature of 7×10^7 G/cm^2, such that the ratio of the transverse to longitudinal field curvatures is 10^6: 1. This large asymmetry in the trapping potential will allow to form a 1D system. When the temperature of the system is low enough, particles are frozen into the quantum mechanical ground state of the transverse dimensions. However, since the ground state energy in the longitudinal direction is roughly 10^3 times smaller than that of the transverse direction (since ground state energy scales as the square root of the field curvature), excited longitudinal states can still be occupied. In this one-dimensional regime, the physics of collisions, thermalization, and quantum degeneracy follow laws which are qualitatively different from those of the typical three-dimensional system.

ACKNOWLEDGMENTS

This work is supported by CNRS, MENRT, Région Ile de France and the European Community. SM acknowledges support from Ministère des Affaires Étrangères.

REFERENCES

1. Anderson, M.H., Ensher, J.R., Matthews, M.R., Wieman, C.E., Cornell, E.A., 1995, Observation of Bose-Einstein Condensation in a Dilute Atomic Vapor, *Science* **269**: 198.
2. Bradley, C.C., Sackett, C.A., Tollett, J.J., Hulet, R.G., 1995, Evidence of Bose-Einstein Condensation in an Atomic Gas with Attractive Interactions, *Phys. Rev. Lett.* **75**: 1687; Bradley C.C, *et al.*, 1997, Bose Einstein Condensation of Lithium: Observation of Limited Condensate Number, *Phys. Rev. Lett.* **78**: 985.
3. Davis, K.B., Mewes, M.-O., Andrews, M.R, van Druten, N.J., Durfee, D.S., Kurn, D.M., and Ketterle, W., 1995, Bose-Einstein condensation in a gas of sodium atoms, *Phys. Rev. Lett.* **75**: 3969.
4. Bloch, I., Hänsch, T.W., and Esslinger, T., 1999, Atom Laser with CW Output Coupler, *Phys. Rev. Lett.* **82**: 3008.
5. Bongs, K., Burger, S., Birkl, G., Sengstock, K., Ertmer, W., Rzazewski, K., Sanpera, A., and Levenstein, M., 1999, Coherent Evolution of Bouncing Bose-Einstein Condensates, *Phys. Rev. Lett.* **83**: 3577.
6. Deng L., *et al.*,1999, 4-wave mixing with matter wave, *Nature*, **398**: 218.
7. Tollett, J.J., Bradley, C.C., Sackett, C.A., and Hulet, R.G., 1995, Permanent magnet trap for cold atoms, *Phys. Rev. A* **51**: R22.
8. Desruelle, B., Boyer, V., Bouyer, P., Birkl, G., Lécrivain, M., Alves, F., Westbrook, C.I., and Aspect, A., 1998, Trapping cold neutral atoms with an iron-core electromagnet, *Eur. Phys. J.* D 1, 255.
9. Jackson, J., 1962, Classical Electrodynamics, Wiley, New York.
10. Vuletic, V., Hänsch, T. W., and Zimmermann, C., 1996, *Europhys. Lett.* **36**: (5) 349.
11. Desruelle, B.,1999, PhD Thesis.
12. Anderson, M.A., Petrich, W., Ensher, J.R, and Cornell, E.A., 1994, Reduction of light-assisted collisional loss rate from a low pressure vapor-cell trap, *Phys. Rev. A* **50**: 83597.
13. Ketterle, W., Davis, K.B., Joffe, M.A., Martin, A., and Pritchard, D.E., 1993, *Phys. Rev. Lett.* **70**: 2253.
14. Ketterle W., and Druten, N.J., 1996, *Advances in Atomic, Molecular and Optics Physics* **37**, (B. Bederson, and H., Walther, eds.)

15. Luiten, O.J., Reynolds, M.W., and Walraven, J.T.M., 1996, Kinetic theory of evaporative cooling of a trapped gas, *Phys. Rev. A* **53**: 381.
16. Walraven, J., 1996, Quantum dynamics of simple systems, *Proceedings of the 44th Scottish University Summer School in Physics*, Stirling.
17. Pakarinen, O.H., and Suominen, K.-A., 1999, Atomic dynamics in evaporative cooling of trapped alkali atoms in strong magnetic field e-print physics/9910043.
18. Desruelle, B., Boyer, V., Murdoch, S.G., Delannoy, G., Bouyer, P., and Aspect, A., 1999, Interrupted evaporative cooling of ^{87}Rb atoms trapped in a high magnetic field, *Phys. Rev. A* **60**: 81759.
19. Suominen, K.-A., Tiesinga, E., and Julienne, P., 1998, *Phys. Rev. A* **58**: 3983.
20. Hall D.S., *et al.,* 1998, *Phys. Rev. Lett.* **81**: 1543.
21. Hagley, E.W., e*t al., 1999,* Measuremcnt of the Coherence of a BoseEinstein Condensate, *Phys. Rev. Lett.* **83**: 3112.

Dissipative Dynamics Of An Open Bose-Einstein Condensate

[1,2]F. T. ARECCHI, [1,3]J. BRAGARD AND [1,4]L. M. CASTELLANO
[1]Istituto Nazionale di Ottica, Largo E. Fermi, 6, 1501 25, Florence, Italy;[2]also at Dept. of Physics, University of Florence, Florence, Italy; [3]also at Dept. of Physics, University of Liege, Liege, Belgium; [4]on leave from Dept. of Physics, University of Antioquia, Medellin, Colombia.

1. INTRODUCTION

The BEC dynamics in an atomic trap is ruled by a Gross-Pitaevskii equation (GP)[1,2] which in fact is a nonlinear Schrödinger equation (NLS) describing a conservative motion. Experimental evidence of BEC in a trap[3-5] confirmed qualitatively a dynamical picture based on a GP description. On the other hand, extraction of BEC-atoms toward an atom laser[6,8] introduces a dissipation which must be compensated for by a transfer from the uncondensed fraction of trapped atoms. Those ones on their turn must be refilled by a pumping process which, in the actual laboratory set ups is a discontinuous process[6,8] but that we here consider as a continuous refilling, even though no working scheme is available yet. In Sect. 2 we describe the addition of dissipative interactions through coupled rate equations, as done by Kneer et al.[9]. In Sect. 3 we provide the physical grounds for an additional space dependent (diffusive) process and introduce an adiabatic elimination procedure, whereby we arrive at a closed equation which in fact is a Complex Ginzburg Landau equation (CGL). In Sect. 4 we re-scale the CGL around threshold for both positive (^{87}Rb) and negative (^{7}Li) scattering lengths, showing that in the first case the BEC is stable against space time variations, whereas in the second case the system can easily cross the instability barrier (so-called Benjamin-Feir line[10-12]). In Sect. 5 we present numerical results showing that in the unstable case, rather than collapsing into one singular spot as in the isolated BEC, the open system presents many un-correlated domains (space-time chaos). In Sect. 6 we compare the strength of the nonlinear dissipative term introduced by us with the 3-body recombination rate.

2. THE DYNAMICS OF AN OPEN BEC

We know[1,2] that a BEC is modeled by the GP

$$i\hbar\frac{\partial\phi}{\partial t}=-\frac{\hbar^2}{2m}\nabla^2\phi+V_{ext}\phi+g|\phi|^2\phi \tag{1}$$

where $\phi=\sqrt{\rho}e^{i\theta}$ is the macroscopic wave function describing the probability amplitude of the condensate, V_{ext} is the trap potential, shaped as a harmonic oscillator with frequency ω, and g is the coupling constant for the nonlinear (density $|\phi|^2$ dependent) self interaction. g is proportional to the s-wave scattering length a_s

$$g=\frac{4\pi\hbar^2}{m}a_s \tag{2}$$

We discuss specific experimental situations concerning ^{87}Rb atoms (m = 1.4×10^{-25} kg, a_s = 5.77nm) and ^{7}Li atoms (m = 0.115×10^{-25} kg, a_s = -1.45nm)[2]. For an anisotropic trap the frequency[2-5] is $\omega=(\omega_x\omega_y\omega_z)^{\frac{1}{3}}$. The above equation is formally a conservative non linear Schrödinger equation (NLS). Thus, it is a straightforward task to attach to a BEC all those space-time features familiar of a NLS as e.g. solitary structures and vortices[14,15], which have been explored in the recent past for a NLS, mainly in connection with pulse propagation in optical fibers[16,17].

On the other hand the idealized picture of a BEC in an isolated system is in contrast with two physical facts, namely:

i) The BEC is made of that fraction of atoms which have collapsed into the ground state (n = 0) of the harmonic oscillator trap potential; these atoms interact via collisions with those ones which are distributed over the excited states (n > 0) of the trap. The uncondensed atomic density n_u, evolves in time not only because of the coupling with the condensed phase described by ϕ, but also because trapping and cooling processes imply a feeding (pumping) at a local rate $R(\vec{r})$, the space dependence accounts for the non uniformities of the pumping process as well as for losses due to escape from the trap, at a rate γ_u; and,

ii) In order to have an atom laser, a radio frequency (RF) field is applied to the trap. The rf changes the magnetic quantum number of the atoms' ground state, thus transforming the trapping potential into a repulsive one and letting atoms escape from the BEC at a rate γ_c[6,8].

Both i) and ii) have been modeled by Kneer et al. in Ref. [9] by adding dissipative terms to Eq. (1) and coupling the resulting equation with a rate equation for n_u (as a fact, rate equation coupling between condensed and uncondensed atoms had already been introduced by Speew et al. in Ref. [18]). In a slightly different formulation, this amounts to the following equations

$$i\hbar\frac{\partial\phi}{\partial t} = -\frac{\hbar^2}{2m}\nabla^2\phi + V_{\text{ext}}\phi + g|\phi|^2\phi - \frac{i}{2}\hbar\gamma_c\phi + \frac{i}{2}\hbar\Gamma n_u\phi \tag{3}$$

and

$$\dot{n}_u = R(\vec{r}) - \gamma_u n_u - \Gamma n_u n_c \tag{4}$$

here Γ, is the rate constant coupling the condensed field ϕ with the uncondensed density n_u, and $n_c = |\phi|^2$ is the local density of the condensed phase. We have modified the model of Ref. [9] as follows. At variance with Ref. [9], where Eqs. (3,4) were written for the overall atomic population N_u, over the whole trap volume V, that is,

$$N_u = \int n_u d^3r \tag{5}$$

which is then coupled to

$$N_c = \int n_c d^3r = \int |\phi|^2(r) d^3r \tag{6}$$

here we prefer to deal with a local coupling. In fact Eqs. (3,4) as written above are more convenient, as they refer to a local interaction. The coupling rate of uncondensed to condensed atoms, $\Gamma(m^3s^{-1})$, is given by the global rate used in Ref. [9] which we call Γ', multiplied by the trap volume V

$$\Gamma' = \frac{\Gamma}{V}\left(s^{-1}\right) \tag{7}$$

Furthermore our local feeding rate *R(r) is* related to the overall rate R_u, of Ref.[9] by

$$R_u = \int R(r)d^3r \tag{8}$$

3. CGL PICTURE OF THE OPEN BEC

The Eqs. (3,4) were the basis of the model reported in Ref. [9]. We wish to improve that picture, based on the following considerations. The uncondensed phase, n_u, is fed by a pumping process *R(r)* which is in general non uniform, and is locally depleted by its coupling with the condensed phase. As a result, n_u has a sensible space dependence and hence it undergoes diffusion processes. Precisely, by the fluctuation-dissipation theorem[19,20], the diffusion in velocity is given by

$$D_v = \gamma_u \frac{k_B T}{m}\left(m^2 s^{-3}\right) \tag{9}$$

The corresponding diffusion constant in real space will be

$$D_r = \frac{D_v}{\gamma_u^2} = \frac{1}{\gamma_u}\frac{k_B T}{m}\left(m^2 s^{-1}\right) \tag{10}$$

For ^{87}Rb at $T = 100$ nK , and for $\gamma_u = 5\times10^2 s^{-1}$ (of the order of the average trap frequency $\tilde{\omega} \approx 415$ Hz)[9] this yields

$$D_r = 2\times10^{-8}\left(m^2 s^{-1}\right) \tag{11}$$

Thus we must add the term $D_r \nabla^2 n_u$, to Eq. (4). Once the BEC has been formed, the escape rate γ_c, in Eq. (3) is compensated for by the feeding rate Γn_u. As we set the BEC close to threshold, because of critical slowing down, the ϕ dynamics will be much slower than the n_u dynamics, thus we can apply an adiabatic elimination procedure[21], find a quasi stationary solution for n_u, in terms of ϕ and replace it into Eq. (3) which then becomes a closed equation for ϕ.

We specify the above procedure by the following steps. First, rewrite Eq. (4) including diffusion

$$\dot{n}_u = R(r) - \gamma_u n_u + D_r \nabla^2 n_u - \Gamma n_u |\phi|^2 \tag{12}$$

Next, we take its space Fourier transform. The linear terms are trivial, whereas the nonlinear term should provide a convolution integral. Even though the condensate is not uniform, we consider only the $k = 0$ component in the nonlinear term which is just a perturbation; then its Fourier transform is $\mathcal{F}|\phi(r)|^2 = |\phi|_0^2 \delta(k)$ and Eq. (12) transforms as

$$\dot{n}_k = R_k - (\gamma_u + D_r k^2) n_k - \Gamma n_k |\phi|_0^2 \tag{13}$$

where $n_k = \mathcal{F} n_u$, $R_k = \mathcal{F} R(r)$. The adiabatic elimination procedure consists in taking the stationary solution of Eq. (13) and replacing into Eq. (3). The stationary solution of Eq. (13) is

$$n_k = \frac{\dfrac{R_k}{\gamma_u}}{1 + \dfrac{D_r}{\gamma_u} k^2 + \dfrac{\Gamma}{\gamma_u} |\phi|^2} \tag{14}$$

For long wavelength perturbations and far from saturation, the two additional terms in the denominator are less than unity. Here we consider[22] a cylindrical volume with $L_x = L_y = 5\ \mu m$, $L_z = 10\ \mu m$ ($V = 250 (\mu m)^3$) containing a condensate of $N_c = 5 \times 10^4$ atoms at a temperature $T = 100$ nK. It follows that $k \approx O(1/L) \approx 0.5 \times 10^5\ m^{-1}$ and hence $D_r k^2 / \gamma_u < 1$. We can then expand Eq. (14) as

$$n_k \approx \frac{R_k}{\gamma_u} \left(1 - \frac{D_r}{\gamma_u} k^2 - \frac{\Gamma}{\gamma_u} |\phi|_0^2 \right) \tag{15}$$

The inverse Fourier transform of Eq. (15) is an operator relation as

$$n_u = \frac{R}{\gamma_u}\left(1+\frac{D_r}{\gamma_u}\nabla^2 - \frac{\Gamma}{\gamma_u}|\phi|^2\right) \tag{16}$$

As we replace this expression into Eq. (3), the operator ∇^2 acts on its right upon the space function ϕ. By doing this, we arrive at a closed equation for ϕ which reads as

$$\frac{\partial\phi}{\partial t} = -\frac{i}{\hbar}[GP] + \frac{1}{2}\left(\frac{R\Gamma}{\gamma_u} - \gamma_c\right)\phi + \frac{RD_r\Gamma}{2\gamma_u^2}\nabla^2\phi - \frac{R\Gamma^2}{2\gamma_u^2}|\phi|^2\phi \tag{17}$$

where the square brackets contain the right hand side of Eq. (1). We now write explicitly` the GP terms thus arriving at the following CGL

$$\dot{\phi} = \left[\frac{1}{2}\left(\frac{R\Gamma}{\gamma_u} - \gamma_c\right) - \frac{iV_{ext}}{\hbar}\right]\phi + \left(\frac{RD_r\Gamma}{2\gamma_u^2} + \frac{i\hbar}{2m}\right)\nabla^2\phi - \left(\frac{R\Gamma^2}{2\gamma_u^2} + i\frac{g}{\hbar}\right)|\phi|^2\phi \tag{18}$$

The dissipative terms of Eq. (18) represent respectively: i) difference between gain and losses, which implies a threshold condition; ii) a real diffusion which implies a spread of any local perturbation; iii) a real saturation term which provides a density dependent gain saturation.

4. RESCALED CGL NEAR THRESHOLD

We herewith list the numerical values as taken from the experiment[22] or from Ref. [9]. We refer to a trap volume $V = 0.25 \times 10^{-15}\text{m}^3$; with a loss rate $\gamma_u = 500\ \text{s}^{-1}$. If we take the value[9] $\Gamma' \approx 7\ \text{s}^{-1}$, then $\Gamma = \Gamma' V \approx 2\times10^{-15}\ \text{m}^3\ \text{s}^{-1}$. Furthermore, a reasonable estimate for the BEC escape rate toward the atom laser[9] is $\gamma_c \approx 50\ \text{s}^{-1}$. Therefore the threshold condition (gain = losses) is fulfilled for $R = \bar{R}$, where

$$\bar{R}\Gamma/\gamma_u = \gamma_c \tag{19}$$

here $\bar{R} \approx 12\times10^{18}\ \text{m}^{-3}\text{s}^{-1}$, corresponding to $\bar{R}_u = \bar{R}\,V\,3000\ \text{s}^{-1}$. Finally we have for $^{87}\text{Rb}^2$, $g/\hbar = (4\pi\hbar a_s)/m = 0.48\times10^{-16}\text{m}^3\text{s}^{-1}$. Furthermore, the treatment here outlined, with the cubic approximation (Eq. (14) to (15)) requires $\left(|\phi|^2\Gamma\right)/\gamma_u \ll 1$, which holds for the Rb condensate up to $N_c \approx 100$.

We can now write the parameterized CGL equation (18) in the dimensionless form

$$\dot{\phi} = \varepsilon\phi + (1 + ic_1)\nabla^2\phi - (1 + ic_2)|\phi|^2\phi \tag{20}$$

where we have used the dimensionless time

$$\tau = \gamma_c t \tag{21}$$

and the dimensionless space coordinates

$$\frac{x}{l_0}, \frac{y}{l_0}, \frac{z}{l_0}, \tag{22}$$

where

$$l_0 = \left(\frac{\overline{R}D_r\Gamma}{2\gamma_u^2\gamma_c}\right)^{1/2} \approx 3\times10^{-6}m \tag{23}$$

is the characteristic length associated with the CGL dissipative dynamics, and the dimensionless condensed wave-function

$$|\tilde{\phi}|^2 = \frac{\overline{R}\Gamma^2}{2\gamma_u^2\gamma_c}|\phi|^2 \tag{24}$$

Note that tilde has been dropped in Eq. (20). It follows from Eq. (18) that

$$\varepsilon = \frac{1}{2}\left(\frac{R\Gamma}{\gamma_u\gamma_c} - 1\right) = \frac{1}{2}\left(\frac{R}{\overline{\overline{R}}} - 1\right) \tag{25}$$

$$c_1 = \frac{\hbar\gamma_u^2}{mRD_r\Gamma} \tag{26}$$

$$c_2 = \frac{2g\gamma_u^2}{\hbar R\Gamma^2} \tag{27}$$

are the significant parameters of Eq. (20). They are pure numbers. The term $-iV_{ext}/\hbar\phi = -ic_0\phi$ in Eq. (18) can be eliminated by a $\phi \to \phi e^{-ic_0 t}$ rotation transformation.

We notice that Eq. (20), derived by sound physical assumptions, is far from being a purely conservative (GP) or purely dissipative (real Ginzburg Landau) equation, but it displays both characters.

However the Benjamin-Feir instability condition[10]

$$c_1(-c_2) > 1 \tag{28}$$

is not met by Rb ($c_1 = 0.52$, $c_2 = 2.05$) and its dissipative CGL is fully inside the stable region. Hence the addition of dissipative terms may add interesting transient effects but does not lead to substantial qualitative changes with respect to the GP equation. Quite different is the case of ^{7}Li ($c_1 = 0.04$, $c_2 = -0.51$). Indeed even though the values of c_1 and c_2 just listed give a stable dynamics, the fact that the scattering length is negative may lead to an instability if the parameters of the open BEC are slightly changed, e.g. if γ_c, is reduced by a factor 10 (which physically corresponds to a Li atom-laser with weaker losses). In such a case, we get $c_1 = 0.4$, $c_2 = -5.1$ and the open BEC is in the unstable region; we will denote this experimental situation by referring to an open Li* BEC.

5. NUMERICAL SIMULATIONS

As we have shown in Sect. 4, the coefficients of the CGL depend on the nature of the atoms forming the open BEC and also depend on the characteristic working parameters of the open BEC. Let us discuss the space tune dynamics of the density of the condensed phase . To do this, we proceed to the numerical integration of Eq. (20). The integration is performed on a $|\phi|^2$ two dimensional domain. This corresponds to a cross section of the 3-D cigar shape where the condensation takes place. This is justified by the fact that $\omega_z << \omega_x$, ω_y. The simulations are done on a 200x200 (Rb) or 256x256 (Li*) grid starting with an initial Gaussian distribution at the center of the domain. The numerical integration code is based on a semi-implicit scheme in time with finite difference in space. The chosen boundary conditions (at $x = \pm L_x/2$ and $y = \pm L_y/2$) are

$$\frac{\partial \phi}{\partial n} = -0.1\phi \tag{29}$$

where n is the normal at the boundary. Eq. (29) expresses the condition of an isotropic output flux of the condensed BEC (in the ideal situation of zero-gravity). The numerical coefficient on the right hand side of Eq. (29) is the dimensionless ratio between γ_c, and the velocity modulus of the condensed atoms, easily evaluated from the ground state solutions of the harmonic oscillator[2]. In fact the ground state of the condensate is not that of the harmonic oscillator, because of the nonlinear term, but for the sake of the computation this is a fair approximation.

In Fig. 1 four snapshots of $|\phi\,(x, y, t)|$ are shown at different times for the Rb case (c_1 = 0.52, c_2 = 2.05 and ε = 0.5). The initial distribution evolves towards a stable quasi-uniform state.

The Fig. 2 displays three cross section of Fig. 1 at different times, the solid line corresponds to the final stationary state and we observe the nearly uniform condensate on the overall domain. For a uniform pump R, the balance between source (uncondensed atom contribution) and sink (boundary escape) eventually yields a uniform condensate profile far from threshold.

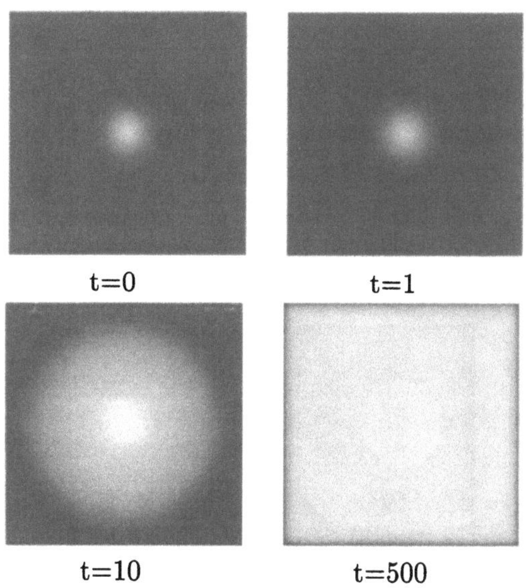

Figure 1. Open BEC of Rb: Time evolution of $|\phi|^2$ using Eq. (20) with c_1 = 0.52, c_2 = 2.05 and ε = 0.5. The time step is (Δ_t = 0.01), the initial condition is a Gaussian. The figures are coded using a grey scale (white corresponds to the maximum value of $|\phi|^2$).

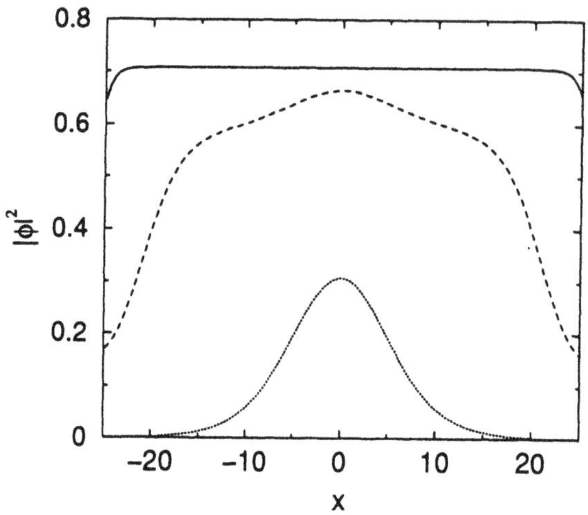

Figure 2. Cuts of $|\phi|^2$ at $y = 0$, same parameters as for Fig. 1. The dotted line corresponds to t = 1, the dashed line is for t = 10 and the solid line (t = 500) corresponds to the final stationary state.

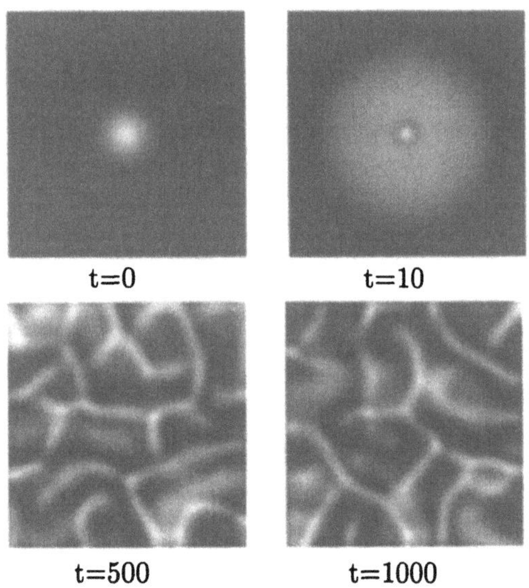

Figure 3. Open BEC of Li* (escape parameters adjusted to be in the unstable regime). Time evolution of $|\phi|^2$ using Eq. (20) with $c_1 = 0.4$, $c_2 = -5.1$ and $\varepsilon = 0.5$.

The Fig. 3 illustrates a quite different situation: The values are now $c_1 = 0.4$, $c_2 = -5.1$ and $\varepsilon = 0.5$ which corresponds to a Li* open BEC in the unstable region of use. The space-time chaotic dynamics emerges after a short transient (t < 10). The Fig. 4 confirms that $|\phi|^2$ is no longer symmetric with respect to $x = 0$ (the same holds for the y-axis). The Fig. 5 is aimed to show the spatial decorrelation of the signal when the condensate has entered the chaotic regime.

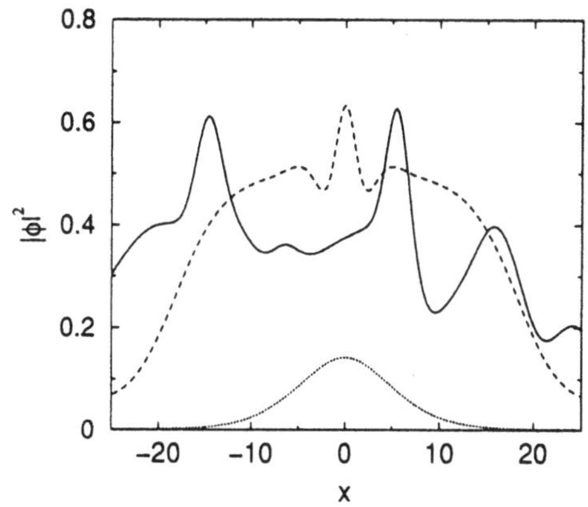

Figure 4. Cuts of $|\phi|^2$ at $y = 0$, same parameters as for Fig. 3. The dotted line corresponds to t = 1, the dashed line is for t = 10 and the solid line (t = 500) corresponds to a chaotic state.

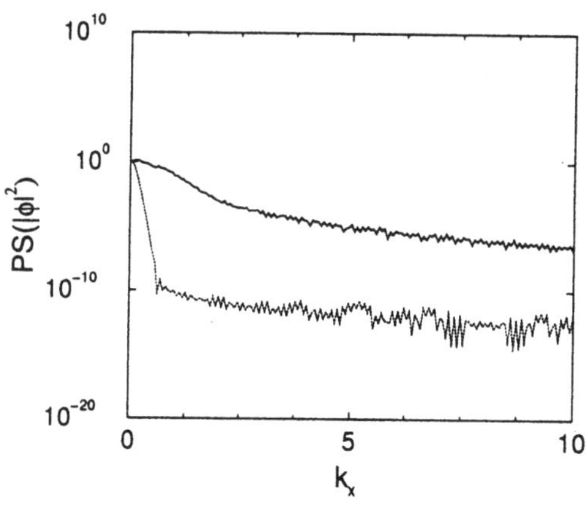

Figure 5. Spatial power spectrum of $|\phi\,(x, y = 0,t)|^2$ for Li*. The lower curve (dotted line) corresponds to t = 1. The upper curve (solid line) is calculated by averaging the spatial power spectrum of $|\phi\,(x, y = 0,t)|^2$ between t = 500 and t = 1000 (by steps of $\delta t = 10$).

The 1D spatial power spectrum of the $|\phi(x)|^2$ function is shown in the lower curve of Fig.5., the upper curve (solid line) is calculated by averaging the power spectrum of the function $|\phi(x)|^2$ over a time interval from $t = 500$ until $t = 1000$ (taking a sampling time $\delta t = 10$) within which the dynamics is statistically stationary. The results clearly indicates the large spatial decorrelation of the signal inside the chaotic regime. Indeed, it is well known that the Fourier transform of a Gaussian function $G \propto e^{-x^2/\sigma_x^2}$ is again a Gaussian function $\widetilde{G} \propto e^{-k^2/\sigma_k^2}$ with $\sigma_k\sigma_x = 1$. On Fig.5 it appears that the bandwidth in the Fourier space is much larger in the chaotic regime than for the initial distribution, which means a decorrelation of $|\phi(x)|^2$ once the system becomes chaotic.

To give a quantitative feeling, in the case of Li*, we have reduced γ_c, by a factor 10, which means that the normalization length is increased by $\sqrt{10}$ with respect to Eq. (23), and it is $l_0^* \approx 10\mu m$. In the numerical calculations we have considered a trap of linear size $L = 50 l_0^*$. Since the ratio of the spectral widths between the chaotic and the initial spectra is about 4 (estimated from Fig.5), it results that the coherence length σ_c, in space time chaos is ~1/4 the length of the initial Gaussian packet σ_0. As seen from Fig.4 $\sigma_0 \cong 5 l_0^* \cong 1/10L$, hence $\sigma c \cong 1.2 l_0^* \cong 1/40L$. These numerical estimates agree with a densitometric analysis of Fig.3.

6. DISCUSSION AND CONCLUSIONS

Kagan et al. in Ref [23] have discussed the collapse of a BEC in ^{7}Li for a number of condensed atoms N_0 larger than the critical value N_{crit}, that is,

$$N_{crit} \approx \frac{a_{h0}}{|a_s|} \tag{30}$$

This relation for N_{crit} is obtained by equating the level spacing $\hbar\omega$ in the trap to the interparticle interaction energy $n_0|g| = N/V|g|$ where $g = 4\pi\hbar^2 a_s/m$ and $V \approx a_{h0}^3$. The Ref. [23] stabilizes the GP via a dissipative term corresponding to 3-body recombination processes. This amounts to a correction corresponding to a 6th power term in ϕ in a free energy potential. The dissipative equation of Ref. [23] is then

$$\dot{\phi} = -i[GP] - \xi|\phi|^4\phi + \text{(pumping from the uncondensed portion)} \tag{31}$$

In Eq. (20), we have already treated the last term, here expressed in words, by the Kneer et al. in Ref. [9] . Let us now compare the 5th power real damping entering Eq. (31) with the 3rd power real damping of Eq. (20). The cubic term is of the form $G_3\phi$, where

$$G_3 = \frac{R\Gamma^2}{2\gamma_u^2}|\phi|^2 = \frac{\gamma_c\Gamma}{\gamma_u}|\phi|^2 \tag{32}$$

Using the numerical values corresponding to the ^{87}Rb we obtain $G_3 \approx 48$ s^{-1}, for ^{7}Li we have $G_3 \approx 1920\ s^{-1}$. The cubic rate (G_3) is a combination of the three characteristic rates of an open BEC. In a similar way we can introduce the rate $G_5 = \xi|\phi|^4$. Taking the numerical values provided in Ref. [23] we have the following ratios between the two dissipation rates

$$\frac{G_3}{G_5} = \begin{cases} 480 & \text{for} \quad ^{87}\text{Rb} \\ 2\times 10^4 & \text{for} \quad ^{7}\text{Li} \end{cases} \tag{33}$$

This result clearly indicates that for an open BEC, the 3-body recombination is negligible with respect to the saturation cubic term that comes from the coupling between the condensed and uncondensed phase of the open BEC.

To summarize, in this Chapter we have shown that in the framework of an atom-laser approach via two coupled equations, one for the uncondensed phase and the other one for the condensed phase, addition of a diffusion term for the uncondensed atoms and application of a proper adiabatic elimination procedure leads to a CGL dynamical equation for an open BEC. In the case of negative scattering length, a suitable adjustment of the escape rate implies entering the unstable regime of the CGL dynamics. Furthermore, within the chosen ranges of the parameters (ε, c_1, c_2), the 3-body recombination processes have a negligible influence.

REFERENCES

1. Gross, E.P., 1961, *Nuovo Cimento* **20**: 454; Pitaevskii, L.P., 1961, *Zh.Eksp.Teor.Fiz.* **40**: 646. [Sov.Phys.JETP 13, 451 (1961)].
2. Dalfovo, F., Giorgini, S., Pitaevskii, L.P., and Stringari, S.,1999, *Rev. Mod. Phys.* **71**: 463.

3. Anderson, M.H., Ensher, J.R., Matthews, M.R., Wieman C.E., and Cornell, E.A.,1995, *Science* **269**: 198.
4. Bradley, C.C. C.A. Sackett, J.J. Tollett and R.G. Hulet, 1995, *Phys. Rev. Lett.* **75**: 1687 .
5. Davis, K.B., Mewes, M.O.,. Andrews, M.R, van Druten, N.J., Diirfec, D.S., Kurn, D.M., and Ketterle, W., 1995, *Phys. Rev. Lett.* **75**: 3969.
6. Mewes, M.O., Andrews, M.R., Kurn, D.M., Durfee, D.S., Townsend C.G., and Ketterle, W., (1997) *Phys.Rev.Lett.* **78**: 852.
7. Hagley, E.W., Deng, L., Kozuma, M., Wen, J. Helmerson, K., Rollston S.L., and Phillips, W.D., 1999, *Science* **283**: 1706.
8. Bloch I., Hansch T.W., and Esslinger, T., 1999, *Phys.Rev.Lett.* **82**: 3003.
9. Kneer, B., Wong, T. Vogel, K. Schleich W.P., and Walls, D.F., 1998, *Phys.Rev. A* **58**: 4841.
10. Benjamin, T.B., and Feir, J.E., 1967, *J. Fluid Mech.* **27**: 417.
11. Shraiman, B., *et al.*, 1992, *Physica* (Amsterdam) **D57**: 241 Chaté, H., 1994, *Nonlinearity* **7**: 185; Chaté, H., 1995. in *Spatiotemporal Patterns in Noneguilibrium Complex Systems,* (P.E. Cladis and P. Palffy-Muhoray, eds.), Addison-Wesley, New York.
12. Chaté, H., and Manneville, P., 1996, *Physica* **A 224**: 348.
13. Arecchi, F.T., Bragard, J., and Castellano, L.M., 2000, *Opt. Comm.*, to be published.
14. Dum, R., Cirac, J.I., Lewenstein, M., and Zoller, P., 1998, *Phys.Rev.Lett.* **80**: 2972.
15. Sammut, R.A., Buryak, A.V., and Kivshar, Yu S., 1998, *J.Opt.Soc.Am.***B 15**: 1488.
16. Hasegawaand, A., and Tappert, F., 1973, *Appl. Phys. Lett.* **23**: 142.
17. Kumar, A., 1990, *Phys. Rep.* **187**: 63.
18. Spreeuw, R.J.C., Pfau, T., Janicke, U., and Wilkens, M.,1995, *Europhys. Lett.* **32**: 469.
19. Kubo, R., 1957, *J.Phys.Soc.Japan* **12**: 570.
20. Callen, T., and Welton, 1951, T.A., *Phys.Rev.* **83**: 34.
21. Haken, H., 1983. *Advanced Synergetics,* Springer-Verlag, Berlin.
22. BEC group at LENS, 1999, private communication, Firenze
23. Kagan, Y., Muryshev, A.E., and Shlyapnikov, G.V., 1998, *Phys. Rev. Lett.* **81** (5): 993.

Non-Ground-State Bose-Einstein Condensation

[1]V. S. BAGNATO, [2]E. P. YUKOLOVA, AND [3]V. I. YUKALOV
[1]Instituto de Física de São Carlos, Universidade de São Paulo, Caixa Postal 369, São Carlos/SP., 13560-970, Brazil; [2]Department of Computational Physics, Laboratory of Computing Techniques and Automation, Joint Institute for Nuclear Research, Dubna, 141980, Rússia; [3]Bogolubov Laboratory of Theoretical Physics, Joint Institute for Nuclear Research, Dubna, 141980, Rússia.

1. INTRODUCTION

The spatial distribution of atoms during the occurence of Bose-Einstein Condensation (BEC) is an important factor which permits both the identification of the phase transition itself and also the reliable production of coherent beams of matter[1,2]. To associate the variations of the spatial distribution with the occurence of BEC, we must be able to follow spatial variations in the density profile as the temperature is lowered towards the critical point. The first part of this Chapter presents, in a tutorial way, how one can identify BEC by following the spatial evolution with temperature[3].

The spatial distribution of BEC in an equilibrium system is basically a picture of the ground state wave function. One could also consider the possibility of realizing such a macroscopic population for some other quantum state rather than the ground state. If that were possible, new applications for quantum degenerated gas could be possible. For example, a non-ground-state BEC could permit the production of various spatial modes in a coherent beam of atoms, with important consequences for the applicability to atom lasers. A non ground-state BEC would create new possibilities for understanding relaxation processes in the quantum degenerate regime. Even more important than these applications, when a new system with unusual features is explored, there is always the chance of finding completely unexpected. phenomena

The main part of this report is devoted to the description of a possible way of producing a non-ground-state BEC and to the study of its behavior in space and time.

2. EQUILIBRIUM SYSTEMS AND THE GROUND STATE BEC[3]

Due to the inhomogeneous character of the confining potential, the atomic cloud undergoes spatial compression during the cooling process. The typical features of this spatial compression have been used as a signature for the occurence of BEC. To associate BEC with the evolution of the spatial profile, we must be able to follow the density $\rho(r,T)$ as the temperature is lowered. One can begin by writing the general spatial density profile as:

$$\rho(\vec{r},T) = \begin{cases} \sum_{\varepsilon} n_\varepsilon \left|\psi_\varepsilon(\bar{r})\right|^2 & T > Tc \\ N_0 \left|\psi_0\right|^2 + \sum_{\varepsilon_1}^{\infty} n_\varepsilon \left|\psi_\varepsilon(\bar{r})\right|^2 & T < Tc \end{cases} \tag{1}$$

where n_ε is the occupation number for the state with energy ε and wave function $\psi_\varepsilon(r)$. To evaluate Eq. (1), one must know μ (T), the dependence of the chemical potential on temperature, because it affects n_ε (T). There is however an alternative way to evaluate $\rho(r,T)$, much simpler than in Eq. (1), without the necessity of knowing ψ_ε. The number of particles occupying a given element of volume in phase space with the same specific momentum and position is given by:

$$dN = \frac{1}{h^3} n_\varepsilon d^3p d^3r \tag{2}$$

and for one specific momentum component, the density may be expressed as:

$$d\rho_P(\vec{r},T) = \frac{4\pi}{h^3} n_\varepsilon(\vec{p},\vec{r},T) p^2 dp \tag{3}$$

Integration over the whole range of available momenta allows one to obtain $\rho(r,T)$. This integration can, in principle, be performed for any confining potential. However for a harmonic oscillator, where the number of particles, the critical temperature and the potential are connected through[4]

$$KTc = 0.941\ k\omega\ N^{1/3}, \tag{4}$$

the spatial density profile can be evaluated as

$$\rho(\vec{r},T) = \begin{cases} \dfrac{1}{\lambda_D^3(T)} \displaystyle\sum_{t=1}^{\infty} \dfrac{Z^t e^{-t\left(\frac{1}{2}M\frac{\omega^2 r^1}{KT}\right)}}{t^{3/2}} & T > T_c \\ \dfrac{1}{\lambda_D^3(T_c)} \displaystyle\sum_{t=1}^{\infty} \dfrac{e^{-t\left(\frac{1}{2}M\frac{\omega^2 r^1}{KT}\right)}}{t^{3/2}} & T = T_c \\ \dfrac{3.099 N^{1/2}}{\lambda_\rho^3(T_c)} \left[1 - \left(\dfrac{T}{T_c}\right)^3\right] e^{-M\frac{\omega^2 r^1}{T}} + \dfrac{1}{\lambda_\rho^3(T)} \displaystyle\sum_{t=1}^{\infty} \dfrac{e^{-t\left(\frac{1}{2}M\frac{\omega^2 r^1}{KT}\right)}}{t^{3/2}} & T < T_c \end{cases} \tag{5}$$

where λ_ρ (T) is the thermal de Broglie wavelength at temperature T. To observe $\rho(\vec{r},T)$ in the whole range of temperature we still need to know the fugacity Z(T) = $e^{\mu(T)\,/\,KT}$ as a function of temperature. To obtain Z (T), consider the case T > Tc, $N = \sum_{\varepsilon} n_\varepsilon$ which can be written as

$$N = \sum_{\varepsilon} \frac{Ze^{-\varepsilon/KT}}{1 - Ze^{-\varepsilon/KT}} = \sum_{\varepsilon} \sum_{j=1}^{\infty} Z^j e^{-j\varepsilon/KT} \tag{6}$$

The sum over all states can be transformed into an integral by introducing the density of states g(ε) (given for an arbitrary potential in Ref. [4]):

$$N = \sum_{j=1}^{\infty} Z^j \int_0^{\infty} g(e) e^{-jE/KT} dE \text{ , or}$$

$$N = \sum_{j=1}^{\infty} D_j Z^j \tag{7}$$

with $D_j = \int_0^\infty g(\varepsilon) e^{-j\varepsilon/KT} d\varepsilon$. Eq. (7) is a series relating the number of particles to the fugacity, and the coefficients cary all the information about the external potential. This series can be inverted, providing Z (N, T). For a harmonic oscillator, we obtain

$$Z(T) = 1.200\left(\frac{Tc}{T}\right)^3 - 0.180\left(\frac{Tc}{T}\right)^6 - 0.01\left(\frac{Tc}{T}\right)^9 + \text{high orders.} \tag{8}$$

Now, Eq. (5) can be evaluted without restriction. Observing only the density at the center of the potential $\rho(0,T)$ we obtain the following behavior. Around Tc a sudden jump in $\rho(0,Tc)$, proportional to $N^{1/2}$, is observed. This behavior can actually be used as an experimental indicator for the occurence of BEC.

3. THE PRODUCTION OF NON-GROUND-STATE BEC[5]

One means of realizing the macroscopic occupation of a non-ground-state of a confining potential would consist of first the production of a conventional ground-state BEC, and then the transfer of the population to a non-ground-state level. This transfer could be achieved by means of an external pumping field.

In the case of a harmonic oscillator type potential, this approach would not work, because of the equal separation between the energy levels. In the harmonic oscillator case, the pumping field would promote atoms from the ground state to the first excited level, but at the same time, transitions from the latter to the second excited state would take place, and so on. In such a case, the effect of the external pumping field would be to disperse the initial ground-state population over a large number of states. Fortunately, because of the nonlinearity resulting from atomic interactions in the system, the energy levels are not equidistant even when the confining potential $U(r)$ is harmonic.

The hamiltonian in this case is of the form

$$H(\varphi)=\frac{-\hbar^2}{2M}\nabla^2+U(\vec{r})+A|\varphi^2| \tag{9}$$

where the atom-atom interaction is modelled by the s-wave scattering length (a) resulting in

$$A=(N-1)4\pi\hbar^2\frac{a}{M} \tag{10}$$

Adding the external pumping field, the total hamiltonian is:

$$\hat{H}=H(\varphi)+Vp(\vec{r,t}) \tag{11}$$

where $V_p(r,t)$ is a time and space dependent external field.

Before dealing with the population transfer problem, let us consider the energy levels of Eq. (9). The stationary states for this nonlinear hamiltonian are defined by $H(\varphi_n)\varphi = E_n\ \varphi_n$, and in the case of a harmonic trap potential we have approximately obtained the energy levels[5].

Consider the confining potential as

$$U(\vec{r})=\frac{1}{2}M\left(\omega_x x^2+\omega\gamma^2 z^2\right) \tag{12}$$

Defining the characteristic frequency and length by

$$\omega_0\equiv(\omega_x\omega_y\omega_z)^{1/2},\qquad l_0\equiv\sqrt{\frac{\hbar}{m\omega_0}} \tag{13}$$

and a dimensionless interaction parameter

$$g=4\pi(N-1)\frac{a}{l_0} \tag{14}$$

the Hamiltonian (9) can be presented as:

$$H=\frac{1}{2}\sum_{i=i}^{3}\left(-\frac{\partial^z}{\partial x_i}+\lambda_i^z x_i^z\right)+g|\psi|^2 \tag{15}$$

where

$$x_i = \frac{x(or\ y\ or\ z)}{l_0}; \quad \lambda_i = \frac{\omega_i}{\omega_0} \quad \text{and} \quad \psi = l^{3/2}\varphi$$

To obtain the energy levels for Eq. (15) we may employ the renormalized perturbation method already successfully used in several examples[6]. This is fully described in Ref [5]. For instance, the ground-state energy, in the case of an isotronic oscillator, is

$$E_0 = \begin{cases} \left(\frac{3}{2} + g_0 - \frac{3}{4} g_0^2\right)\omega_0\hbar & g_0 \ll 1 \\ \frac{5}{4}(2g_0)^{2/5}\omega_0\hbar & g_0 \gg 1 \end{cases} \tag{16}$$

where $g_o \equiv g / 2(\pi)^{3/2}$.

For atoms with a negative scattering length, ($g < 0$), the real solution for the spectrum exists till some critical value g_c after which the solution becames complex.

For an isotropic potential, $g_c \sim -4.213$, which defines the critical number of atoms

$$N_c = 1 + \frac{g_c l_0}{4\pi_a} \tag{17}$$

that can condensate in a stable state. After this critical valve, the interaction is strong enough to cause a collapse of the system. This type of behavior has already been reported[7].

The energies of arbitrary excited levels can also be calculated. Thus, for the first excited levels E_{100}, E_{110}, and E_{200} we obtain

$$E_{100} = \begin{cases} \left(\frac{5}{2} + 0.04762g\right)\hbar\omega_0 & g \ll 1 \\ 0.607943g^{2/5}\hbar\omega_0 & g \gg 1 \end{cases}$$

$$E_{110} = \begin{cases} \frac{7}{2} + 0.035715 g\hbar\omega_0 & g >> 1 \\ 0.675012 g^{2/5}\hbar\omega_0, & g >> 1 \end{cases} \tag{18}$$

$$E_{200} = \begin{cases} \left(\frac{7}{2} + 0.040676 g\right)\hbar\omega_0 & g << 1 \\ 0.632193 g^{2/5}\hbar\omega_0 & g >> 1 \end{cases}$$

where the indices correspond to the quantum number of the nonlinear oscillator and the top value refers to the weak-coupling limit while the bottom value refers to the strong-coupling limit.

The interlevel separation is characterized by the transition frequency $\omega_{m_x m_y m_z} = \left(E_{m_x m_y m_z} - E_0\right)/\hbar$ which is shown below.

	Weak-Coupling	Strong-Coupling
ω_{100}	$1 - 0.015874g\ \omega_0$	$0.060405g^{2/5}\ \omega_0$
ω_{110}	$2 - 0.027779g\ \omega_0$	$0.084655g^{2/5}\ \omega_0$
ω_{200}	$2 - 0.022818g\ \omega_0$	$0.1274745^{2/5}\ \omega_0$

As mentioned above, the interaction makes the energy levels no longer equidistant, making possible population transfer from the ground to an excited level without a dispersion of the population.

Consider now the problem of resonant pumping by an external field. Start with the Hamiltonian (11), having as an external field

$$V_p(\vec{r},T) = V(\vec{r})\cos\omega t \tag{19}$$

where the pumping field frequency is detuned by $\Delta\omega = \omega - \omega_p$ from the transition frequency between E_p and E_0. We consider the case of $\Delta\omega$ small, such that the pumping would not influence the neighboring states, but mainly the E_p-state.

We can look for a solution of the time-dependent nonlinear Schrödinger equation in the form of a sum,

$$\varphi(\vec{r},t) = \sum_n (C_n(t)\varphi_n(\vec{r})\exp\left(\frac{-i}{\hbar}E_n t\right) \tag{20}$$

over the stationary states discussed previously. Substituting this expansion into the Schrödinger equation and employing the guiding-center approach involving averaging over fast oscillations, we obtain the following system of equations:

$$\begin{cases} \dfrac{dC_0}{dt} = -i\alpha npC_0 - \dfrac{i}{2}\beta e^{+is\alpha t}Cp \\ \dfrac{dCp}{dt} = -i\alpha n_o Cp - \dfrac{i}{2}\beta^* e^{-i\Delta\alpha t}C_0 \\ \dfrac{dC_x}{dt} = 0(K \neq 0, K \neq p) \end{cases} \tag{21}$$

where we have defined the following quantities:

$$\begin{aligned} \alpha &\equiv \frac{A}{\hbar}\int\left|\varphi_0(\vec{r})\right|^2\left|\varphi_p(\vec{r})\right|^2 d\vec{r} \\ \beta &\equiv \frac{1}{\hbar}\int\varphi_0^K(\vec{r})\,\mathrm{V}(\vec{r})\,\varphi(\vec{r})\,d\vec{r} \\ n_j &= \left|Cj\right|^2 \end{aligned} \tag{22}$$

α represents the interaction between different levels, while β is the external pumping strength. The amplitude n_j represents the population fraction for the j-level.

The solution of Eqs. (21), with the initial conditions $C_0(0) = 1$, $C_p(0) = 0$, and $C_k(0) = 0$, provides us with the time evolution for the population fraction of each of the levels: ground state $n_0 = |C_o(t)|^2$ and excited-state $n_p = |C_p(t)|^2$. For a spatially non-homogenious pumping field $\beta \neq 0$. The system of Eqs. (21) can be solved by means of the averaging method, which results in:

$$\begin{cases} n_o = 1 - \dfrac{|\beta|^2}{\Omega^2} sin^2 \dfrac{\Omega t}{2} \\ n_p = \dfrac{|\beta|^2}{\Omega^2} sin^2 \dfrac{\Omega t}{2} \end{cases} \tag{23}$$

with $\Omega^2 = [\alpha(n_o - n_p) - \Delta\omega]^2 + |\beta|^2$. The functions in Eq. (23) describle the time evolution of the fractional populations of the ground state and of the excited state which has been chosen to be connected to the ground state via the pumping field. The oscillatory behaviour is predicted to be nonsinusoidal because of the dependence of the effective Rabi frequency (Ω) on the populations. It is possible to choose the transition amplitude β and detuning $\Delta\omega$ so that to realize, at some instants of time, the complete transfer from the ground to an excited state, producing a non-ground-state macroscopic occupation. If the pumping field is suddenly removed when $n_p \rightarrow 1$ the system will remain in that state, and its decay may take a considerably long time.

4. CRITICAL EFFECTS IN POPULATION DYNAMICS

Once we have choosen the excited state and a convenient pumping field, we can take a closer look at the population transfer dynamics as different parameters of the system are allowed to vary. It is convenient to analyse the system in terms of dimensionless parameters:

$$\begin{aligned} b &\equiv |\beta| / \alpha \\ \delta &\equiv \Delta\omega / \alpha \end{aligned} \tag{24}$$

As before, we keep $\delta \ll 1$. An accurate analysis of the behavior of the solutions of Eqs. (21) was done numerically for small b; the fractional populations oscillate according to the sine-squared law. When b increases, the amplitude of oscillations also increases. The overall behavior stays the same until we reach a critical value b_c, when the dynamics of the system change dramatically. Around the critical value $b_c \cong 1/2 - \delta$, the system experiences a sharp change. The system evolves from the normal oscillatory behavior for $\delta = 0$ to a period doubling and a flattening out of n_p for $\delta = 10^{-4}$. An additional change in δ on the order of 10^{-7} promotes the appearance of an upward cusp in n_p. Increasing the detuning further results in the squeezing of the period of oscillation. The same phenomena occur when other values of parameters are chosen, such that $b+\delta \sim 1/2$.

This unusual behaviour of the fractional populations is certainly due to the nonlinearity present in the Hamiltonian of the system. In such a system, infinitesimal changes of parameters can produce drastic change in behavior. The change of one type of behavior to another is normally refered to as a bifurcation. At a bifurcation point the dynamical system is structurally unstable, and effects similar to phase transitions and critical phenomena can occur. To elucidate these effects for a nonequilibrium system, we have to consider its time-averaged behavior. Defining an effective hamiltonian from the equations of motion

$$i\frac{dC_0}{dt} = \frac{\partial Heff}{\partial C_o^*}$$
$$i\frac{dCp}{dt} = \frac{\partial Heff}{\partial C_p^*} \qquad (25)$$

we find

$$Heff = \alpha n_o n_p + \frac{1}{2}\left(\beta e^{i\Delta\omega t} C_0^* Cp + \beta^* e^{-i\Delta\omega t} C_0 C_p^*\right)$$

Taking into account the condition $n_0 + n_p = 1$, yields

$$Heff = \alpha n_p^2 + n_p \Delta\omega \qquad (26)$$

Averaging the fractional population n_p over the time, we get the average population $\overline{n_p} = b/2\varepsilon$ where $\varepsilon = \overline{\Omega}/\alpha$ and $\overline{\Omega}$ is the average Rabi frequency.

Substituting the average population n_p into the Hamiltonian (26) we obtain the average effective energy

$$Eeff = \frac{\alpha b^2}{2E^2}\left(\frac{b^2}{2E^2} + \delta\right) \qquad (27)$$

from which some thermodynamical parameters can be derived.

A "heat" capacity of the system can be defined as the variation of Eeff with $|\beta|$. We also define an order parameter $\eta = n_0 - n_p$ and a susceptibility χ_δ, for which we have:

$$\eta = 1 - \frac{b^2}{E^2}$$
$$C_\beta = \frac{\partial Eeff}{\partial|\beta|} \tag{28}$$
$$\chi_\delta = \left|\frac{\partial\eta}{\partial\delta}\right|$$

Note that C_β is actually the capacity of the system to incorporate the pumping energy. The quantities defined in Eq. (28) show discontinuities around the critical line $b+\delta \sim 1/2$, similar to the jumps observed during a phase transition.

The asymptotic behaviour of the characteristic quantities defined in Eq. (28), around the vicinity of the critical line, defines the corresponding critical indices. For small relative deviation $\tau \equiv |b - b_c|/b_c \to 0$, we have the following behaviour:

$$\eta - \eta_c \cong \frac{\sqrt{2}}{2}(1 - 2\delta)\tau^{1/2}$$
$$C_\beta \cong \frac{\sqrt{2}}{8}\tau^{1/2}$$
$$\chi_\delta \cong \frac{1}{\sqrt{2}}\tau^{-1/2}$$

The related critical indices are 1/2 for all quantities.

It is worth noting that the method considered for resonant excitation of coherent modes can be employed for exciting varios modes, including vortex states. To create a vortex it is necessary to have a specific dependence on spatial coordinates for the pumping field. For instance, if $V(\vec{r}) \sim \cos k\varphi$, where φ is the polar coordinate of a cylindrically symmetric trap field, vortex states with the winding number can be excited.

5. CONCLUSION

In conclusion, we can say that having a critical effect within a degenerate quantum gas may bring interesting new physics, as well as better understanding of macroscopic manifestations of quantum mechanics.

ACKNOWLEDGMENT

We would like to acknowledge FAPESP (Fundação de Amparo à Pesquisa do Estado de São Paulo) and Programa PRONEX.

REFERENCES

1. Andrews, M.R., Towsend, C.G., Miesner, H.J., Durfee, D.S., Kurn D.M., and Ketterle W., 1997, *Science* **275**: 637.
2. Deng, L., Hagley E., Wen J., Trippenbach M., Band Y., Julienne P.S., Simsarian J.E., Helmerson K., Rolston S., and PhillipsW.,1999, *Nature* **398**: 218.
3. BagnatoV.S., 1997, *Phys. Rev. A* **56**: 4845.
   Napolitano R., Deluca J., Bagnato V.S., and Marques G.C., 1997, *Phys. Rev A* **55**: 3954.
   Bagnato V.S., Marcassa L.G., Zilio S.C., Napolitano R., Deluca J. and, Weiner J.,1997, *Laser Phys.* **7**: 40.
4. Bagnato V.S., Pritchard D.E., and Kleppner D., 1987, *Phys. Rev. A* **35**: 4354.
5. Yukalov V.I., Yukalova E.P., and Bagnato V.S., 1997, *Phys. Rev A* **56**: 4845.
6. Yukalov V.I., 1995, *Phys. Rev. Lett.* **75**: 3000.
7. Bradley C.C., Sackett C.A., and Hulet. R.G., 1997, *Phys. Rev. Lett.* **78**: 985.

Towards A Two-Species Bose-Einstein Condensate

E. ARIMONDO
Dipartimento di Fisica and Istituto Nazionale per la Fisica della Materia, Università di Pisa, Via F. Buonarroti 2, I-56127 Pisa, Italy

1. INTRODUCTION

In the framework of Bose-Einstein condensation, the observation of a condensate mixture composed of two spin states of ^{87}Rb atoms[1] has prompted a significant interest into the two-species Bose-Einstein condensates (BECs). In these systems two different atomic species are simultaneously confined in a magnetic trap and cooled to very low temperatures. Two different macroscopic wavefunctions describe the two condensate species. In the condensation phase, the atomic properties are determined by the confining potential of the magnetic trap, but also by the collisions within each atomic species, and, for a two-species condensate, by the collisions with the other species. Thus for the two-species condensate the interatomic collisions play an additional role. The presence of an additional term in the condensate interaction energy produces an extra freedom in the construction of the Bose-Einstein condensate.

An apparatus for the simultaneous cooling and confinement of two species, rubidium and cesium, has been built in Pisa. The apparatus is based on a double-magneto optical trap (MOT) in a vertical geometry. From the lower MOT the atoms are transferred into a time-orbiting-potential (TOP) magnetic trap, more precisely into a TOP triaxial anisotropic trap with a geometry similar to that operating at NIST[2-4]. The future aim of the experiment is to probe heteronuclear cold collisions, sympathetic cooling, two-species relative scattering length, interaction between one condensate species and a non-condensed one. Up to now the apparatus has been used to measure the light-assisted cold collisions between rubidium and cesium cold atoms contained in the upper MOT. These results complement those obtained by previous authors on different atomic species[5-10]. Loading with rubidium atoms into the TOP magnetic trap has been achieved. These experimental progresses are here presented.

Furthermore, as a theoretical test of the properties associated to the preparation of a rubidium and cesium condensate system, we have solved the coupled Gross-Pitaevskii equations for a two-species condensate confined within a fully anisotropic triaxial magnetic trap. The calculation of the two-species excitation spectrum has shown that the mode frequencies of the individual condensates are modified in a two-species condensate interatomic interaction, with new excitation modes present in the mixed system[11]. The spatial distributions of the two condensates show deformations produced by the interatomic interactions, with either compression of the atoms in the central region of the trap, or with a decrease of the atomic density in that region.

In Sect. 2 the experimental apparatus is presented. The following Section examines the harmonic potential produced in our magnetic trap. The experimental results on the rubidium-cesium light assisted cold collisions will be presented in Sect. 4. In Sect. 5 the results of the theoretical analysis for the Gross-Pitaevskii equation will be presented. Sect. 6 concludes the presentation.

2. APPARATUS

The apparatus, presented in Fig. 1, is based on two MOT's in a vertical configuration connected by a narrow graphite tube creating the pressure gradient between the low vacuum upper chamber and the high vacuum lower chamber. The lower quartz cell dimensions are 18 x 24 x 70 mm^3. Alkali dispensers produce the Rb and Cs background atoms collected by the laser cooling process in the upper MOT.

In the laser system, based on master-slave configuration, the Rb cooling slave is a tapered amplifier with 300 mW output, the Cs cooling a 200 mW diode laser. Radiation from both cooling lasers is combined within a single mode optical fiber, with 65% transmission at both wavelengths. All the optical components at the fiber output required for the laser cooling of the two species operate at both 780 and 850 nm wavelengths.

The transfer from the upper MOT to the lower one is produced by a push beam entering into the apparatus from the top part of the vacuum system and acting on the atoms for 1 ms. The upper MOT is loaded with atoms for a maximum time of 100 ms, in order to produce a MOT with a contained transverse size, and to realize a transfer into the lower MOT not perturbed by the graphite narrowing. The transfer efficiency with the push beam reaches the value of 90 percent.

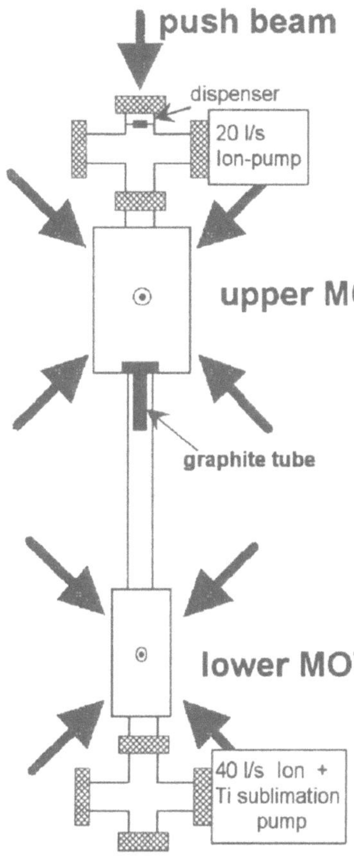

Figure 1. Schematic representation of the apparatus with upper and lower vacuum systems, the alkali dispensers in the top part, the graphite narrowing producing the vacuum gradient, and the direction of the push beam transferring atoms from the upper MOT to the lower one.

The magnetic trap in the lower chamber is based on the TOP configuration with the symmetry axis of quadrupole field in the horizontal plane and the rotating field, at 10 kHz, also contained in the horizontal plane. The harmonic potential associated to this configuration, similar to that applied in the experiments at NIST[2-4], produces a very symmetric condensate wavefunction. Because the rotating field lies in the horizontal plane, also the circle-of-death is in the horizontal plane, whence the circle-of-death evaporation is not influenced by the gravity. The quadrupole field is generated by 28 windings of 3 mm copper tubing cooled by 5 bar pressured water. With a maximum current of 240 A, the 2b' gradient of the quadrupole field along the symmetry axis reaches a value of 550 G cm^{-1}, the dissipated power being 2.4 kW. A maximum TOP rotating field of 20 G is applied between 5 and 10 kHz, in the frequency range where the TOP coils were resonantly tuned.

The preparation of the Rb atoms in the magnetic trap takes place through a standard sequence of steps: loading the MOT, transferring the cold atoms into a compressed magneto-optical trap (CMOT), decreasing of the atomic temperature through an optical molasses stage, and finally transferring into the TOP magnetic trap.

3. TOP TRAP

The magnetic field composed by the quadrupole field, with gradient 2b' along the horizontal x axis, and by the field B_0 rotating at angular velocity ω_0 in the horizontal plane is given by:

$$\vec{B}_{TOP} = \begin{pmatrix} 2b'x \\ -b'y \\ -b'z \end{pmatrix} + \begin{pmatrix} B_o \cos\omega_o t \\ B_o sin\omega_o t \\ 0 \end{pmatrix} \tag{1}$$

The magnetic energy U_M for the atomic momentum μ coupled to the TOP magnetic field is:

$$U_M = -\vec{\mu}\cdot\vec{B}_{TOP} = -\mu B_o \cdot \sqrt{1+\frac{4b'^2x^2}{B_o^2}+\frac{b'^2y^2}{B_o^2}+\frac{4b'^2z^2}{B_o^2}+\frac{4b'x\cos\omega_o t}{B_o}-\frac{2b'ysin\omega_o t}{B_o}} \tag{2}$$

The average of the magnetic energy over the rotation period results:

$$\langle U_M \rangle_{\omega_o} = -\mu B_o - \frac{1}{2} M\omega^2\left(4x^2+y^2+2z^2\right) \tag{3}$$

with:

$$\omega^2 = \frac{\mu b'^2}{2MB_o} \tag{4}$$

M being the atomic mass, whence the harmonic frequencies along the three axis are in the ratio $2{:}1{:}2^{1/2}$. The total potential U_{tot} is composed by that of Eq. (3) and by the gravitational potential:

$$U_{tot} = \langle U_M \rangle_{\omega_o} + Mgz \tag{5}$$

From this potential the equilibrium position of the atoms is derived, leading to a gravitational sag δz from the z = 0 position TOP center:

$$\delta z = -\frac{g}{2\omega^2}\frac{1}{\sqrt{1-\eta^2}} \tag{6}$$

where the parameter η is given by:

$$\eta = \frac{Mg}{\mu b'} \tag{7}$$

Eqs. (4) and (7) provide the conditions for generating same oscillation frequencies and same gravitational sag if two different atomic species are loaded into the magnetic trap. The two species should have the same value of the ratio M/μ, where for an atomic state (F,m_F) the magnetic moment μ is given by:

$$\mu = g_F m_F \mu_B \tag{8}$$

with g_F the gyromagnetic ratio and μ_B the Bohr magneton. If we consider the state ($F=3,m_F=-3$) for ^{133}Cs and the state ($F=1,m_F=-1$) for ^{87}Rb, it results:

$$(M/\mu)_{Cs} = 177.3\, a.u. \tag{9a}$$

$$(M/\mu)_{Rb} = 174\, a.u. \tag{9b}$$

Whence the difference between the oscillation frequencies and the gravitational sag for the two atoms is less than 2%.

4. TOP LOADING

We have examined the loading of the TOP magnetic trap with a cloud of Rb atoms. With the operational parameters of our magnetic trap b' = 47 G cm^{-1} and B_o = 19.5 G. the gravitational sag δz_g of Eq. (6) is quite large for the Rb atoms, roughly -1.5 mm. Thus it is of paramount importance that the spatial center of the CMOT is identical to the operational point of the magnetic trap. In absence of additional magnetic fields, the CMOT center is the zero for the quadrupole field and lies $-\delta z_g$ above the TOP operation point, so the two traps do not share a common center. A set of three coils has been used to shim by the quantity Δz the vertical CMOT position in order to match that of the TOP trap. This shimming is required to reduce the sloshing motion of the atom cloud after loading into the magnetic trap.

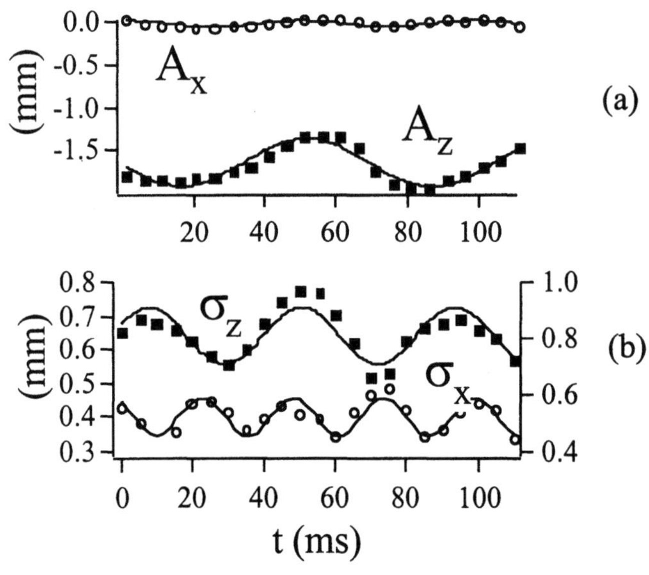

Figure 2. Experimental results for the time dependence of the amplitudes in the sloshing motion, in (a), and in the breathing motion, in (b), along two axes, for the Rb cloud after loading into the TOP magnetic trap. For the breathing motion amplitude σ_z left scale, for the σ_x amplitude, right scale.

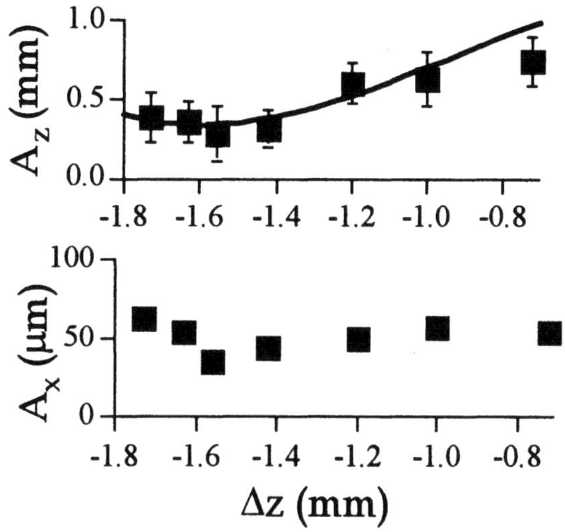

Figure 3. Data for the amplitude of the Rb cold cloud after loading of the TOP trap. In (a) amplitude A_z for the motion along the *z* vertical axis, and in (b) amplitude A_x for the motion along the magnetic trap *x* axis versus the Δz mismatch. Squares represent the experimental results. The continuous line in (a) has been obtained from a Monte Carlo simulation supposing the Rb atoms to be loaded with initial vertical velocity of 4 cm/s.

Fig. 2 reports experimental results for the sloshing and breathing motion of the atomic cloud produced after loading the magnetic trap from the CMOT, in the operation conditions where a good mode matching was supposed to be realized[12]. Motion along two different axes, *x* and *z*, was monitored. Small oscillations are produced along the *x* axis, while large amplitude oscillations are produced along the z axis. The amplitudes of those oscillations versus the Δz shimming is shown in Fig. 3. The A_z amplitude reaches a minimum when the Δz matches the δz_g gravitational sag. However that minimum is not zero, and the residual minimum in the oscillation amplitude was produced by a momentum transfer to the atoms in the optical molasses stage. The continuous line in Fig. 3(a) has been obtained from a Monte-Carlo simulation analysis supposing that an initial velocity of 4 cm/s has been applied to the Rb atoms. A similar dependence of A_z or Δz has been obtained on the basis of the conservation for the sum of kinetic and potential energies.

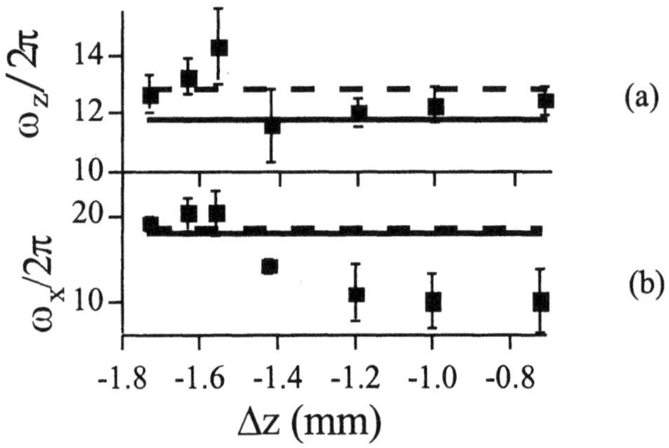

Figure 4. Frequencies $\omega_x/2\pi$ and $\omega_z/2\pi$, in Hz, of the oscillations along the x and z axes, versus the Δz mismatch. Lines represent theoretical predictions for those frequencies with measured trap parameters using an averaged potential (dashed line) and a non averaged one (continuous line).

It appears from Fig. 2 that the oscillations frequencies of the breathing modes are twice those of the corresponding sloshing modes. The different amplitudes σ_x and σ_z for the breathing motions along x and z axes arise because their potential energies are equal but the oscillation frequencies are different. The oscillation frequencies for the sloshing motion along the x and z axes have been measured versus Δz and results are shown in Fig. 4. Within the accuracy of the measurements the data are nearly fitted by the predicted constant value (continuous and dashed lines), even if some systematic deviation takes place when Δz matches the value of the gravitational sag. The dashed lines represent the sloshing frequencies predicted for atomic motion in the harmonic time averaged potential of Eq. (3). The continuous lines report the predicted values derived from a Monte Carlo simulation using the exact potential of Eq. (2). It appears that the adiabatic elimination is not exact, but a test of those non-adiabatic corrections would require a more precise determination of the magnetic trap parameters.

5. Rb-Cs COLD COLLISIONS

Two different alkali species, Rb and Cs, have been simultaneously cooled within the upper MOT of the apparatus described above. Thus the modifications produced by the light assisted cold heteronuclear collisions on the operation of the two species MOT (2MOT) have been examined.

For the light assisted cold collisions, the main difference between homonuclear and heteronuclear cold species is that the interatomic potential between a ground state atom and an excited one, in the homonuclear case depends on R^{-3} and in the heteronuclear case depends on R^{-6}. Furthermore in the homonuclear case the potential is always attractive, while, in the heteronuclear case, the potential can be either attractive or repulsive. More precisely the attraction or repulsion depends on which atomic species has the lower energy excited state. For instance in the case of the rubidium and cesium, where the cesium 6P excited state is lower in energy than the rubidium 5P excited state, it results that all electronic molecular states asymptotically connected to the Rb(5P)+Cs(6S) are repulsive and all those connected to the Rb(5s)+Cs(6P) are attractive[13]. Thus a collision between a cold excited Rb atom and a cold ground Cs atom is characterised by a repulsive potential, i.e., the atoms will not be able to reach small internuclear distances where the loss processes of radiative escape and fine structure changing collisions may take place. On the contrary a cold collision between a ground Rb atom and an excited Cs atom attracts the atoms to the small internuclear distances where loss processes take place. Another characteristic of the heteronuclear collisions is that, owing to the R^{-6} potential, at the laser detunings usually applied in the cooling experiments, the heteronuclear atomic excitation takes place at internuclear distances smaller than in the homonuclear case[8,10]. As a consequence the survival factor, i.e., the probability of reaching the very small internuclear distances where the crossing between the molecular potential leading to losses occurs, is nearly unity. Thus all collisions taking place along the attractive potential produce a trap loss.

As final point on the heteronuclear loss processes, the kinetic energy gained in the collision process is shared between the two colliding species, whence the lighter species gets most of that energy and escapes from the trap[8]. In the case of Rb-Cs 2MOT, Rb is the loser. The experimental configuration used to measure the Rb-Cs loss rate constants is presented in Fig. 5 with the photodiode monitoring the light emitted from the 2MOT on the Rb resonance line. A decrease in the number of Rb atoms contained in the 2MOT as consequence of the Rb-Cs collisions appears as a decrease on the emitted Rb fluorescence light.

The fluorescence of each trapped atomic species was measured using narrow-band interference filters and calibrated photodiodes. Cold cloud sizes and shapes were measured with charged-coupled-devices (CCD) cameras. The number of atoms was derived from the fluorescence rate using a steady state rate calculation. Individual trapped densities were ~ 5.10^{10} cm^{-3}, with ~ 2.10^6 Rb atoms and ~ 5.10^6 Cs atoms. The trap parameters were chosen to produce a r.m.s diameter of the Rb cloud (0.4 mm) smaller than that of the Cs cloud (0.65 mm). Because the larger loss rate is expected for the lighter atom, the loading rate of the Rb atoms was investigated.

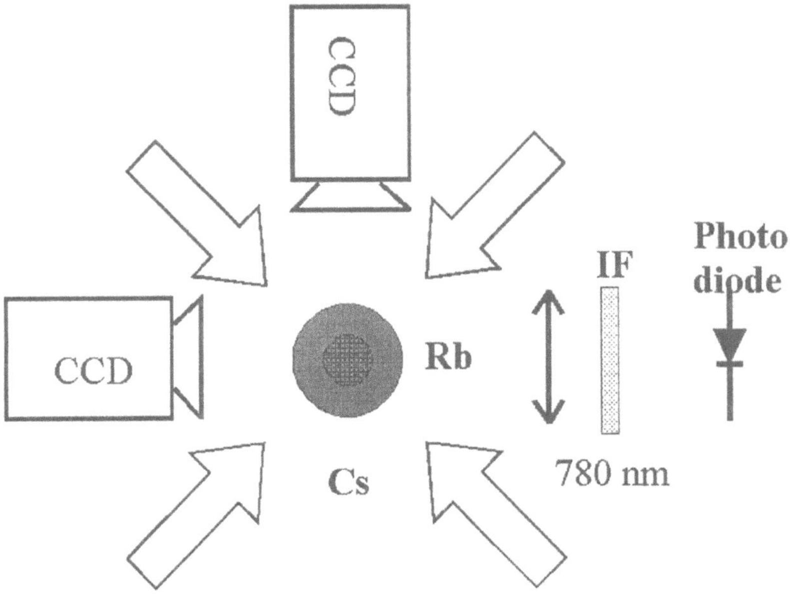

Figure 5. Experimental configuration for monitoring the rubidium loss processes from the 2MOT system, with a 780 nm interference filter and a photodiode recording the light emitted by Rb atoms. The cold Rb cloud is at the center of the cold Cs cloud. The CCD's were oriented so that complete three dimensional overlap of the Rb and Cs clouds could be assured.

The collision rates in the 2MOT were derived from an analysis of the trap loading, following a switch-on of the cooling lasers. The loading curves for both Rb and Cs atoms were well fitted by exponential functions. The loading rate for the number N_i of the species i in the presence of the number N_j of the species j is described by the following equation:

$$\frac{dN_i}{dt} = L - \alpha_i N_i - \beta_i \int n_i^2 dV - \gamma_{ij} \int n_i n_j dV \qquad (10)$$

Here L is the trap loading rate, α_i represents the loss rate due to background collisions, β_i is the single species cold collision rate, γ_{ij} is the loss rate from cold collisions between species i and species j. Because both traps operate in the density limited regime, and the Rb cloud is immersed into the region of the Cs cloud where its density can be assumed constant, the equation for N_{Rb} may be written:

$$\frac{dN_{Rb}}{dt} = L - \alpha_{Rb} N_{Rb} - \beta_{Rb} f(N_{Rb}) n_{Rb} N_{Rb} - \gamma_{RbCs} n_{Cs} N_{Rb} = \\ = L - R_P N_{Rb} \tag{11}$$

where the exponential rate R_P for the Rb loading has been introduced:

$$R_P = \alpha_{Rb} + \beta_{Rb} f(N_{Rb}) n_{Rb} + \gamma_{RbCs} n_{Cs} \tag{12}$$

$f(N_{Rb})$ is a factor that accounts for the deviation of the Rb cloud density distribution from the uniform density n_{Rb} value[14]. Experimental results for R_P vs the total intensity I_{tot}^{Rb} of the Rb trap lasers are shown in Fig. 6(a), for the case of Rb only and Rb in presence of Cs atoms. R_P appears nearly independent of the laser intensity for the case of Rb only, and instead decreases when cold collisions with Cs take place. That behaviour arises because the absolute value of γ_{RbCs} is large compared to the typical value, $< 10^{-11}$ cm^3 s^{-1}, for the β_{Rb} rate coefficient of Rb+Rb* collisions. Using the definition of Eq. (12), the value of γ_{RbCs} has been derived from the R_P data, as shown in Fig. 6(b). γ_{RbCs} presents a dependence on I_{tot}^{Rb} that was fitted with the following linear slope:

$$\gamma_{RbCs} = \gamma_{RbCs}^{0} + \gamma_{RbCs}^{I} \cdot I_{tot}^{Rb} \tag{13}$$

A fit of the data of Fig. 6(b) leads to the following values:

$\gamma^0_{RbCs} = 2.1(4)10^{-10}$ cm^{3}s^{-1} ,

$\gamma^I_{RbCs} = 0.04(1)10^{-10}$ cm^{3}s^{-1}.

The dependence of γ_{RbCs} on I_{tot}^{Rb} may be produced by different mechanisms. One mechanism is the change in the Rb trap depth with I_{tot}^{Rb}. Another mechanism is the modification of the ground state occupation by I_{tot}^{Rb}. Because the Rb-Cs trap losses require a ground state Rb atom colliding with a Cs excited atom, the loss parameter γ_{RbCs} should follow the ground state occupation Π_{Rb}. In Fig. 6(c) the loss rate γ_{RbCs} is shown versus Π_{Rb}, calculated using the equation for the excited state fraction of ref.15. It appears that γ_{RbCs} does follow the variation of Π_{Rb}. Thus both mechanisms may contribute to the dependence shown in Fig. 6(b). The value of γ_{RbCs} derived from our analysis has the same order of magnitude as that derived by previous authors[8-10] for other heteronuclear collision rates.

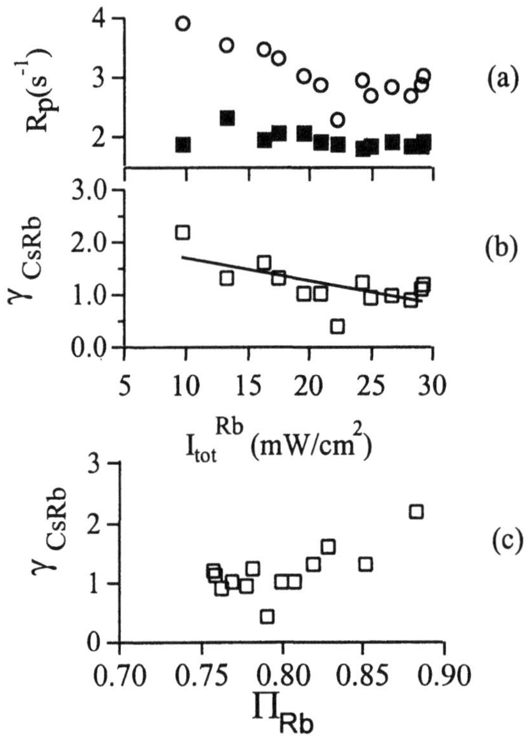

Figure 6. Experimental results for the Rb losses produced by the collisions with Cs in a 2MOT, for I_{tot}^{Cs} = 40 mW cm^{-2}, δ_{Rb} = -2Γ_{Rb}, and δ_{Cs} = -1.4Γ_{Cs}. In (a) MOT loading rate R_p, in units of s^{-1}, vs the total intensity I_{Rb} of Rb pumping lasers, with closed squares for the case of Rb only, and the open circles when Cs cooling takes place. In (b) γ_{RbCs} values, in units of 10^{-10} cm^{3}s^{-1}, derived from data in (a) using n_{Cs} = 7.2x10^9 cm^{-3}. In (c) γ_{RbCs} versus ground state occupation Π_g, modified by the Rb trap laser intensity.

6. EQUATIONS FOR 2-CONDENSATE

Several important properties of the two-species Bose condensate (2-condensate) have been previously examined[16-26], the most important one being the spatial symmetry breaking associated to the macroscopic wavefunction of the condensate. Thus in a symmetric magnetic potential the 2-condensate wavefunction does not match to the trap symmetry[19,20]. Also, the phenomenon of density compression takes place, with one atomic species squeezed out and forced to form a shell around the trap center which holds the second atomic species, receiving extra confinement from the first atomic species[18,21].

The system composed by two Bose-Einstein condensates, Rb and Cs, confined in a triaxial fully anisotropic magnetic trap has been investigated[11]. In the analysis the well known Rb-Rb scattering length, and the current best estimates for the Cs-Cs have been inserted into the numerical calculations. The estimated large negative scattering length of cesium implies that a condensate of Cs atoms is stable only for a very small number of bosons. Aim of the investigation was to verify whether, using the density compression provided by the Rb condensate with a large number of bosons, it is possible to realize a stable condensate of Cs atoms with a sizable number of atoms, i.e., larger than that achievable in presence of single species. This should give some indications whether, as a final product of the sympathetic cooling between two different alkali species, as opposed to two isotopes of the same species, 2-condensates could be created even in very unfavourable cases.

To deal with a condensate containing a small number of atoms, like Cs with a negative scattering length, the Thomas-Fermi approximation cannot be used, and both a variational approach based on Gaussian trial functions and a direct numerical integration of the tridimensional time dependent Gross-Pitaevskii equation (GPE) was used. From the numerical solution, the deviation in the spatial distribution from a Gaussian shape for both atomic species has been derived, with either a compression or a depression at the center of the magnetic trap.

In mean field approximation the 2-condensate is described by two macroscopic wave functions, that evolve in time according to the self-consistent GPE's. The coupling constants in the GPE's are given by the s-wave scattering lengths between atoms of the same species, and between different species. Like in our experiments in magnetic traps for single atomic species, the two condensates have been supposed confined inside a magnetic trap with the triaxial harmonic potential of Eq. (3). Moreover as for the cesium and rubidium states of Eqs. (9) the gravity sag of the two condensates has been supposed to be exactly compensated in order to produce the two harmonic potentials centered at the same point in space.

To solve the GPE's in the variational analysis, Gaussian trail functions have been used, with variables the widths and positions of the two Gaussians in the three spatial directions and their derivatives. For the numerical integration of the time dependent GPE equation, a modified split operator technique, adapted to the integration of a Schrödinger equation has been used. This algorithm had already briefly accounted in Cerboneschi et al.[27], but it has been opportunely extended to the two species case[11]. The oscillation frequencies of the condensates were found applying a small perturbation to the steady state of the GPE, evolving the condensate in real time, and looking at the frequency spectrum of the widths and positions of the two condensates.

In the numerical simulations we have considered atoms of ^{133}Cs and ^{87}Rb, trapped respectively in the $|F=3,m_F=-3\rangle$ and $|F=1,m_F=-1\rangle$ hyperfine sublevels. For the Rb-Rb scattering length we used $a_{Rb\text{-}Rb}$ = 109.1 a.u. for the $|F=1,m_F=-1\rangle$ hyperfine sublevel. For the Cs-Cs scattering length $a_{Cs\text{-}Cs}$ we have used the value of -400 a.u., in agreement with different recent estimates[28]. Finally, for the Rb-Cs collisions an estimate $a_{Rb\text{-}Cs}$ = 150 a.u. has been used. This value[i] is smaller than the critical value $|a^c_{Rb\text{-}Cs}| = (|a_{Rb\text{-}Rb}\, a_{Cs\text{-}Cs}|)^{1/2}$ above which in the Thomas-Fermi regime the two-species condensate cannot exist[18,24,29,30].

If rubidium is not present, there exists a critical number N_C for the bosons number above which Cs collapses because of the attractive term of auto interaction within the GPE[31]. That critical number can be increased reducing the spring constants of the harmonic magnetic confining potential. Such dependence suggests that condensate stability with a number of bosons greater than N_C can be realized adding a second condensate that interacts repulsively with cesium. This configuration is realized for a negative Cs-Cs scattering length and a positive Rb-Cs scattering length.

Using the variational method the steady state of the GPE's has been derived by varying N_{Cs} and N_{Rb}. For N_{Rb} different from zero it was found that, increasing N_{Cs}, a steady state exists until a critical value of the Cs atomic number is reached where the lowest oscillation frequency of the 2-condensate goes to zero. This particular frequency is related to the vibration of the center of mass along one of the axes, and its change in sign indicates that the state has become unstable by broken symmetry of the overall GPE

[i] The rubidium-cesium scattering lenght has been derived on the basis of published spectroscopy data for the ground state of the RbCs molecule (A. Bambini, private communication).

solution[ii]. In these conditions the two condensates separate spatially moving into two distinct regions, so that they are no longer at the center of the trap. The broken symmetry value coincides with the number of cesium atoms for which the cesium condensate collapses: as soon as the two-species condensate separates in space, the cesium condensate is no longer sustained against collapse by the other condensate species. Whence for a broken symmetry a condensate collapse occurs with a condensate cloud having a number of bosons larger than the one-species critical value.

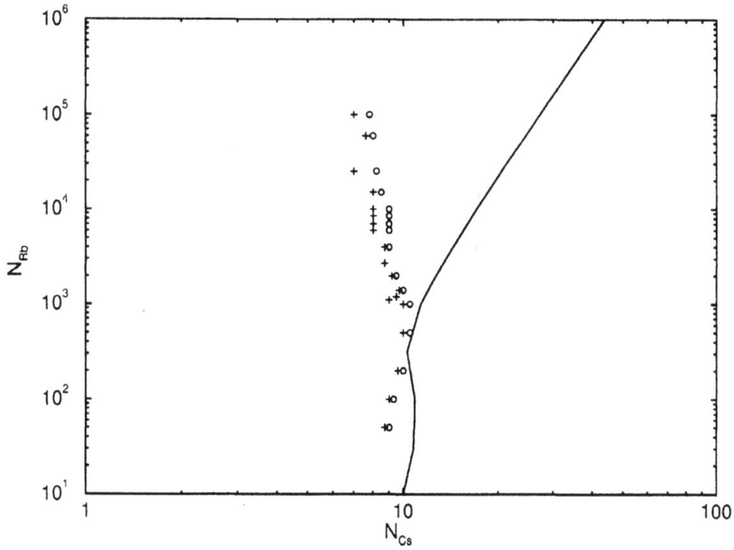

Figure 7. Stability region for a Cs-Rb mixture. To the left of the solid line (obtained from the variational approach) there is the region where the Cs condensate is stable. The circles (daggers) mark the values of the Cs and Rb atom numbers yielding an unstable (stable) cesium condensate: the condition of stability is derived numerically.

Fig. 7 shows the stability region of the two-species condensate on the (N_{Cs}, N_{Rb}) plane. The solid line divides the stable region, at lower atom number values, from the unstable region, at larger atom number values, as computed using the variational method. Increasing N_{Cs}, the critical number of Cs atoms of the broken symmetry point is larger than the critical N_C value for a single species. Points computed using the numerical integration on the lattice are shown by symbols: values of N_{Cs}, N_{Rb} supporting a stable or unstable mixture are shown as circles or daggers, respectively. The transition

[ii] For a fully anisotropic trap, in the presence of a broken symmetry, the wavefunction invariance under reflection on each of the (x,y,z) planes no longer applies.

line between stable and unstable mixtures lays between the circles and the daggers of the figure. It is clear from the figure that the variational method overestimates the region of stability. The reason of this discrepancy between the two methods is linked to the spatial distribution of the condensate.

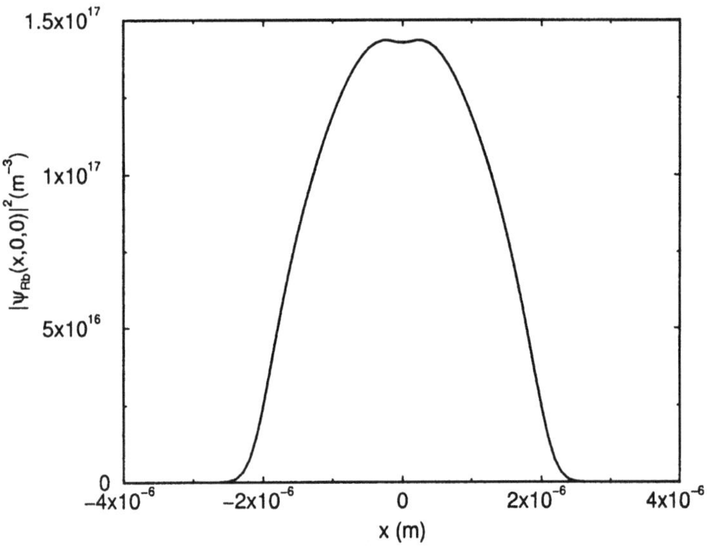

Figure 8. Probability density distribution $|\psi_{Rb}(x,0,0)|^2$ of the Rb condensate in the presence of a Cs condensate with negative scattering length, for the following values of atom numbers, $N_{Rb} = 6000$ and $N_{Cs} = 8$. The line shows the results from the numerical approach.

Fig. 8 shows the probability density distribution for the Rb atoms along the x axis, $|\psi_{Rb}(x,0,0)|^2$, obtained by the numerical method in the case of negative $a_{Cs,Cs}$: at larger N_{Cs}, at the trap center, instead of a maximum the Rb distribution has a depression caused by the presence of cesium. This deformation of the Rb cloud produces an additional squeezing of the cesium distribution towards the trap center. The numerical analysis allows us to conclude that this change in the distribution at the trap center has two important effects: first, one of the hydrodynamics frequencies connected to the motion of the Gaussian width goes to zero; second, the collapse of the Cs condensate does not happen via broken symmetry, as for the case of the variational analysis.

7. CONCLUSIONS

The experimental progress towards the realization of BEC in Rb and the measurements of cold collisons between Rb and Cs have been examined.

Moreover numerical analyses for a double Bose-Einstein condensate confined within a triaxial magnetic trap, using both a variational approach and the numerical solution of the two-species GPE have been performed. Several characteristic BEC quantities, to be tested in experiments on two-species, have been examined. The most important features are associated to the spatial separation of the two atomic species, produced by the positive interatomic scattering length. For instance at large number of rubidium atoms the interatomic repulsion is so large that the potential experienced by the cesium atoms is no more harmonic. Viceversa the cesium atoms, even at small number, produce a large deformation in the rubidium cloud. It may be imagined that the presence of the small condensate cesium could be detected not directly on the cesium absorption, but on the spatial deformation of the rubidium cloud. Actually for the values of scattering lengths introduced in the analysis, the modifications in the rubidium cloud are very small and not easily monitored on the integrated absorption profile. However for other scattering lengths, and whence for different atomic species, larger deformations may be produced and detected. Up to now double species condensates have not yet been produced, so that the phenomena here predicted could not be yet tested.

Note: after the School, while these lecture notes were prepared, BEC in rubidium was observed in the apparatus here described.

ACKNOWLEDGMENTS

This work has been supported by the INFM through the PRA on BEC, by the Consiglio Nazionale delle Ricerche through a Progetto Integrato in collaboration with the IFAM of Pisa, and by the MURST. The present research has been performed in collaboration with D. Ciampini, M. Fazzi, F. Fuso, J.H. Müller, O. Morsch, G. Smirne, P. Verkerk, D. Wilkowski. I want to thank F. Cervelli for help in preparing these notes.

REFERENCES

1. Myatt, C.J., Burt, E.A., Ghrist, R.W., Cornell, E.A., and Wieman, C.E., 1997, *Phys. Rev. Lett.* **78**: 586-589.
2. Kozuma, M., Deng, L., Hagley, E.W., Wen, J., Lutwak, R., Helmerson, K., Rolston, S.L., and Phillips, W.D., 1999, *Phys. Rev. Lett.* **82**: 871-875.
3. Deng, L., Hagley, E.W., Wen, J., Trippenach, M., Band, Y., Jiulienne, P.S., Simsairan, J.E., Helmerson, K., Rolston, S.L., and Phillips, W.D., 1999, *Nature* **398**: 218-220.
4. Hagley, E.W., Deng, L., Kozuma, M., Wen, J., Helmerson, K., Rolston, S.L., and Phillips, W.D., 1999, *Science* **283**: 1706-1709.

5. Süptitz, W., Wokurka, G., Strauch, F., Kohns, P., and Ertmer, W., 1994, *Opt. Lett.* **19**: 1571-1573.
6. Santos, M.S., Nussenzveig, P., Marcassa, L.G., Helmerson, K., Fleing, J., Zilio, S.C., and Bagnato, V.S., 1995, *Phys. Rev. A* **52**: R4340-R4343.
7. Scherf, W., Wippel, V., Fritz, T., Gruber, D., and Windholz, L., 1998, *Europhys.* Conf. Abstr. **22** D: 6-31.
8. Telles, G.D., Marcassa, L.G., Muniz, Miranda, S.G., Antunes, A., Westbrook, C., and Bagnato, V.S., 1999, *Phys. Rev. A* **59**: R23-R26.
9. Shaffer, J.P., Chapulpczak, W., and Bigelow, N.P., 1999, *Phys. Rev. Lett.* **82**: 1124-1127; *Phys. Rev. A* **60**: R3365-R3368.
10. Schlöder, U., Engler, H., Schünemann, U., Grimm, R., and Weidemüller, M., 1999, *Eur. Phys. J.* D **7**: 331-340.
11. Montina, A., Mannella, R., and Arimondo, E., 1999, *Phys. Lett. A* **261**: 337-344.
12. Ketterle, W., 1999, in *Bose-Einstein Condensation in Atomic Gases* (M. Inguscio, S. Stringari and C. Wieman, eds), IOS, Amsterdam, pp. 67-176.
13. Marinescu, M., and Sadeghpour, H.R., 1999, *Phys. Rev. A* **59**: 390-404.
14. Feng, P,. Hoffman, D., and Walker, T., 1993, *Phys. Rev. A* **47**: R3495-R3498.
15. Grego, S., Colla, M., Fioretti, A., Müller, J.H., Verkerk, P., and Arimondo, E., 1996, *Opt. Commun.* **132**: 519-526.
16. Ho, Tin-Lun, and Shenoy, V.B., 1996, *Phys. Rev. Lett.* **77**: 3276-3279.
17. Goldstein, E., and Meystre, P., 1997, *Phys. Rev. A* **55**: 2935-2940.
18. Esry, B.D., Greene, Chris, H., Burke. James, P., Jr., and Bohn, J.L., 1997, *Phys. Rev. Lett.* **78**: 3594-3597.
19. Law, C.K., Pu, H., Bigelow, N.P., and Eberly, J., 1997, *Phys. Rev. Lett.* **79**: 3105-3108.
20. Esry, B.D., and Greene, C.H., 1998, *Phys. Rev. A* **57**: 1265-1271; Esry, B.D., 1998, *Phys. Rev. A* **58**: R3399-R3402.
21. Burke, J.P., Jr., Bohn, J.L., Esry, B.D., and Green, C.H., 1998, *Phys. Rev. Lett.* **80**: 2097-2100; 1999, *Phys. Rev. A* **59**: 1457-1460.
22. Pu, H., and Bigelow, N.P., 1998, *Phys. Rev. Lett.* **80**: 1130-1133, and 1134-1137.
23. Öhberg, P,. and Stenholm, S., 1997, *Phys. Rev. A* **57**: 1272-1279.
24. Gordon, D., and Savage, C.M., 1998, *Phys. Rev. A* **58**: 1440-1444.
25. Ao, P., and Chui, S.T., 1999, *Phys. Rev. A* **58**: 4836-4840.
26. Sinatra, A., Fedichev, P.O., Castin, Y., Dalibard, J., and Shlyapnikov, G.V., 1999, *Phys. Rev. Lett.* **82**: 251-254.
27. Cerboneschi, E., Mannella, R., Arimondo, E., and Salasnich, L., 1998, *Phys. Lett. A* **249**: 495-500.
28. Vuletic, V., Kerman, A.J., Chin, C., and Chu, S., 1999, *Phys. Rev. Lett.* **82**: 1406-1409, and references therein.
29. Andreev, A. F., and Bashkin, E. P., 1957, *Zh. Eksp. Teor. Fiz.*, **69**: 319 [*Sov. Phys. JETP* **7**: 698]
30. Sigga, E.D., and Ruckenstein, A.E., 1980, *Phys. Rev. Lett.* **44**: 1423-1426.
31. Dalfovo, F., Giorgini, S., Pitaevskii, L.P., and Stringari, S., 1999, *Rev. Mod. Phys.* **71**: 463-512.

Atom Interferometry With Ultra-Cold Atoms

M. KASEVICH
Sloane Physics Laboratory, Yale University, New Haven, CT 06520-8120

1. INTRODUCTION

Atom interferometer inertial force sensors offer the prospect of unprecedented performance for rotation, acceleration and gravity gradient measurements. Applications range from inertial navigation and oil/mineral exploration, to tests of General Relativity and measurements of G (the gravitational constant).

The current generation of laboratory instruments, based on single-particle atom interference, have reached sensitivity levels where they now compete favorably with state-of-the-art sensors. The recent demonstration of Bose-Einstein condensed atomic sources[1] and the progress in atom interferometry offer the potential of a new class of sensors drawing on advances in both areas. For example, the novel coherence properties of quantum degenerate atomic sources might be exploited to operate interferometers below the shot-noise limit[2]. Since the present generation instruments run at, or near, the atomic shot-noise limit, these techniques might ultimately provide for a significant increase in sensor performance. As another example, the use of these sources allows, in principle, for realisation of large spatial separations (> 1 cm) of atomic wavepackets. Since inertial sensor sensitivity scales, in general, with this spatial separation, Bose-Einstein condensed sources may enable new interferometer geometries with substantially enhanced intrinsic sensitivity.

The organisation of this Chapter is as follows. Sect. 2 presents a brief summary of the current status of single-particle atom interferometer sensors. Our intent is to provide core information which enables evaluation of basic sensor characteristics, and to allow for comparison with possible next-generation instruments based on degenerate atomic sources. Sect. 3 discusses the prospects of this latter class of instruments.

2. SINGLE PARTICLE LIGHT-PULSE ATOM INTERFEROMETRY

We first summarise basic principle underlying the operation of single-particle atom interferometers force sensors based on pulses of light (Sect. 2.1). We then specialise these ideas to the demonstration of a gyroscope[3] and gravity gradiometer[4] in Sects. 2.2 and 2.3.

2.1 Overview

We begin with a classical analogy for atom interferometer force measurements, which is illustrated in Fig. 1a. The presence of an inertial force results in the deflection of the trajectory (solid line) of a proof-mass from the unperturbed trajectory (dashed line). The resulting acceleration can be extracted from measurement of the curvature of the perturbed trajectory. This curvature can be determined, for example, by measuring the distance of the proof-mass from reference platforms (shaded gray) at three equally spaced points in time. The resulting distance measurements are used to infer acceleration by $a = [\ell(t_1) - 2\ell(t_2) + \ell(t_3)]/T^2$. In a classical measurement, these distance measurements could be made, for example, using optical interferometry techniques (by reflecting a laser from the proof-mass).

The atom interference method essentially replaces the macroscopic proof-mass with an atom, and is illustrated in Fig 1b. The curvature of the atomic trajectory is measured through three successive interactions with laser beams. The laser beams are tuned to be resonant with a transition between two of the atom's internal quantum states (levels $|1\rangle$ and $|2\rangle$). Under appropriate conditions, the atom records the phase of the driving electromagnetic field during each resonant interaction. This phase is directly proportional to the distance of the atom from the reference platform to which the laser field is anchored (*ie.* $\phi = 2\pi\ell/\lambda$, where λ is the wavelength of the laser). If the atom is initially prepared in state $|1\rangle$, the probability of finding the atom in state $|2\rangle$ following the three interactions is $P_2 = 1/2[1-\cos\Delta\phi]$ where $\Delta\phi = \phi(t_1) - 2\phi(t_2) + \phi(t_3)$ This transition probability follows from standard application the Schrödinger equation. Measurement of the number of atoms in state $|2\rangle$ following the interaction sequence allows determination of $\Delta\phi$. This measurement is accomplished by standard resonance fluorescence methods. The atom's acceleration is extracted from $\Delta\phi$ according to the relation $a = \Delta\phi\ \lambda/2\pi T^2$. In practice, the acceleration of an ensemble of atoms is recorded to improve signal to noise. This is possible since the acceleration induced phase shift $\Delta\phi$ is independent of an atom's initial position and velocity.

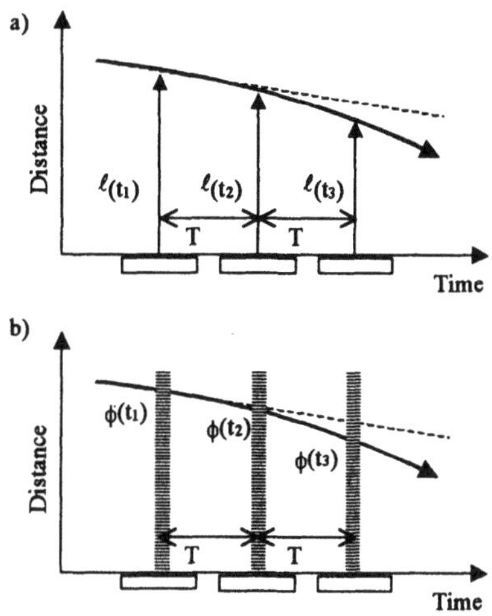

Figure 1. Single particle atom interferometry. (a) Generic model for acceleration measurements. (b) Schematic illustration of atom interferometer acceleration measurement technique.

A complete treatment requires consideration of the momentum exchange between the atom and the laser field since the momentum of the atom is comparable with the momentum associated with the photon used to interrogate the atom's position (a Cs atom's velocity changes by 3 mm/sec when it absorbs a single photon). Consequently, an atom which emits (absorbs) a photon of momentum $\hbar\mathbf{k}$ will receive a momentum impulse of $\Delta\mathbf{p}=-\hbar\mathbf{k}$ ($+\hbar\mathbf{k}$). Under the resonant optical excitation described above, the internal state of the atom becomes correlated with its momentum: an atom in its ground state $|1\rangle$ with mean momentum $\mathbf{p}$ (labeled $|1,\mathbf{p}\rangle$) is coupled to an excited state $|2\rangle$ of momentum $\mathbf{p}+\hbar\mathbf{k}$ ($|2,\mathbf{p}+\hbar\mathbf{k}\rangle$).

The conditions required to measure the curvature of the atomic trajectory also lead to the coherent division and recombination of an atomic wavepacket in a manner which is analogous to an optical Mach-Zehnder interferometer. The first interaction [which records $\phi(t_1)$] puts an atom initially in the $|1,\mathbf{p}\rangle$ state in a coherent superposition of states $|1,\mathbf{p}\rangle$ and $|2,\mathbf{p}+\hbar\mathbf{k}\rangle$, and is analogous to the input beamsplitter in the optical interferometer. As time passes, these two states separate spatially due to their differing mean momenta, so that after a time T the wavepackets will have separated by an amount $2\hbar\mathbf{k}T/m$, where m is the mass of the atom. The next interaction [which records $\phi(t_2)$] induces the transitions $|1,\mathbf{p}\rangle \rightarrow |2,\mathbf{p}+\hbar\mathbf{k}\rangle$ and $|2,\mathbf{p}+\hbar\mathbf{k}\rangle \rightarrow |1,\mathbf{p}\rangle$ so that after another interval T the two wavepackets again merge. These interactions are analogous to the mirrors in the optical interferometer. The final interaction results in the interference of the two wavepacket trajectories [and records $\phi(t_3)$]. This interference determines the detection probability P_2 described above, and is analogous to the exit beamsplitter. Note that the simple semi-classical argument given in the previous paragraph accurately predicts the acceleration induced phase shifts $\Delta\phi$.

In practice, we use 2-photon stimulated Raman transitions[5] between the F=3, m_f=0 and F=4, m_f=0 groundstate hyperfine levels of atomic Cs to realize the above system. The laser beams are nearly resonant with the 850 nm optical resonance in Cs, and are aligned to counter-propagate.

We now briefly describe application of the above techniques to the detection of rotations and gravitationally induced accelerations. *Gyroscope*: A rotation induces a Coriolis acceleration $2\Omega\times\mathbf{v}$ (Ω is the rotation rate and $\mathbf{v}$ the particle velocity) which deflects the particle trajectory. High sensitivity is achieved in a geometry where a thermal atomic beam (v ~ 200 m/sec for Cs) propagates through three spatially separated laser beams whose propagation axes are oriented perpendicular to the direction of the atomic velocity[3]. *Gravimeter*: High sensitivity measurements of gravitationally induced accelerations are obtained by launching ensembles of atoms on vertical ballistic trajectories. A travelling wave laser beam, whose propagation axis is also vertically oriented, is pulsed on three times in order to measure the gravitationally induced acceleration. Long measurement times T are obtained by working with laser cooled atoms[6]. *Gravity Gradient*: Gravitational gradient measurements are made by simultaneously launching two independent, vertically separated, ensembles of atoms on vertical ballistic trajectories. A travelling wave laser beam, whose propagation vector passes through both ensembles, simultaneously records the acceleration of each ensemble as well as their differential acceleration[4].

2.2 Gyroscope

A schematic illustration our prototype gyroscope apparatus is shown in Fig. 2. The overall apparatus length is 2 m. A UHV vacuum system which contains counter-propagating high-flux Cs atomic beams is supported on a separate frame located just above the table. Note that technical vibrations of the support structure do not effect gyroscope stability, since the atomic beams, which form the inertial reference, are not coupled to this structure.

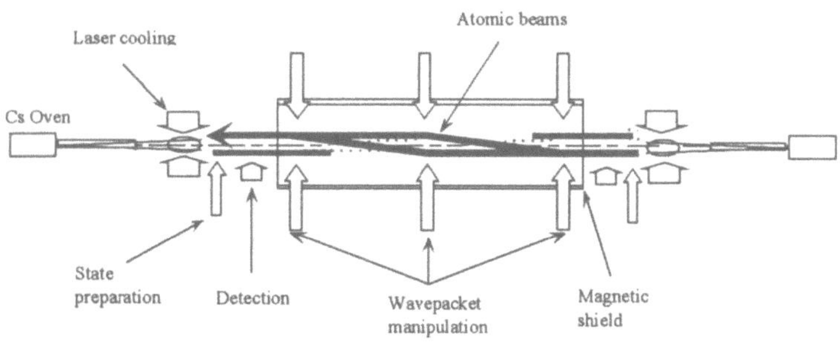

Figure 2. Schematic illustration of the atom interferometer gyroscope.

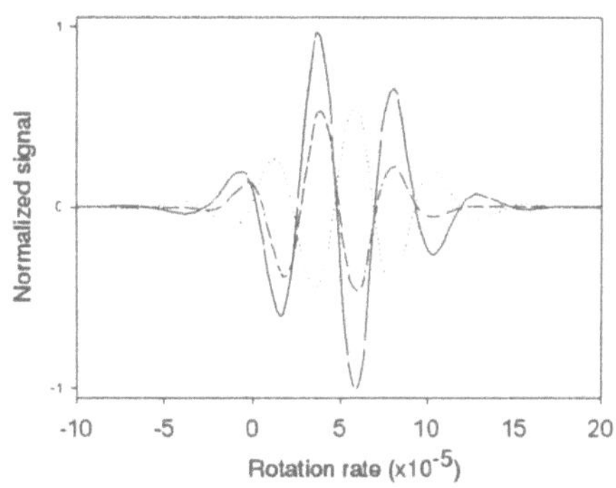

Figure 3. Gyroscope interference fringes. Solid line indicates the normalised difference signal. Dot/Dash lines indicate the rotation signal from individual atomic beams.

Typical interference signals are shown in Fig. 3. Counter-propagating (atomic) beams are used to suppress systematic phase shifts common to each beam. This figure shows normalised difference between the rotation signal from each beam (solid curve) as well as the rotation signals from each beam (dotted/dashed). The central zero crossing indicates an absolute rotation zero. The scale factor for this geometry is 8 rad/(Earth rotation rate), and is comparable to the scale factors used in optical gyroscopes. The signal-to-noise indicates a rotation sensitivity of $2\text{x}10^{-6}$ deg/hr$^{1/2}$, and is limited by atom shot-noise (~10^{10} atom/sec contribute to the interference signal).

An essential feature of this work is the use of an all-optical technique to provide a rotation bias to the gyroscope. In previous work, we scanned the interference fringes through application of a mechanical dither to the optical table. In this work, we acousto-optically frequency shift the Raman difference frequency for each of the three Raman beam pairs. The acousto-optic shifts are chosen to compensate the Doppler shift (~50 Hz) on the Raman difference frequency which is induced by the Earth's rotation. The rotation read-out, then, is given by the applied frequency shift which just compensates for the table rotation.

2.3 Gravity Gradiometer

Established atom interferometry techniques are used to measure the relative accelerations of two ensembles of laser cooled atoms. Each ensemble of ultra-cold atoms is trapped and cooled in its own UHV vacuum system. The laser light needed for the cooling and trapping is generated using an all-diode laser system and delivered through optical fibers to each apparatus. A second laser system is used to generate the light used to measure the atom accelerations. Vertically oriented laser beams derived from this system are used to track the trajectories of atoms in each ensemble using a sequence of optical pulses. The acceleration of the ensembles is measured using a three pulse excitation sequence, and is inferred from the measured populations of the atomic states following the sequence. Finally, the gravitational gradient is determined by taking the difference between the measured acceleration of each ensemble and dividing by the ensemble separation.

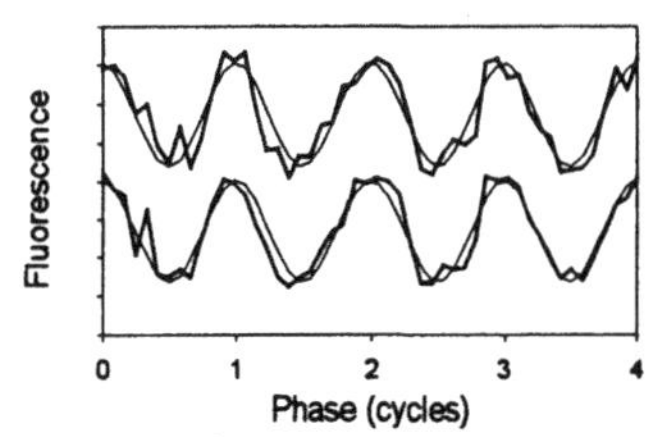

Figure 4. Raw gravity gradiometer interference fringes.

Fig. 4 shows characteristic interference fringes from which gravity gradient information is extracted. The upper trace corresponds to data taken from the upper chamber, and the lower trace to data from the lower chamber. The gradient extracted from these data is in excellent agreement with the known gradient. The signal-to-noise ratio indicates a 30 $E/Hz^{1/2}$ sensitivity for a 10 m instrument (1 E $\equiv$ 10^{-9} sec^{-2}). We expect straighforward improvements in the apparatus to bring the sensitivity to below 1 $E/Hz^{1/2}$ in the coming year.

2.4 Future Performance

Tables 1 and 2 summarise anticipated performance characteristics of mature laboratory gyroscopes and gravity gradiometers.

Table 1. Anticipated performance for laboratory gyroscope.

Attribute	
Sensitivity	10^{-7} deg/hr$^{1/2}$
Accuracy	10^{-7} deg/hr
Scale factor stability/linearity	Exceptional
Integrated Accelerometer	10^{-8} g/Hz$^{1/2}$
Prototype size	2m x 0.3m x 0.3m

Table 2. Anticipated performance for laboratory gravity gradiometer.

Attribute	
Sensitivity	0.1 $E/Hz^{1/2}$
Accuracy	< 0.1 E
Prototype size	2m x 0.3m x 0.3m

Key characteristics of the light-pulse approach include exceptional long-term stability, intrinsic calibration, immunity to environmental perturbations, high sensitivity, conceptual simplicity and robust operation. These characteristics emerge from the following attributes: (1) *Atomic-proof mass.* This insures that the material properties of the proof-mass will be identical from one instrument to another. It allows for straight-forward characterisation of environmental perturbations. For example, sensitivity to external magnetic field and electric field gradients can be directly estimated from calculable atomic properties (polarizability and magnetic moment). The designs are immune to variation in temperature. In comparison, the material properties of macroscopic proof-masses change with temperature. (2) *Laser distance measurements.* Since distances are measured in terms of the wavelength of a laser (whose frequency is stabilised to an atomic resonance), the acceleration measurements are intrinsically calibrated. This guarantees long term stability. (3) *No moving parts.* With the exception of the atoms, there are no moving parts in these designs. We anticipate robust, long-lived, low maintenance, and low cost instruments.

3. INTERFEROMETRY WITH BOSE-EINSTEIN CONDENSED ATOMIC SOURCES

In this section we explore the potential for Bose-Einstein condensed atomic sources. We first focus on a novel interferometer geometery available with Bose-Einstein condensed atoms, then discuss the possibility of exploiting the coherence properties of the condensate to demonstrate interferometry below the shot-noise limit.

3.1 Atom heterodyne measurements with BEC

Atom-laser heterodyne experiments, illustrated in Fig. 5, are sensitive probes of the gravitational potential. In this class of measurements, two atom laser sources, located at different vertical positions, are aligned to interfere at a detector sensitive to atom density.

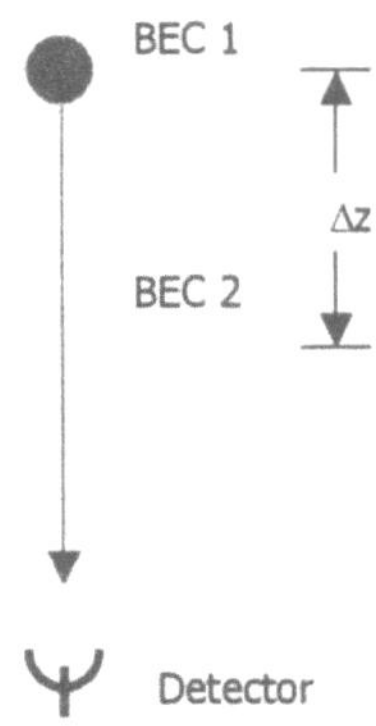

Figure 5. Atom heterodyne measurement of g with two atom laser sources.

The interfering beams lead to a time-dependent modulation in the atomic density at a frequency proportional to the gravitational potential difference between the two atom sources. Our optical lattice work with Bose-Einstein condensed samples, described below, can be considered a proof-of-principle for the proposed work at the 10^{-5} g/Hz$^{1/2}$ level. These techniques could lead to a new class of ultra-sensitive inertial force sensing instruments.

We estimate the potential sensitivity as follows. If two coherent sources of atoms are separated by a distance of Δz and acted upon by the acceleration due to gravity g, the energy difference between the two sources is $\Delta E = mg\Delta z$ per atom. The corresponding beatnote frequency is $\omega = mg\Delta z/\hbar$. For a source separation of $\Delta z = 1$ mm, this frequency is 2 MHz. If 10^6 atoms interfere over an observation time of T = 100 sec, an acceleration of $7x10^{-10}$ g produces a 1 rad phase shift over the observation interval. Assuming shot-noise limited detection, this amounts to an acceleration sensitivity of $7x10^{-13}$ g.

It is interesting to compare this approach with the single-particle methods described in Sect. 2. In either case, sensitivity scale linearly with the distance scale for separation of interfering wavepackets. For the single-particle interferometers, this distance scale is limited by the momentum transfer associated with the initial atom beamsplitter, and the interrogation time T. In the atom heterodyne instrument, this distance scale is determined by the separation of the atomic sources, and can, in principle, be much larger. In the single particle instrument, coherence between wavepackets is guaranteed by the beamsplitting process. In the atom heterodyne instrument, relative coherence derives from the coherence properties of Bose-Einstein condensed atomic sources.

One possible demonstration of the heterodyne concept would be to create a single BEC, divide it in two using optical dipole force induced potentials, spatially separate the two condensate halves, then allow atoms to tunnel from each trap and interfere. The beatnote could be picked up by detecting the temporal modulation of a probe laser beam. A related implementation would be to separate and recombine the two halves by manipulating them with optical dipole force potentials. In this case the interference would be read out as an overall shift between the relative phase of the two condensates. Finally, a straight forward extension of our previous lattice work to larger period lattices will lead to substantially enhanced sensitivities.

The key barrier to achieving ultra-high performance levels is controlling mean field shifts due to atom-atom interactions. In practice, this means working at very low atomic densities – which could be achieved through adiabatic relaxation of the trapping potential strength. On the other hand, it is likely that techniques to manipulate the s-wave scattering length, which controls the interaction strength, will mature to the point where they can be used to suppress these shifts.

3.2 Atomic Tunnel Arrays

We made a first demonstration of the above idea with Bose-Einstein condensed atoms confined in the nodes of an optical lattice[7]. We demonstrated that careful measurement of the beatnote frequency associated with atom-tunnelling from these traps could lead to determination of g, the acceleration due to gravity. This work can also be interpreted as a first demonstration of a mode-locked atom laser, or as a demonstration of the AC Josephson effect.

The experimental system is illustrated in Fig. 6. Bose-Einstein condensed atoms are initially loaded into a vertically oriented optical lattice. The lattice was created by a retroreflected laser beams detuned ~70 nm from the optical resonance. The depth of the lattice was chosen so that atoms could tunnel from the individual lattice traps into unbound states. Once in the unbound continuum, they accelerate due to gravity. Since the atoms are loaded into the traps from the same initial condensate, the de Broglie waves tunnelling from each lattice site have well defined relative macroscopic quantum phases.

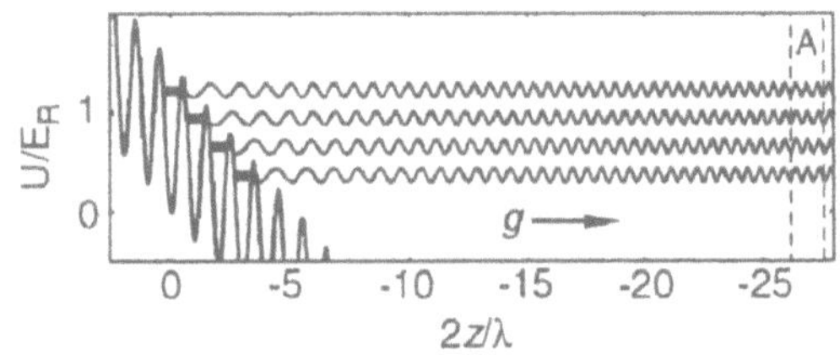

Figure 6. Atomic tunnel array. De Broglie waves tunnelling out of individual traps formed from the potential U interfere constructively in region A to form coherent pulses of atoms. ER is the recoil energy.

The combination of well defined initial phases and the gravitationally induced energy offset between adjacent traps leads to a periodic interference of the array output, analogous to the interference of adjacent cavity modes in a mode-locked laser. The pulse output frequency is set by the gravitational energy difference between adjacent sites $\Delta E_g = mg\lambda/2$ (corresponding to an output frequency $\omega = mg\lambda/2\hbar$), where λ is the wavelength of the laser used to make the optical lattice. These output pulses are illustrated in Fig. 7, which shows array output after a 10 msec holding times in the lattice.

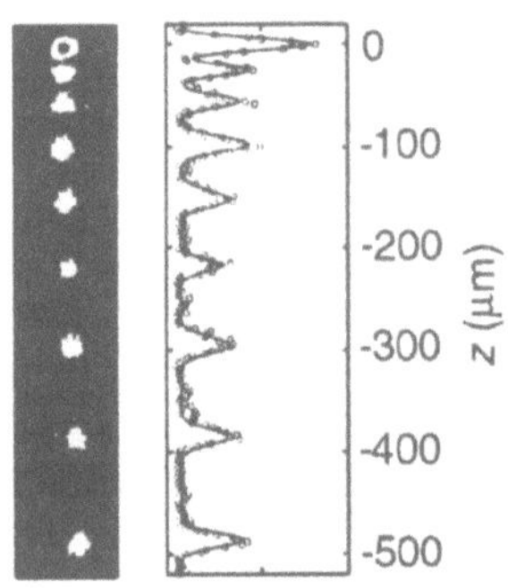

Figure 7. Atomic tunnel array output.

We extracted the acceleration due to gravity from the measured pulse period, which was obtained by fitting a series of Gaussians to the pulse profile (as shown in Fig. 7). Our accuracy for this measurement was limited by our knowledge of the magnification of the imaging system (~ 5%). The sensitivity to changes in g was $\sim 10^{-5}\ g/Hz^{1/2}$. This is the first measurement of g using a macroscopic quantum, atomic system.

This system is directly analogous to the AC Josephson effect in superconducting electronic systems, where a chemical potential difference between two superconducting reservoirs separated by an insulating barrier leads to a time dependent current at a frequency determined by the chemical potential difference[8]. In our experiment, this difference is provided by the gravitational potential and the tunnel barriers are created by the optical potential. Since we operated in a regime where the mean field interactions were negligible, one also expects to be able to describe the physics in terms of single particle quantum states (*eg.* solutions to the Schrödinger equation). From this perspective, the observed atom pulsing is a direct observation of Bloch oscillations (the Bloch frequency is just the Josephson frequency identified above)[9]. The two views are equivalent and complementary.

3.3 Correlated-state Interferometry

We have proposed a Heisenberg-limited (sub-shot noise) measurement scheme using correlated atomic Fock states[2]. A key step in the experimental demonstration of this method is the preparation of the required initial state.

One possibility is to exploit mean-field interactions to prepare the required entangled quantum states. To illustrate the method, we first consider a paradigm double-well system. In the following, we treat this as a two-mode quantum system, which can be described in terms of raising/lowering operators for each mode. There are two physical processes: tunnelling between wells and mean-field interactions. The Hamiltonian governing this system is

$$H = -\delta/2\ (a_L^+ a_R + a_R^+ a_L) - 4\kappa\ a_L^+ a_L a_R^+ a_R,$$

where δ characterises the tunnelling rate between well and κ parameterises the strength of the mean-field interaction[10]. We now find the ground states of this Hamiltonian. In the limit where mean-field interactions dominate, we can neglect the first term on the right. In this limit, the eigenstate is a dual Fock state: $|\psi\rangle \sim \{a_L^+\}^{N/2}\{a_R^+\}^{N/2}|vac\rangle$, since the Hamiltonian reduces to a product of number operators. In the other limit, where tunnelling dominates, the ground-state is then described by macroscopic occupation of the single-particle ground state: $|\psi\rangle \sim \{(a_L^+ + a_R^+)/\sqrt{2}\}^N\ |vac\rangle$. Physically, we can understand these results from the following naïve picture. Strong interactions suppress tunnelling since the hopping atom spoils the energy resonance between wells which exists when the atoms are uniformly distributed between the wells.

We can exploit the fluctuation properties of the ground state to prepare non-trivial entangled quantum states by adiabatically changing the ratio of the tunnelling to interactions strength. For example, adiabatic manipulation of the ratio of the tunnelling time to the mean field interaction can be used to transform the ground state from a coherent state to a squeezed state - and ultimately to a Fock state.

3.4 Squeezed States

We can use the optical lattice system to realise a system qualitatively similar to the double well system. The proposed experiments involve creation of a chain of traps in a harmonic potential - achieved simply by slowly turning on a far-detuned standing wave of light while atoms are confined in a (magnetic) harmonic trap. When the interactions are strong, the variance of the many-body quantum state describing each well is reduced below its value for a non-interacting ensemble. Numerical variational calculations for the ground state of the system show that the atom number variance at each lattice site can be reduced by a factor of 10 for experimentally realisable parameters.

We can create an array of squeezed states by adiabatically raising the strength of the tunnel barrier to a regime where the variance is significantly reduced from its classical value. A key question is whether we can satisfy the adiabaticity requirement. For the two well system, the time scale for adiabaticity is given by the frequency $\omega_a = (N\kappa\delta)^{1/2}$ (corresponding to the energy difference between the ground state and first excited state). For a lattice system, this characteristic frequency is reduced by the number of lattice sites in the chain of traps. For our lattice parameters $N\kappa/2\pi \sim 1$ kHz, $\delta/2\pi \sim 0.01$ Hz - 1 kHz (by changing the lattice strength), while the trap lifetime is in excess of 4 seconds and the number of traps is 10 - 30 (depending on the initial loading conditions). Thus we expect to be able to be safely in the adiabatic regime.

Our experimental signature for preparation of the squeezed state is by interferometric readout of the relative phase of adjacent wells. For this measurement, we suddenly turn off the lattice+harmonic potential, allowing atoms to ballistically expand from each lattice site. After and expansion time of ~10 msec, we detect the atom density distribution. In the non-interacting limit, this experiment results in the observation of sharp interference fringes, since a well defined relative phase exists between the lattice sites. For a squeezed state, however, these fringes will tend to wash out, since the relative phase between lattice sites is no longer sharp. (Note that with only two lattice sites, interference contrast would not wash out, but fluctuate from one experiment to the next). Since we are using phase coherence as our read-out, we need to be able to experimentally discriminate between the dephasing mechanism described above, and other mechanisms. We believe this is possible by adiabatically cycling the tunnelling rate from values corresponding to squeezed states, then coherent states, and back. In this case we expect to be able to recover sharp interference fringes, since they depend only on final ratio of tunnelling rate to mean field interaction strength. On the other hand, competing dephasing mechanisms will not allow the recovery of phase coherence.

3.5 Heisenberg-limited Atom Interferometry

The resolution of a Heisenberg-limited phase measurement is $\Delta\phi \sim 1/N$, where N is the number of particles in the measurement and $\Delta\phi$ is the phase uncertainty of the measurement. In comparison, the resolution of a shot-noise limited method scales inversely with the square root of N. Precision phase measurements may have application to time standards, parity non-conservation measurements or searches for a permanent electric dipole moment, in addition to next generation atom interferometer sensors.

We will first review our original proposal, then describe a possible generalisation to the optical lattice system. The basic idea draws on earlier work in the quantum optics community, and is closely related to a proposal by Holland and Burnett for optical interferometry at the Heisenberg limit[11]. The method requires the creation of two degenerate atomic ensembles which are coupled via an interaction potential. In the proposed experiment, these ensembles are atoms in differing hyperfine ground states prepared in overlapping spatial regions. At least one of the ensembles needs to be in a state characterised by a known number of atoms (Fock state). The two ensembles are coupled via magnetic dipole transitions between the two states. The coupling potential is switched on in a time sequence analogous to a Ramsey interference experiment: a first $\pi/2$ pulse, followed by an interrogation time T, and then another $\pi/2$ pulse. The Heisenberg limit is obtained by observing the fluctuations in the number particles in one of the two hyperfine states following the second $\pi/2$ pulse. When no phase imbalance is present, there is no fluctuation in the number at the output port. On the other hand, the presence of a (small) phase shift in one of the interferometer arms introduces relatively large (compared to traditional methods) fluctuations in the number of atoms after the two $\pi/2$ pulses. The minimum resolvable phase shift $\Delta\phi$ for a single experiment is determined by a shift which produces a final distribution with a variance of ~ 1 atom. For this dual-Fock state, we have shown numerically that $\Delta\phi$ N ~ 1. In subsequent (unpublished) work, we have shown that the constraints on the initial state can be further relaxed and that the Heisenberg limit is obtained when just one of the two states is in a Fock state. Physically, we can understand the high sensitivity to phase shifts as resulting from the conversion of Fock states into state of well defined relative phase after the first $\pi/2$ pulse (beamsplitter). The corresponding number variance in each interferometer arm following the initial $\pi/2$ pulses is ~N.

We now consider a specialisation to the two-well system, which is directly analogous to the coupled two-state system described in the previous section. The first step is to prepare a dual Fock state, which we accomplish by adiabatically raising the tunnelling barrier. We then implement the analogue of a $\pi/2$ pulse by suddenly lowering the barrier height to a value where tunnelling dominates the mean field interaction for a fixed time $t_{\pi/2}$. Following this time, we suddenly raise the barrier back up to its initial value, thus locking in the new many-body state, which we expect to have large number fluctuations (an thus, by the number-phase uncertainty relation, a well defined relative phase). We then hold the system in this new state for a time T before implementing the second $\pi/2$ pulse by suddenly lowering the barrier a second time for duration $t_{\pi/2}$. Finally, we adiabatically reduce the tunnel barrier height, and read-out the final state phase coherence by suddenly turning off the double-well potential and allowing the atoms to ballistically expand. Small phase shifts acquired during the interrogation time T will substantially alter the contrast of the detected interference.

One expects that a similar combination of adiabatic/non-adiabatic manipulations in the optical lattice system will yield a sub-shot noise phase shift sensitivity. We have recently verified this numerically through solution of the equations of motion for the evolution of the lattice states using a time-dependent variational technique[12]. Experimentally, the key step will be creation of phase states from the array of Fock states with a non-adiabatic manipulation of the tunnelling time. We can use this system to probe the gravitational potential by turning off the harmonic potential during the interrogation time T, then turning it back on just before the final non-adiabatic barrier manipulation.

We do not expect the above systems to immediately produce an improvement over the current state-of-the-art for precision gravity measurements. Rather, we view this system as a vehicle for proof-of-principle studies of the efficacy of squeezed atomic state interferometry, and as a trigger for future work. With currently available parameters, we expect an order of magnitude reduction below the shot-noise limit for signals of 10^4 atoms.

3.6 Attractive Interactions

If we consider the two-well system described with attractive, rather than repulsive, atom-atom interactions, we discover that the ground state of the system has remarkable properties. Intuitively, an ensemble of atoms with attractive interactions minimises its energy by forming clusters of atoms. For the double well system, in the limit where tunnelling is dominated by interactions, the ground state is a coherent superposition of all atoms in the left well and all atoms in the right well. In other words, under appropriate conditions, adiabatically raising a tunnel barrier will produce the Schrödinger cat state $|\psi\rangle \sim [\{a_L^+\}^N + \{a_R^+\}^N]|vac\rangle$[13].

What is perhaps even more remarkable is the condition required for the formation of this state. A signature of the formation of this state is the development of a degeneracy between the ground state and first excited state of the system. Numerical calculations of this energy difference show a sharp transition to the Schrödinger cat state when $N\kappa/\delta = 1$. This threshold is a characteristic zero-temperature, second-order quantum phase transition. Quantum phase transitions have received significant attention from the condensed matter theory community. However, there have been relatively few experimental realisations of these systems.

A fascinating feature of this ground state is that it is not described by a single condensed wave-function, but a superposition two macroscopically populated states. Formally, this ground state is fragmented, meaning that the single particle density matrix has more than one macroscopically populated eigenvalue. Recent theoretical work in this area indicates that finite angular momentum, negative scattering length systems, will also fragment[14].

We expect to be able to explore this physics with the negative scattering length ^{7}Li system loaded into a far detuned optical lattice at 780 nm (where we currently have a 0.5 W laser system). Production of pure Schrödinger cat state requires the initial energy offset between adjacent wells to be essentially zero (otherwise the system would energetically favour one well over the other) – a condition we probably cannot satisfy experimentally. However, as in the case of positive scattering length (repulsive interactions), we expect an intermediate range of parameters which will produce non-trivial quantum states with properties similar to the ideal case identified above. Finally, we note that the state described above is equivalent to the spin-squeezed state identified in Ref. [15] for use in sub-shot noise interferometry.

The relevance of this work to sensor development is that it could provide an alternative mechanism for achieving Heisenberg limited, sub-shot noise readout: platform rotation breaks the symmetry between two otherwise degenerate modes. The mean field interaction forces all atoms to collapse into the lowest energy mode. Therefore, by monitoring population of the mode we can determine whether the platform is rotating left or right. The platform could be served to a zero rotation condition, where the populations would balance between the states.

4. CONCLUSIONS

The current generation of single-particle force sensors have achieved performance levels where they now compete with state-of-the-art instruments. Whether Bose-condensed atomic sources will enable a further performance gain remains to be seen. However, there are a number of tantalising possibilities which are emerging from current studies of condensate cohenerence properties.

ACKNOWLEDGMENTS

This work was supported be grants from the Office of Naval Research, National Science Foundation and NASA.

REFERENCES

1. Anderson, M. H., Ensher, J., Matthews, M., Wieman, C., and Cornell, E., 1995, *Science* **269**: 198; Bradley, C. C., Sackett, C. A., Tollett, J. J., and Hulet, R. G., 1995, *Phys. Rev. Lett.* **75**: 1687; Davis, K., Mewes, M., Andrews, M., van Druten, N., Durfee, D., Kurn, D., and Ketterle, W., 1995, *Phys. Rev. Lett.* **75**: 3969.
2. Bouyer, P. and Kasevich, M., 1997, *Phys. Rev. A* **56**: R1083.
3. Gustavson, T., Bouyer, P., and Kasevich, M., 1998, *Proc. SPIE*, **3270**: 62; Gustavson, T., Bouyer, P. and Kasevich, M., 1997, *Phys. Rev. Lett.* **78**: 2046.
4. Snadden, M., McGuirk, J., Bouyer, P., Haritos, K. and Kasevich, M., 1998, *Phys. Rev Lett.* **81**: 971.
5. Kasevich, M. and Chu, S., 1991, *Phys. Rev. Lett.* **67**: 181.
6. Kasevich, M. and Chu, S., 1992, *Appl. Phys B.* **54**: 321.
7. Anderson, B. and Kasevich, M., 1998, *Science* **282**: 1686.
8. Barone, A. and Paternò, G., 1982. *Physics and Applications of the Josephson Effect.* John Wiley & Sons, New York.
9. Observation of Bloch oscillations in non-degenerate ensembles of atoms appears in Wilkenson, S., Bharucha, C., Madison, K., Niu, Q., and Raizen, M., 1996, *Phys. Rev. Lett.* **76**: 4512 and Dahan, M., Peik, E., Reichel, J., Castin, Y., and Salomon, C., 1996, *Phys Rev. Lett.* **76**: 4508.
10. See, for example, Smerzi, A, Fantoni, S., Giovanazzi, S., and Shenoy, S., 1997, *Phys. Rev. Lett.* **79**: 3164; Zapata, I., Sols, F., and Leggett, A., 1998, *Phys. Rev. A* **57**: R28.
11. Holland, M., and Burnett, K., 1993, *Phys. Rev. Lett.* **71**: 1355.

12. Fenseleau, M. and Kasevich, M., in preparation.
13. A related proposal, based on repulsive interactions between two condensed species, appears in Cirac, I., Lewenstein, M., Mølmer, and Zoller, P., 1998, *Phys. Rev. A* **57**: 1208.
14. Wilken, N., Gunn, J., and Smith, R., 1998, *Phys. Rev. Lett.* **80**: 2265.
15. Bollinger, J., Itano, W., Wineland, D., and Heinzen, D., 1996, *Phys. Rev. A* **54**: R4649.
16. Anderson, M. H., Ensher, J., Matthews, M., Wieman, C., and Cornell, E., 1995, *Science* **269**: 198; Bradley, C. C., Sackett, C. A., Tollett, J. J., and Hulet, R. G., 1995, *Phys. Rev. Lett.* **75**: 1687; Davis, K., Mewes, M., Andrews, M., van Druten, N., Durfee, D., Kurn, D., and Ketterle, W., 1995, *Phys. Rev. Lett.* **75**: 3969.
17. Bouyer, P. and Kasevich, M., 1997, *Phys. Rev. A* **56**: R1083, (1997).
18. Gustavson, T., Bouyer, P., and Kasevich, M., 1998, *Proc. SPIE*, **3270**: 62; Gustavson, T., Bouyer, P. and Kasevich, M., 1997, *Phys. Rev. Lett.* **78**: 2046.
19. Snadden, M., McGuirk, J., Bouyer, P., Haritos, K. and Kasevich, M., 1998, *Phys. Rev Lett.* **81**: 971.
20. Kasevich, M. and Chu, S., 1991, *Phys. Rev. Lett.* **67**: 181.
21. Kasevich, M. and Chu, S., 1992, *Appl. Phys B.* **54**: 321.
22. Anderson, B. and Kasevich, M., 1998, *Science* **282**: 1686.

Classical And Quantum Josephson Effects With Bose-Einstein Condensates

A. SMERZI
Istituto Nazionale di Fisica della Materia and International School for Advanced Studies, via Beirut 2/4, I-34014, Trieste, Italy

1. INTRODUCTION

The Josephson effects are a paradigm of the physical manifestation of phase coherence on a macroscopic scale. These phenomena were predicted in the early '60 by Brian Josephson[1] who considered two superconducting metals separated by a thin oxide barrier, and connected with an external current/voltage source. The key achievement was to realize that the current flowing through the junction was proportional to the sine of the relative collective phase of the two bulk systems[2]. The concept of a (macroscopic) phase coherence was introduced quite early, in the '38 by London[3], and in the '50, by Ginzburg and Landau[3], who described the superconducting state in terms of a one-body macroscopic wave-function obeying a Schroedinger-like equation. Ten years later, this approach was generalized by Gross and Pitaevskii to superfluid systems[4]. Driven by such ideas, Feynman reformulated the Josephson problem as coherent oscillations in a two-mode quantum system[5]. His analysis, however, have to be considered with some warning: the superconducting Josephson effects come out from the interplay between a classical piece (the external circuit) and the quantum component (the two superconducting grains)[6,7]. The interparticle interaction plays another crucial role: it is the presence of the "charging energies" that makes such oscillations highly non trivial, respect to the simple Rabi regime (governing, for instance, single atom oscillations among two quantum levels).

Bose-Einstein Condensates and Atom Lasers
Edited by Martellucci *et al.*, Kluwer Academic/Plenum Publishers, 2000

Trapped bosons are neutral, and external circuits are obviously absent. This raises immediately the suspect that, exception made of some analogies, the physics of a boson Josephson junction (BJJ), regardless how it might be created, would be different from the physics of a superconducting Josephson Junction (SJJ). Clearly, I'm not trivially referring to the fact that in one case we have bosons, while in the other Cooper-pairs, but to the fact that the dynamical regimes accessible with a BJJ could be essentially different from those accessible with a (superconducting) SJJ (or with superfluid Helium systems[8]). This is of paramount importance, since it can drive the search of new macroscopic quantum phenomena: and such query is one of the main reasons of the interest on the BEC physics.

Two aspects identify the fingerprint of the Josephson effects[7,9]. The first requires a coherent transfer of matter between two bulk systems connected by a "weak link". The transfer can be through tunneling, contact, or coupling with external fields, the precise mechanism being irrelevant. The second aspect relies on the existence of a macroscopic phase in each bulk. The role of the weak link is to make the energetic cost for the variations of the relative phase of the two bulks, cheaper than the cost of varying the phase inside the bulk itself.

So far, two different experimental schemes have been proposed to observe the Josephson effects with BECs. In both cases, the junction is provided by a double well potential created, say, by an harmonic trap cutted by a blue-detuned laser sheet[10], through which the atoms can tunnel. In the first proposal[11-16], a chemical potential difference is created by an initial population/phase imbalance. This induces coherent condensate oscillations whose dynamics is governed, in a mechanical analogy, by a "non-rigid pendulum" Hamiltonian (i.e. by a pendulum having a momentum dependent length) with the relative population playing the role of the momentum, and the relative phase that of the angle respect to the vertical axis[14,15]. This is a major difference respect to traditional SJJ, whose dynamics is described by a "rigid" pendulum equation[7]. As a consequence, the boson dynamics is richer than the corresponding superconducting analog. New regimes include the "macroscopic quantum self-trapping" (MQST), a self-maintained population imbalance in a symmetric double well potential, and "π-states", oscillations having an average phase-difference across the junction equal to π. There is a further, deeper physical difference between SJJ and BJJ. In SJJ the Cooper-pair population imbalance remains essentially locked to zero (considering two equal-volume superconducting grains) due to the presence of the external circuit that suppresses charge imbalances[6]. In BJJ, on the other hand, the (non-rigid) pendulum dynamics is associated with the superfluid density oscillations.

A closer analogy with a current driven SJJ is provided by a different kind of boson junction[17,18]. A laser barrier moves adiabatically across the trapping potential (with the condensate initially in equilibrium). Below a critical barrier velocity, the atoms tunnel through the barrier trying to keep the chemical potential difference across the junction locked to zero. However, the tunneling flow is bounded by the Josephson critical current: above a critical value of the laser velocity a chemical potential difference (i.e. a macroscopic population difference between the two wells) develops. Within such scheme, the Josephson effect is evidenciated by a sudden jump in the chemical potential/laser velocity phase diagram, a sharp transition that can be easily monitored experimentally with destructive or non destructive techniques. I would like to remark that the two boson junctions I have described differ in an important aspect: in the first case the Josephson effects manifest dynamically as coherent oscillations between the two trap, while in the second case they do a sharp transition between two clearly distinct quasi-stationary regimes (with a closer analogy with the "dc" and "ac" effects observed in a current driven SJJ[7]).

So far, I have implicitly assumed the "classical" nature of the Josephson effects, meaning, in our context, that they can be understood in the framework of the Gross-Pitaevskii theory[19]. Quantum corrections (that, generally speaking, arise from the many-body nature of the problem) take into account the impossibility to define with arbitrary precision the number/phase observables. A cheap way to include such many-body corrections is to quantize classically conjugate observables. In the context of BJJ, this problem has been studied both analytically[20] and numerically[13,21,22]. Typical effects include the collapses and revivals of the population/phase oscillations, and a coherent destruction of the tunneling in the MQST regime.

At this point, it is important to make clear that it is not relevant, for the present discussion, the distinction between the "internal" and the "external" Josephson effects (JE). In the" internal" JE the oscillations between two condensates, trapped in different hyperfine levels, are induced by the coupling with an external electromagnetic field[25] and not by the atomic tunneling through a potential barrier, as in the "external" JE. Since the mathematics describing the "internal" and the "external" JJ is, for most practical purposes, identical, all the effects we will explore here can be immediately applied for the internal JE.

The following sections are devoted to a brief analysis of the different aspects I have mentioned, while I refer to re/preprints for more detailed discussions. I'll consider a condensate at zero temperature, in order to ignore damping effects. The analysis of the SJJ physics can be found in Refs. [7, 9].

Experimental aspects are presented in the chapters of D. Hall and M. Kasevich.

2. COHERENT CONDENSATE OSCILLATIONS IN A DOUBLE WELL POTENTIAL

The coherent oscillations between two condensates trapped in a double-well potential have been widely studied in the limit of non-interacting atoms [11], in the interacting case for small amplitude oscillations[12] and including finite temperature (damping) effects[16,21,23].

I should first notice that the classical Josephson equations should be retrieved, in some limit, from the dynamical Gross-Pitaevskii equation. At a variational level, this has been done in[14,15], where the results presented in this section are illustrated in detail. A Bose condensate in a double-well trap is governed by a nonlinear, two-mode equation for the time-dependent amplitudes $\psi_{1,2}(t)=\sqrt{N_{1,2}(t)}\,e^{i\theta_{1,2}(t)}$ where $N_{1,2}(t)$ and $\theta_{1,2}(t)$ are the number of atoms and the phases of the condensate in the trap 1, 2 respectively, with the total number of atoms being conserved. These amplitudes are coupled by a tunneling matrix element between the two traps:

$$i\hbar\frac{\partial\psi_1}{\partial t}=\left(E_1^0+U_1N_1\right)\psi_1-K\psi_2 \tag{1a}$$

$$i\hbar\frac{\partial\psi_2}{\partial t}=\left(E_2^0+U_2N_2\right)\psi_2-K\psi_1, \tag{1b}$$

Here $E_{1,2}^0$ are the zero-point energies in each well, $U_{1,2}N_{1,2}$ are proportional to the atomic self-interaction energies, and K to the hopping amplitude. The fractional population imbalances, $z(t)=\frac{N_1(t)-N_2(t)}{N_T}$, and the relative phases $\phi(t)=\theta_2(t)-\theta_1(t)$ are classically conjugate dynamical variables:

$$\dot{z}(t)=-\frac{\partial H}{\partial\phi}=-\sqrt{1-z^2(t)}\sin(\phi(\mathrm{t})), \tag{2a}$$

$$\dot{\phi}(t) = \frac{\vartheta H}{\vartheta z} = \Delta E + \Lambda\, z(t) + \frac{z(t)}{\sqrt{1 - z^2(t)}} \cos(\phi(t)), \tag{2b}$$

respect to the effective Hamiltonian:

$$H = \frac{\Lambda\, z^2}{2} + \Delta\, Ez - \sqrt{1 - z^2}\,\cos\phi \tag{3}$$

with $t2K/\hbar \to t$, $\Delta E = \left(E_1^0 - E_2^0\right)/2K + \left(U_1 - U_2\right)/2K$, $\Lambda = UN_T/2K$, $U = \left(U_1 + U_2\right)/2$. Eqs. (2) are the Josephson equations of two weakly-linked, dilute Bose-Einstein condensates. Such equations are integrable in terms of elliptic Jacobian functions[15]. The Hamiltonian Eq. (3) can be interpreted, in a mechanical analogy, as the equation of a pendulum, with z being the momentum and ϕ the angle respect to the horizontal axis. The peculiarity of such pendulum is to have a length that depends on the momentum, through the "contraction factor" $\sqrt{1 - z^2}$. This analogy will greatly help to visualize the dynamical Josephson regimes.

There is a further aspect that deserves mentioning. So far (and in the following), I'm considering an atomic condensate with positive scattering length. This means having a positive "charging energy" U, as in superconducting systems. However, condensates with a negative scattering length do exist in nature, and a Boson Josephson Hamiltonian with a negative "charging energy" is allowed. The BJJ equations have a nice symmetry respect to the change of sign of the scattering length. In particular, the dynamics remain unchanged under the formal replacement $\Lambda \to -\Lambda, \Delta E \to -\Delta E, \phi \to -\phi + \pi$. It is interesting to note that, with a negative scattering length, and for $|\Lambda > 1|$, the ground state of the BJJ is fragmented, as shown right below.

Stationary solutions. A first peculiar consequence of the "non-rigidity" of the BJJ pendulum equations is the existence of z-symmetry breaking stationary states, i.e., states that, even in a symmetric double-well potential, $(\Delta E = 0)$ have a condensate in one well in equilibrium with a larger condensate in the second well, with their relative phase equal to $\pi : \phi_0 = (2n + 1)\pi, z_0 = \pm\sqrt{1 - \frac{1}{\Lambda^2}}$ provided that $|\Lambda > 1|$. The energy of these two degenerate states is $E_{sb} = \frac{1}{2}\left(\frac{1 + \Lambda^2}{\Lambda}\right)$, which can be respectively greater or smaller than the energies of the symmetric and antisymmetric eigenstates of the GPE, depending on the sign of the scattering length.

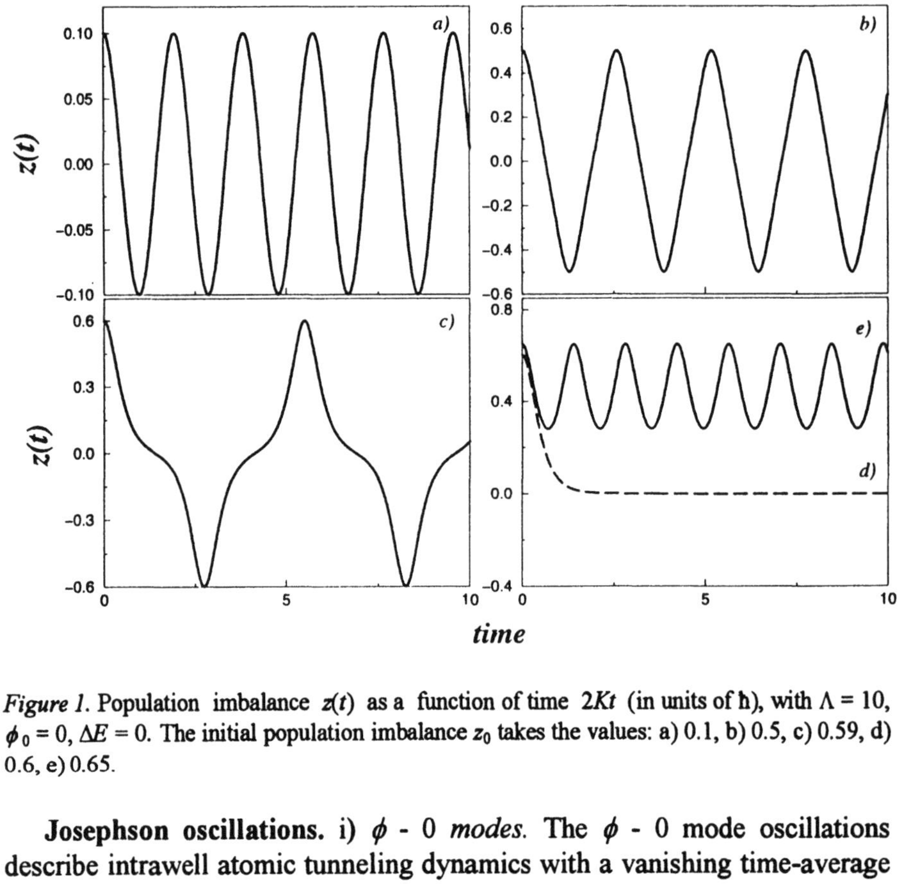

Figure 1. Population imbalance $z(t)$ as a function of time $2Kt$ (in units of $\hbar$), with $\Lambda = 10$, $\phi_0 = 0$, $\Delta E = 0$. The initial population imbalance z_0 takes the values: a) 0.1, b) 0.5, c) 0.59, d) 0.6, e) 0.65.

Josephson oscillations. i) ϕ - 0 *modes.* The ϕ - 0 mode oscillations describe intrawell atomic tunneling dynamics with a vanishing time-average value of the phase across the junction $\langle\phi(t)\rangle = 0$. The frequency of the small amplitude oscillations, $\omega_0 = \sqrt{2UN_T K + (2K)^2}/\hbar$ is of the order of 100 Hz for typical trap parameters which is much smaller than the SJJ plasma frequencies (of the order of GHz). In Fig. 1 it is shown $z = \dfrac{N_1 - N_2}{N_T}$ as a function of time, with the initial value of the phase difference $\phi(0) = 0, \Lambda = 10$ and for increasing values of the initial population imbalance $z(0)$ = 0.1, 0.5, 0.59, 0.6, 0.65 from a) through e), respectively. Increasing $z(0)$ adds higher harmonics to the sinusoidal oscillations, corresponding, in the mechanical analogy, to large amplitude oscillations of the pendulum bob. This is shown in Fig. 1(b)(c). The period of such oscillations increases with $z(0)$, then decreases, undergoing a critical slowing down, Fig. 1(d), dashed line, with a logarithmic divergence. The singularity in the period corresponds to the bob staying on top and the pendulum in a vertically upright position.

ii) ϕ - running modes. In addition to an harmonic Josephson oscillations, other novel effects occur in BJJ. For an initial population imbalance greater than a critical value, the populations become macroscopically self-trapped, corresponding to the pendulum running about the fixed center Fig. 1(e). There are different ways in which this state can be achieved. All of them correspond to the MQST condition $H_0 \equiv H(z(0), \phi(0)) = \frac{\Lambda}{2}z(0)^2 - \sqrt{1-z(0)^2}\cos(\phi(0)) > 1$. MQST is a genuine nonlinear effect arising from the nonlinear self-interaction of the atoms.

iii)ϕ *-π modes.* The non-rigidity of the BJJ pendulum equations allows oscillations with a time-average value of the phase across the junction $\langle\phi(t)\rangle = 0$. This corresponds to the pendulum oscillating around the top position. Such states are not observable with SJJ, which are described by a rigid pendulum equation. The small amplitude oscillation frequency is: $\omega_\pi = \sqrt{(2K)^2 - 2UN_TK/\hbar}$. Notice that the ratio of the frequency of the small amplitude ϕ - 0 and ϕ - π mode oscillations is $\frac{\omega_\pi}{\omega_0} = \sqrt{\frac{1-\Lambda}{1+\Lambda}} < 1$. Large amplitude $z(t)$ π-oscillations are also tolerated, as well as π-MQST states.

The dynamical phase diagrams of the BJJ equations in a symmetric $(\Delta E = 0)$ double-well potential, are shown in Fig. 2. The vertical axis reports the values of Λ, while the horizontal axis shows initial values of the population imbalance z_0 (the figures are symmetric with respect to the vertical $z_0 = 0$ axis). Fig. 2(a) corresponds to a dynamical BJJ regimes with $\phi(t = 0) = 0$. The solid line, given by $\Lambda_r = \frac{2}{z_0^2}\left(1 + \sqrt{1 - z_0^2}\right)$, separates the MQST (running phase) from the 0 - *phase* oscillations. For an initial value of the phase difference ϕ $(t = 0) = \pi$, Fig. 2(b), the phase diagram is manifestly richer. There are four different regions: the solid line, $\Lambda_{st} = \frac{2}{z_0^2}\left(1 - \sqrt{1 - z_0^2}\right)$ divides the self-trapping regimes (above the solid line) from those untrapped. The long-dashed line corresponds to the (z-symmetry breaking) stationary values $\Lambda_s = \frac{1}{\sqrt{1 - z_0^2}}$ which separates two further regions, having the time averaged population imbalance

$\langle z \rangle < |z_0^s| \neq 0$, and $\langle z \rangle > |z_0^s| \neq 0$, respectively. These three regimes have a time averaged value of the phase $\langle \phi \rangle = \pi$. Above the dashed-line, $\Lambda = \frac{2}{\sqrt{1 - z_0^2}} = 2\Lambda_s$, the system falls again in the running-phase regime (equivalent to that of Fig. 2(a)).

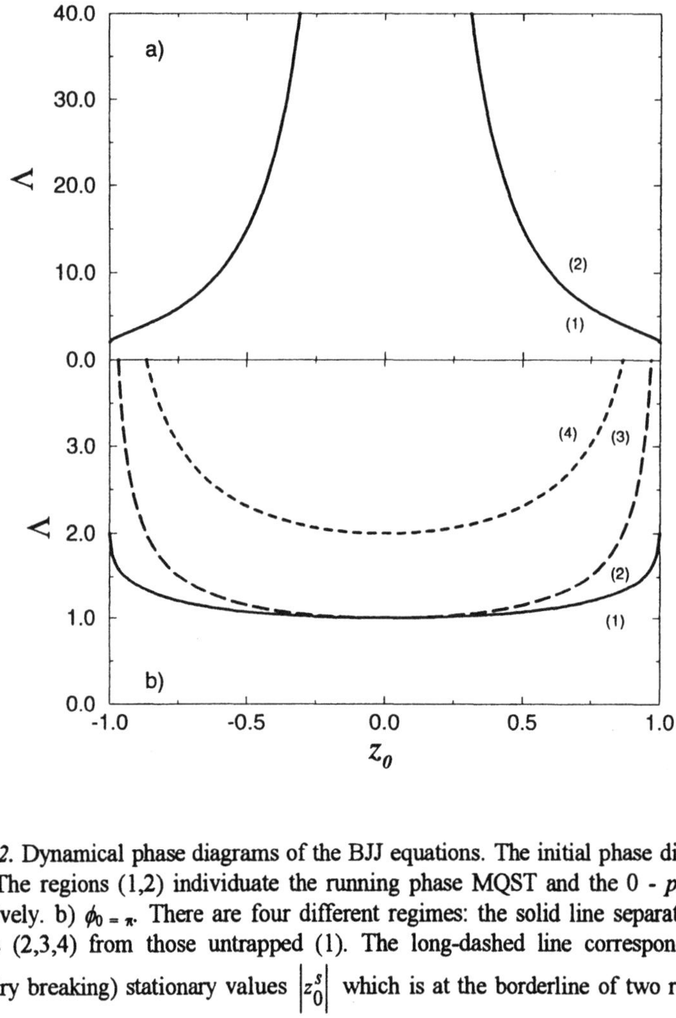

Figure 2. Dynamical phase diagrams of the BJJ equations. The initial phase difference is: a) $\phi_{0} = 0$. The regions (1,2) individuate the running phase MQST and the 0 - *phase* regimes, respectively. b) $\phi_{0} = \pi$. There are four different regimes: the solid line separates the MQST regimes (2,3,4) from those untrapped (1). The long-dashed line corresponds to the (z-symmetry breaking) stationary values $|z_0^s|$ which is at the borderline of two regions having the time averaged population imbalance $\langle z \rangle < |z_0^s| \neq 0$, (2), and $\langle z \rangle > |z_0^s| \neq 0$, (3), respectively. The regions (1,2,3) correspond to a time averaged value of the phase $\langle \phi \rangle = \pi$ Above the dashed line, region (4), the system falls again in the running phase regime.

3. THE BJJ "DC" AND "AC" EFFECTS

A superconcucting Josephson junction is usually biased by an external circuit that typically includes a current drive I_{ext}. The signatures of the Josephson effects are contained in the voltage-current characteristic (V-I_{ext}), which contains the "dc"-branch ($V = 0$, $I_{ext} \neq 0$), and "ac"-branch ($V \approx RI_{ext}$)[7]. Although external circuits and current sources are absent in two weakly linked Bose condensates, it is still possible to design an experiment to observe the analog of such effects with BEC[17,18]. A current-biased SJJ can be simulated by a tunneling barrier moving with constant velocity across the trap. The equations of motion for the relative population and phase are still given by Eq. (2), but with a time-dependent zero point energy (due to the dynamical change of the effective volumes in which the two condensates are confined): $\Delta E(t) = Fl_z(t)$ where F is the average force per particle exerted by the barrier and $l_z(t)$ is its center position. At low barrier velocities $v = \frac{dl_z}{dt}$ the two condensates remain in equilibrium (i.e. in their instantaneous ground state), thanks to a tunneling current sustained by a constants non-zero relative phase between the two condensates. The flow keeps the chemical potential difference between the two subsystems locked to zero (as in the SJJ dc-branch): $\dot{\phi} = \Delta\mu = -Fvt + U(N_2 - N_1) = 0$.

However, the superfluid component of the current flowing through the barrier is up-bounded by a critical value $I_c = \frac{d}{dt}(N_2 - N_1) = KN_T$, see Eq. (2). As a consequence, there is a critical barrier velocity v_c, above which a non-zero chemical potential difference develops across the junction. This regime is characterized by a running-phase mode, $\phi \approx \Delta\mu_0 t$, and provides the analog of the ac-branch in SJJ's. In Fig. (3) the analytical results (full line) are compared with a numerical integration of the time-dependent GPE (squares) in an experimentally realistic geometry[17]. It has been considered a typical JILA setup, with $N = 5 \times 10^4$ Rb atoms in a cylindrically symmetric harmonic trap, having the longitudinal frequency $\omega_0 = 50\ s^{-1}$ and the radial frequency $\omega_r = 17.68\ s^{-1}$. The value of the scattering length is $a = 58.19\ \mathring{A}$. A Gaussian shaped laser sheet is focused in the center of the trap, cutting it into two parts. The (longitudinal) $1/e^2$ half-width of the laser barrier is 3.5 μm and the barrier height $V_0/\hbar = 650\ s^{-1}$. At $t = 0$ the laser is at rest in the middle of the trap, $l_z=0$ and the two condensates are in equilibrium. For $t > 0$ the laser moves across the trap, with constant velocity, and the relative atomic population is observed at $t_f = 1\ s : N_2 - N_1 = \frac{F_v}{U} t_f$. The experiment is repeated increasing the velocity, but still measuring the relative population after one second, (Fig. 3). The critical velocity is $v_c = 0.4$. Generally speaking, the motion of the laser sheet with respect to the magnetic trap or, viceversa, the motion of the magnetic trap, with opposite velocity, are equivalent.

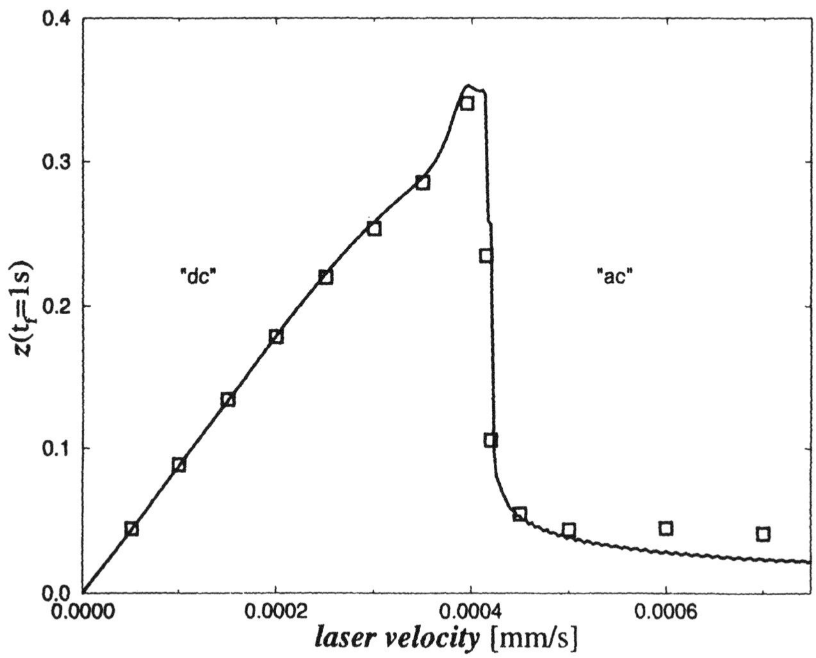

Figure 3. The fractional population imbalance $z(t_f)$ calculated at $t_f = 1s$ versus the velocity of the laser barrier. A sharp transition between the "dc" and the "ac" branches occurs at the barrier critical velocity $v_c = 0.4\ \mu m\ s^{-1}$. The solid line and the squares are the analytical and the numerical calculations, respectively.

4. QUANTUM JOSEPHSON EFFECTS

The Josephson theory we have described so far has a classical nature. By classical, in this context, is intended that the conjugate dynamical observables number/phase have well defined values. The uncertainty principle, on the other hand, tells us that such variables cannot be defined with arbitrary precision. This raises two problems: one is the study of the quantum corrections of the classical Josephson equations, Eq. (2), and the second is to understand the conditions for the classicity to emerge. The results presented in this section together with the details of the calculations, have been worked out in Ref. [20]; numerical analysis can be found in Refs. [13, 22].

The borderline between the classical and the quantum dynamics is characterized by the ratio of the "Josephson coupling energy" $E_J = 2KN_T$ and the "charging energy" U[7]. In the limit $E_j >> U$ both the phase difference and the relative number of condensate atoms are well defined. In this case the classical Josephson equations can be "microscopically" derived by the GPE[14,15]. On the other hand, quantum corrections can significantly modify the classical dynamics even for $E_j/U \sim 10^2$, a regime accessible in current BEC experiments[24]. As we have previously seen, the classical boson Josephson junction (BJJ) equations, can be cast in terms of the relative population $N = \frac{1}{2}(N_1 - N_2)$ and phase $\phi = \phi_1 - \phi_2$ between the two traps. Quantizing BJJ, the c-numbers N and ϕ are replaced by the corresponding operators, satisfying the commutation relation[26] $[\hat{\phi}, \hat{N}] = i$. Then the Hamiltonian of two weakly coupled condensates reads[i]

$$\hat{H} = \frac{U}{2}\hat{N}^2 - E_j \cos\hat{\phi} + \Delta E\hat{N} \tag{4}$$

The Eq. (4) has been solved within a time-dependent variational approach[20]. The quantum Josephson equations are:

$$\dot{N} = -\Gamma \sin\phi\, e^{-\frac{\sigma_\phi^2}{2}} \tag{5a}$$

$$\dot{\phi} = 2N + E_0 \tag{5b}$$

$$\dot{\sigma}_\phi = 2\sigma_\phi \delta \tag{5c}$$

$$\dot{\delta} = -2\delta^2 + \frac{1}{2\sigma_\phi^4} - \Gamma\cos\phi\, e^{-\frac{\sigma_\phi^2}{2}} \tag{5d}$$

[i] Eq. (4) holds for $N/N_T << 1$ and $U >> E_J/N_T^2$ where N_T is the total number of condensate atoms/Cooper-pairs.

with the effective Hamiltonian:

$$H = N^2 + \sigma_N^2 - \Gamma \cos\phi \, e^{-\frac{\sigma_\phi^2}{2}} + E_0 N \tag{6}$$

where $\sigma_N = \frac{1}{2\sigma_\phi}\sqrt{1+4\sigma_\phi^4\delta^2}$ is the relative population dispersion, $\Gamma = \frac{2E_j}{U}$, $E_0 = \frac{2\Delta E_0}{U}$ and the time has been rescaled as $\frac{U}{2}t \to t$. The canonically conjugate dynamical variables are N, Gras in the classical Josephson Hamiltonian, and the pair $\frac{\sigma_\phi^2}{2}, \delta = \frac{1}{\sigma_\phi}\sqrt{\sigma_N^2 - \frac{1}{4\sigma_\phi^2}}$ which characterizes the quantum fluctuations, with, as expected, $\sigma_N\sigma_\phi \geq \frac{1}{2}$.

The variational ground state energy is given by: $E_{gs} = \frac{1}{4\sigma_{\phi,s}^2} - \Gamma e^{-\sigma_{\phi,s}^2/2} - \frac{E_0^2}{4} = \sigma_{N,s}^2 - \Gamma e^{-1/8\sigma_{N,s}^2} - \frac{E_0^2}{4}$ where $\sigma_{\phi,s}, \sigma_{n,s}$ are the solution of: $2\sigma_{\phi,s}^4 \, e^{-\sigma_{\phi,s}^2/2} = e^{-\frac{1}{8\sigma_{N,s}^2}} \Big/ 8\sigma_{N,s}^2 = \frac{1}{\Gamma}$.

Linearizing Eq. (5) for small amplitude ϕ oscillations, we have: $\ddot{\phi} = -2\Gamma e^{-\sigma_{\phi,2}^2/2}\phi$. The condensate atoms oscillate coherently with a frequency (unscaled): $\omega_q = \sqrt{UE_j e^{-\sigma_{\phi,s}^2/4}}$, while the classical Josephson relation gives $\omega_c = \sqrt{UE_j}$. Thus, the quantum fluctuations renormalize the classical frequency, with $\frac{\omega_q}{\omega_c} = \exp\left(-\sigma_{\phi,s}^2/4\right) = \left(\sigma_{\phi,s}^2\sqrt{2\Gamma}\right)^{-1}$. Notice that in the linear regime, the current-phase Eqs. (5a) (5b) are effectively decoupled from the dynamics of the respective fluctuations Eqs. (5c) (5d). On the contrary, for large amplitude ϕ oscillations, Eqs. (5) cannot be decoupled. In this case the exponential factor modulates the amplitude and the frequencies of the oscillations, inducing partial collapses and revivals both in the relative population and phase, as can be seen in Fig. 3(a,b). Thus, quantum corrections introduce a new time scale on the Josephson problem, namely that associated with the envelope of the oscillations.

Above the critical point (N = 0, $\phi = \pi/2$), the phase $\phi(t)$ starts running, and the system is set into MQST mode Fig. (4c). The width of the wave function grows and the amplitude of oscillations 'collapses', inducing a coherent destruction of tunneling, Fig. (4c). In the deep MQST regime, when $N(t) \cong N(t=0)$, the phase diffuses as $\sigma_\phi^2(t) \cong \sigma_{\phi,s}^2 + \frac{E_c^2}{4\sigma_{\phi,s}^2} t^2$, regardless of the initial value of N $(t = 0)$. The relative population oscillations collapse with a life time $\tau \cong 2 E_j^{-\frac{1}{4}} E_c^{-\frac{3}{4}}$, while the $\sigma_N^2(t)$ tends to a constant value. However, since the total number of condensate atoms is finite, the phase can eventually revive partially or completely on a much larger time-scale.

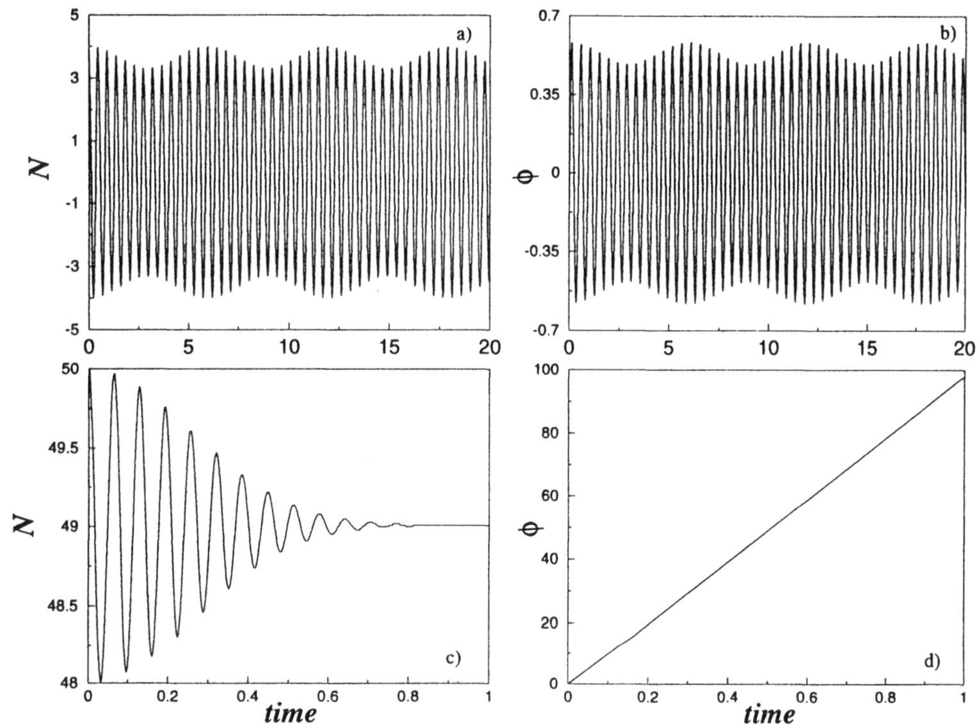

Figure 4. The population imbalance N and the relative phase ϕ as a function of time. The initial conditions are: a,b) $N_0 = 4$, $\phi = 0$, $\Gamma = 100$, $\sigma_N = 0$, $\sigma_\phi = 0.26$ and $E_0 = 0$; c,d) $N_0 = 50$.

Classical limit. The classical Josephson equations are retrieved in the limit $\sigma_\phi \to 0, \sigma_N / N_T \to 0$. Increasing the number of atoms, $\sigma_\phi \approx \Gamma^{-\frac{1}{4}} \approx N_T^{-\frac{1}{2}}$ and $\sigma_N / N_T \approx \Gamma^{\frac{1}{2}} / N_T \approx N_T^{-\frac{1}{5}}$ for typical trap geometries[20]. Thus, in the large condensates limit, Eqs. (5a) (5b) decouple from Eqs. (5c) (5d), and the time evolution of the mean values of current and phase become independent of the corresponding dispersions. In the MQST regime, the collapse time (and, consequently, the time over which the semiclassical predictions are reliable), increases as $\tau \approx N_T^{1/5}$.

Numerical estimates. Following the analytical estimations of the Josephson coupling energy and the on-site energy given for two weakly coupled condensates in Ref. [16], we have: $\Gamma \cong 1.7 N_T \frac{a_0}{a_s} \frac{\exp(-S)}{\tanh(S/2)}$, with a_0, a_s the trap length and the scattering length, respectively, and with $S \approx \frac{1}{\hbar}\sqrt{2m\sigma_B^2(V_0 - \mu)}$. σ_B is the width of the barrier, V_0 its height, and μ the chemical potential. For typical traps and condensates, $a_0 \sim 10^4$ Å, $a_s \sim 50$ Å and $\sigma_B \sim \mu$m. With a height of the barrier such that $(V_0 - \mu) \sim 10$ nK, we have $\Gamma \sim 10$ - 100 for $N_T \sim 1000$. Varying the width and/or the height of the barrier, and the total number of condensate atoms, the system can span from the $\Gamma << 1$ to the $\Gamma >> 1$ limits. The temperature should be small compared to the Josephson coupling energy[21,16] to avoid destroying the quantum fluctuations. Damping effects are also reduced by decreasing the total number of atoms. The Eq. (5) can be easily generalized to describe interwell tunneling in an array or in a torus of coupled trapped condensates. Recently, an array of condensates has been created in [25], with an average population per site of the order of thousand atoms: a case where quantum fluctuations can play an important role.

ACKNOWLEDGMENTS

It is a pleasure to thank my colleagues and friends with whom I collaborated on developing the themes illustrated here: S. Fantoni, S. Giovanazzi, S. Raghavan and S. Shenoy. I also would like to thank the participants of the Erice school for several and stimulating discussions, and for the pleasant time I spent with them.

REFERENCES

1. Josephson, B. D., 1962, *Phys. Lett.* **1**: 251.
2. Anderson, P.W., 1966, *Rev. of Mod. Phys.* **38**: 298.
3. London F., *Nature*, 1938, **141**: 643; Ginzburg, V.L., and Landau, L.D., 1950, *Zh. Eksperim. i Teor. Fiz.* **20**: 1064.
4. Pitaevskii, L. P., *Sov. Phys.*, 1961, JETP, **13**: 451; Gross, E. P., 1961, *Nuovo Cimento* **20**: 454; *J. Math. Phys.* **4**: 195 (1963).
5. Feynman, R.P., Leighton, R.B., and Sands, M., 1964, "*The Feynman lectures on Physics*", Addison-Wesley Publishing Company.

6. Ohta, H., 1977, in SQUID: *Superconducting Quantum Devices and their A*pplications, (H. D. Hahlbohm and H. Lubbig, eds.) Walter de Gruyter, Berlin.
7. Barone, A., and Paternò, G., 1982, *Physics and Applications of the Josephson Effect* (Wiley, New York).
8. Avenel, O., and Varoquaux, E., 1985, *Phys. Rev. Lett.*, **55**: 2704; Pereverzev S. V., et al., 1997, *Nature* **388**: 449; Backhaus, S., et al., 1998, *Science*, **278**:1435; Backhaus, S., et al., 1998, *Science* **392**: 687.
9. Barone, A., NATO ASI Series Quantum *Mesoscopic Phenomena and Mesoscopic Devices* in Microelectronics, Ankara June 1999 (I.O. Kulik and R. Ellialtioglu, Eds.) Kluwer (in press).
10. Andrews, M. R., et al., 1997, *Science* **275**: 637.
11. Javanainen, J., 1986, *Phys. Rev. Lett,*. **57**: 3164.
12. Dalfovo, F., Pitaevskii, L., and Stringari, S., 1996, *Phys. Rev. A* **54**: 4213.
13. Milburn, C. J., Corney, J., Wright, E. M., and Walls, D. F., 1997, *Phys. Rev. A* **55**: 4318.
14. Smerzi, A., Fantoni, S., Giovanazzi, S., and Shenoy, S. R., 1997, *Phys. Rev. Lett.* **79**: 4950.
15. Raghavan, S., Smerzi, A., Fantoni, S., Giovanazzi, S., and Shenoy, S. R., 1999, *Phys. Rev. A* **59**: 620.
16. Zapata, I., Sols, F., and Leggett, A., 1998, *Phys. Rev. A* **57**: R28.
17. Giovanazzi, S., Smerzi A., and Fantoni, S., *Phys. Rev. Lett.*, in press.
18. Giovanazzi, S.,1998, Ph.D. Thesis. SISSA Trieste Italy. Unpublished.
19. Dalfovo, F., Giorgini, S., Pitaevskii L. P., and Stringari S., 1999, *Rev. Mod. Phys.* **71**: 463.
20. Smerzi, A., and Raghavan, S., *Phys. Rev. A,* in press.
21. Ruostekoski, J., and Walls, D. F., 1998, *Phys. Rev. A* **58**: R50.
22. Raghavan, S., Smerzi, A., and Kenkre, V. M., 1999, *Phys. Rev. A* **60**: R1787.
23. Marino, I., Raghavan, S., Fantoni, S.,. Shenoy, S. R, and Smerzi, A., 1999,*Phys. Rev. A* **60**: 487.
24. Anderson B. P., and Kasevich, M. A., 1998, *Science* **282**: 1686.
25. Williams, J., Walser, R., Cooper, J., Cornell, E., and Holland, M., 1999, *Phys. Rev. A* **59**: R31.
26. This quantization scheme suffers from some well known problems: Carruthers, P., and Nieto, M. M., 1984, *Phys. Rev. Lett.* **21**: 353; Pegg D. T., and Barnett, S. M., 1989, *Phys. Rev. A* **39**: 1665.

Josephson Qubits For Quantum Computation

[1]G. FALCI, [1]R. FAZIO, [1,2]E. PALADINO AND [3]U. WEISS
[1]Dipartimento di Metodologie Fisiche e Chimiche per l'Ingegneria and Istituto Nazionale per la Fisica della Materia, Catania (Italy); [2]Consorzio Ennese Universitario, Cittadella degli Studi, Enna (Italy); [3]II.Institut für Theoretische Physik, Universität Stuttgart, D-70550 Stuttgart.

1. INTRODUCTION

The study of quantum dynamics of nanofabricated devices is a fertile meeting point between fundamental and applied physics. Fundamental aspects concern for instance the existence of states of a solid state device which display coherent dynamics and entanglement or the problem of Macroscopic Quantum Coherence[1] which addresses the question of the progressive emergence of classical behaviour due to the interaction with an environment. On the other hand the progress in fabrication techniques which may allow miniaturization down to nanoscale of solid state devices has opened the new field of Mesoscopic Electronics. In this regime the dynamics of the devices is qualitatively new and offers the possibility of new applications, as Quantum Computation (QC)[2].

The great interest the field of QC is related to the fact that it is possible to solve some problems which are intractable with classical algorithms, as the factorization problem[3]. There have been various recent proposal of implementation of a quantum computer using systems of mesoscopic Josephson junctions[4,5]. In this contribution we will focus on Charge-Josephson (CJ) devices[4] which have been recently observed to display coherent dynamics[6]. These devices offer two possible advantages namely the tunability of couplings, which allows to control the coherent dynamics, and

Bose-Einstein Condensates and Atom Lasers
Edited by Martellucci *et al.*, Kluwer Academic/Plenum Publishers, 2000

the possibility of fabricating large integrated networks which are required for realistic applications of QC.

Any realistic quantum system is coupled to the environment and is subject to the ubiquitous phenomenon of dephasing and decoherence spoiling quantum behaviour. Decoherence is the major obstacle for any implementation of QC. Sources of decoherence in CJ devices are for instance linear quantum noise coming from the electromagnetic environment, shot noise coming from impurities in the substrate, quantum leakage. Moreover the dynamical control of couplings may become complicated when picosecond time scales are involved. This means that on the theoretical side realistic models of complicated devices are needed and they have to be studied using accurated techniques. The aim of this work is to show how this is done using techniques of dissipative quantum mechanics[7].

This contribution is organized as follows: first the basic principles of QC (Sect. 2) and the CJ-Qubit (Sect. 3) are described; then a more technical part comes where a model for linear dissipation is derived (Sect. 4) and its dynamics is studied by techniques of dissipative quantum mechanics (Sect. 5). Some result is discussed in the conclusions.

2. QUANTUM COMPUTATION

QC is based on the controlled unitary evolution of elementary units called *Qubits*, a quantum Two-State System (TSS) which replaces the two values of the bit of classical logic devices[2]. Already at this stage a fundamental difference between classical and quantum bits emerges: information is stored either in 0 or in 1 in the former while in the latter any state $|\psi\rangle = a|0\rangle + b|1\rangle$ can be used. *Quantum gates* perform operations on Qubits, i.e. unitary transformations of some initial state where information is stored, $|\psi_f\rangle = U\,|\psi_i\rangle$. If they are realized by time evolution, $U = \exp(-i\,H\,t)$ is the evolution operation. For instance the NOT gate, is defined as

$$a|0\rangle + b|1\rangle \xrightarrow{U_{NOT}} a|1\rangle + b|0\rangle \quad \text{i.e} \quad U_{NOT} = |0\rangle\langle 1| + |1\rangle\langle 0|.$$

Let $\{|0\rangle, |1\rangle\}$ be spin ½ states relative to the z axis. U_{NOT} is then realised by letting the system evolve for a time $t=\pi/\Delta$ under the action of the Hamiltonian $H=\Delta\,\sigma_x$. It is then clear that the ability in performing operations is related to the possibility of manipulating the Hamiltonian of the system.

A quantum computer is a n-Qubit array and operations are unitary transformations in its 2^n-dimensional Hilbert space. It is important to notice that any of these operations can be decomposed in a sequence of one-bit gates (e.g. the Hadamard gates) and two-bit gates (e.g. the control-NOT gate). This implies one has to design one and two-coupled Qubits, as well as operational procedures (i.e. tune the parameters of a TSS Hamiltonian) which produce controlled time evolution. The quantum dynamics has then to be studied in detail, accounting also for preparation effects, for sources of depahsing and for the backaction of the measurement apparatus.

3. MESOSCOPIC JOSEPHSON DEVICES

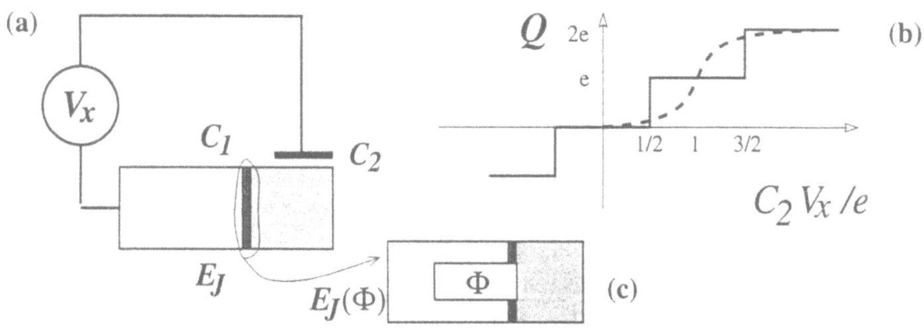

Figure 1. (a) The single-electron box: a normal metal island is connected to the voltage source V_x via a capacitor (C_2) and a tunnel injection (C_1); (b) electrons can be injected one by one in the island (solid line) by increasing V_x; if the electrodes are superconducting Cooper pairs charge states are mixed close to degeneracy(dashed line); (c) a tuneable E_J is obtained by substituting a SQUID to the Josephson junction.

Charging effects A nanofabricated single-electron box[8] (Fig.1(a)) may behave as a TSS due to charging effects. Suppose that electron tunneling through the junction is extremely weak. Then the (discrete) charge on the metal island is a good quantum number and the basis of charge states {|q>, q= 0, ±1,...}, (the extra charge Q in the island is Q=eq) can be used to describe the device. By increasing the external voltage V_x charge is pushed into the island. If the electrostatic energy for adding an electron is large enough, electrons are injected *one by one* (see Fig.1(b)). This means that the energy of the charge states is modulated by V_x: simple electrostatics[8] gives $E_q = E_C (q - q_x)^2$ where $E_C = e^2/(2\,C)$ ($C=C_1+C_2$ is the total capacitance) is the charging energy and $q_x = C_2 V_x /e$. The steplike curve in Fig.1(b) reflects the change of the ground state |q> for increasing q_x. This phenomenology, has been observed[8] in systems with $C \cong 10^{-15}$F with junction resistance $R_t >> R_Q = h/(4e)^2 \cong 6.5k\Omega$, at temperatures $T << E_X \cong 1K$. It is important to notice that if $T << E_C$ and if $0 < V_x < e/C$ only two charge states are important and that the level splitting of this TSS can be modulated with the source V_x.

Charge-Josephson Qubits Although the single electron box can be a TSS this is not enough for QC. Indeed the TSS hamiltonian is diagonal while nondiagonal elements (possibly tunable) are needed for QC. This requirement can be accomplished if we use superconducting electrodes. The key ingredient is Josephson tunneling[9]. The superconducting island is still described by the basis of charge states $\{|q\rangle, q=0,\pm1,\dots\}$, where now q refers to Cooper pairs, so we redefine $Q=2eq$ and $\varphi_x=C_2V_x/2e$. Josephson coupling between the electrodes determines the coherent transfer of Cooper pairs, $|q\rangle \rightarrow |q\pm1\rangle$. The hamiltonian of the device can be written as

$$H_J = \sum_{q=-\infty}^{\infty} |q\rangle 4E_C(q-q_x)^2 \langle q| - \frac{E_J}{2} \sum_{q=-\infty}^{\infty} |q+1\rangle\langle q| + |q\rangle\langle q+1|$$

To make connection with the usual form of the Josephson term[9] notice that the difference φ of the phases of the order parameters of the two superconducting electrodes is canonically conjugated to the number of Cooper pairs passing the junction, thus also to the excess charge in the box, $[\hbar\varphi/(2e), Q] = i\,\hbar$. This implies that $\exp(\pm i\varphi)=\Sigma_q|q\pm1\rangle\langle q|$ and the standard form $-E_J \cos\varphi$ of the Josephson term is recovered. Under suitable conditions again the system is practically a TSS whose Hilbert space is spanned by the states $|0\rangle$ and $|1\rangle$. The hamiltonian is obtained by projecting H_J onto this subspace and can be finally expressed using Pauli matrices

$$H = (\varepsilon/2)\,\sigma_z - (E_J/2)\,\sigma_x . \qquad (1)$$

The "diagonal coupling" is tuneable, $\varepsilon=8E_C - 2(C_2/C)eV_x$. The diagonal term would determine a stepwise behavior of the number of Cooper pairs in the island vs V_x . Nondiagonal matrix elements due to the Josephson coupling E_J mix charge states. As recently observed[10], steps get smeared close to the degeneracy points, indicating the presence of superpositions of states $|q\rangle$.

In order to modulate nondiagonal elements[4] in Eq. (1) one may substitute the Josephson junction with a SQUID ring (see Fig. 1(c)). In a SQUID the critical current I_J and also the associated effective coupling $E_J = \hbar I_J/(2e)$ can be modulated by varying the flux Φ of the magnetic field threading the ring[9].

Real CJ-Qubits CJ-Qubit controlled dynamics has been observed in a recent experiment by Nakamura et al.[6]. The two charge states in were brought to degeneracy for a controlled time by a V_x pulse. The charge Q was detected by quasiparticle current in a subsidiary circuit (see Fig. 2(a)). Coherent oscillations of Q were observed up to 2ns. Times >100 times larger are required for typical QC applications so possible sources of dephasing have to be analysed. While the major problem in the set-up of Ref. [6] is probably connected to imperfect Josephson-quasiparticle measurement cycles, there are other problems, common also to other implementations, as linear noise (which here is due to thermal and quantum fluctuations of the circuit), shot noise (due moving background charges trapped in the insulating substrate) and quantum leakage[11] (due to the fact that the real CJ-Qubit is a multistate system and not a TSS). In what follows we will focus on linear noise, presenting the main lines of the detailed theoretical analysis.

4. MODEL FOR LINEAR DISSIPATION

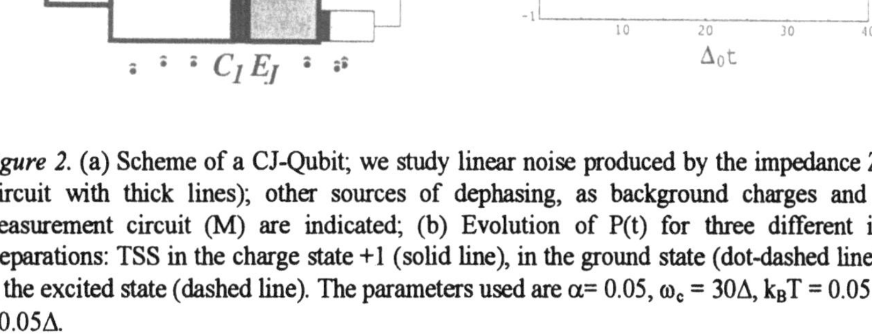

Figure 2. (a) Scheme of a CJ-Qubit; we study linear noise produced by the impedance $Z^0(\omega)$ (circuit with thick lines); other sources of dephasing, as background charges and the measurement circuit (M) are indicated; (b) Evolution of P(t) for three different initial preparations: TSS in the charge state +1 (solid line), in the ground state (dot-dashed line) and in the excited state (dashed line). The parameters used are $\alpha = 0.05$, $\omega_c = 30\Delta$, $k_B T = 0.05\,\Delta$, $\varepsilon_0 = 0.05\Delta$.

In order to discuss the effect of linear dissipation due to the impedance $Z^0(\omega)$ of the circuit we introduce the imaginary time path-integral generating functional[8] associated to the circuit of Fig. 2(a). Phases across each element are related to voltages via the Josephson relation[9]. The effective action is[8]

$$S = \int_0^\beta d\tau \frac{C_1}{2}\left(\frac{\hbar\dot{\varphi}_1}{2e}\right)^2 - \int_0^\beta d\tau\, E_J \cos(\varphi_1) + \int_0^\beta d\tau \frac{C_2}{2}\left(\frac{\hbar\varphi_2}{2e}\right)^2 + \\ + \frac{1}{2}\int_0^\beta d\tau\, d\tau' \dot{\varphi}_z(\tau) D_z(\tau-\tau') \dot{\varphi}_z(\tau') \tag{2}$$

The first two terms are charging and Josephson coupling for the junction, the third is charging energy for the capacitor, the last is the Caldeira-Leggett[8,7] dissipative action for the impedance, the kernel being defined as $D_Z(\omega) = [2\,\pi\,|\omega|\,\zeta^0(\omega)]^{-1}$, where $\zeta^0(\omega)=Z^0(-i|\omega|)/R_Q$, and $\omega=2\pi n/\beta$. The voltage source $V_x(t) = 2e/\hbar\,(d\varphi_x/dt)$ enters via the circuit equation, $d(\varphi_1+\varphi_2+\varphi_Z+\varphi_x)/dt=0$, as a constraint for the path-integral. The generating functional is

$$Z = \int D\varphi_1 D\varphi_2 D\varphi_z \delta\,[\dot{\varphi}_1 + \dot{\varphi}_2 + \dot{\varphi}_z + \dot{\varphi}_x]\exp(-S[\varphi_1,\varphi_2,\varphi_z]) \tag{3}$$

Charge discreteness in the box enters via the non trivial boundary condition $[\varphi_1(\beta) - \varphi_2(\beta)]/2=[\varphi_1(0) - \varphi_2(0)]/2+2\pi m$. Summation over the winding number m is implicitly included in Eq. (3).

At this stage we can eliminate φ_2 and integrate out φ_Z via standard Gaussian integration which yields an effective model in the variable ϕ_1 only. This latter can be reexpressed in the quantum-dual representation, that is in terms of the "momentum" conjugated with φ_1 which is a discrete quantity associated with the charge in the box. The transformation to this "charge representation"[8] for general dissipation and source can be exactly performed starting from the full perturbation series[7] in E_J. The result is

$$Z[J_x(\tau)] = \sum_{p=0}^{\infty} \frac{[E_J(\Phi)/2]^{2p}}{(p!)^2} \int_0^\beta \left\{\prod_{i=1}^{2p} d\tau_i\right\} \sum_{\{q(\tau)\}} \exp\left\{-\int_0^\beta d\tau 4E_C q^2(\tau)\right\} \times \\ \exp\left\{-\frac{1}{4}\int_0^\beta d\tau d\tau' K(\tau-\tau')\,[q(\tau)-q(\tau')]^2 + i\int_0^\beta d\tau J_x(\tau) q(\tau)\right\} \tag{4}$$

where all possible paths of the kind $q(\tau)=q+\sum_{i=1}^{2p} n_i \Theta(\tau-\tau_i)$ with q integer and $n_i=\pm1$ are included. The kernel is given by

$$K(\omega) = -\left(\frac{C_2}{C}\right)^2 \frac{2\pi\zeta^0(\omega)}{1 + R_Q C_{eff}\zeta^0(\omega)|\omega|}|\omega| \qquad (5)$$

where $C_{eff} = C_1\ C_2\ /C$. The source $J_x(\tau)$ is related to $V_x(\tau)$. For small enough V_x and large E_C only two states are relevant. Truncation of Eq. (4) to a TSS is performed by writing $\sigma(\tau)=2\ [q(\tau)-\ 1/2]$ and retaining only $\sigma=\pm 1$ ("leakage" from the TSS Hilbert space is discussed in Ref. [11]). We obtain

$$Z[\varepsilon(\tau)] = \sum_{p=0}^{\infty}\left[\frac{E_J(\Phi)}{2}\right]^{2p}\int_0^{\beta} d\tau_1 \ldots \int_0^{\tau_{2p-1}} d\tau_{2p} \sum_{\{\sigma(\tau)\}} F^{(E)} \exp\left\{-\frac{1}{2}\int_0^{\beta} d\tau\varepsilon(\tau)\sigma(\tau)\right\}$$

$$F^{(E)}\{\sigma(\tau)\} = \exp\left\{-\frac{1}{16}\int_0^{\beta} d\tau\, d\tau' K(\tau - \tau')[\sigma(\tau) - \sigma(\tau')]^2\right\} \qquad (6)$$

$$\varepsilon(\tau) = 4E_C[1 - C_2 V_x(\tau)/e] - \int_0^{\beta} d\tau' K(\tau - \tau') C_1 V_x(\tau)/e \qquad (7)$$

Eqs. (5,6,7) allow to identify the dissipative TSS whose quantum dynamics can be studied with the methods of Ref. [7]. For instance an interesting special case is $Z^0(\omega) = R$. The bath, identified by its spectral density $G(\omega)=-\mathrm{Im}[K(|\omega| \to i\,\omega)]/\pi$, is ohmic and has a Drude cutoff[7]

$$G(\omega) = \frac{2\alpha\omega}{1 + (\omega/\omega_c)^2}, \quad \alpha = \left(\frac{C_2}{C}\right)^2 \frac{R}{R_Q}, \quad \omega_c = \frac{1}{C_{eff} R} \qquad (8)$$

This model has been considered[4] for studying the dephasing in a CJ-Qubit for static V_x. For general bias we need the analytic continuation of eq.(7). For $V_x(t) = V_{dc}+V_{ac}\cos(\Omega t)$ we get $\varepsilon(t) = \varepsilon_0 + \varepsilon_1(\Omega)\cos(\Omega t+\gamma)$ where

$$\varepsilon_0 = 8E_C\left[1 - \frac{C_2 V_{dc}}{2e}\right]; \qquad \varepsilon_1(\Omega) = -2e\frac{C_2}{C}V_{ac}\left|1 + i\Omega C_{eff} Z(\Omega)\right|^{-1}$$

5. DISSIPATIVE TSS DYNAMICS

Eqs. (5, 6, 7) express the imaginary time reduced generating functional of the driven spin-boson model[7,12,13]

$$H_{SB}(t)=\frac{1}{2}\varepsilon(t)\sigma_z-\frac{1}{2}\Delta\sigma_x+H_B-\frac{1}{2}\sigma_z\sum_i c_i x_i, \tag{9}$$

which represents a driven TSS coupled to a bath of harmonic oscillators, $H_B=\Sigma_i[p_i^2/(2\,m_i)+m_i\,\omega_i^2\,x_i^2/2]$. The connection with the CJ-Qubit is made by identifying $\Delta=E_J$ and the spectral density, which entirely captures the influence of the bath on the TSS dynamics, $G(\omega)=1/(2\hbar)\Sigma_i(c_i^2/m_i\;\omega_i)\;\delta(\omega-\omega_i)$.

We now study the dynamics of the reduced density matrix (RDM) $\rho(t)=\mathrm{Tr}_B\{W(t)\}$, $W(t)$ being the density matrix of the global system. We assume the initial state $W(t_0)=\rho(t_0)\,W_B$, where $\rho(t_0)$ is a general TSS initial state

$$\rho(t_0)=\begin{pmatrix} p_R & a+ib \\ a-ib & p_L \end{pmatrix} \tag{10}$$

and $W_B=\exp\{-\beta\,H_B\}/\,\mathrm{Tr}\{e^{-\beta\,H_B}\}$ is the equilibrium density matrix of the bath. The trace over the bath degrees of freedom can be performed exactly and the time evolution of the RDM is then expressed by a double path functional integral with spin paths from time t_0 until time t,

$$\rho_{\sigma,\sigma'}(t)=\sum_{\sigma_0,\sigma_0'}\int \mathrm{D}\sigma\int \mathrm{D}\sigma'\mathrm{A}[\sigma]\mathrm{A}^*[\sigma']\mathrm{F}[\sigma,\sigma']\rho_{\sigma,\sigma'}(t_0)$$

Here, $\mathrm{A}[\sigma]$ is the probability amplitude of the TSS to follow the path $\sigma(t')$ in the absence of the bath coupling. The real time influence functional reads

$$F[\sigma,\sigma']=e^{\left\{\int_{t_0}^{t}dt'\int_{t_0}^{t'}dt''\{\dot\sigma(t')-\dot\sigma'(t')\}\{Q(t'-t'')\sigma(t'')-Q^*(t'-t'')\sigma'(t'')\}\right\}}.$$

The kernel Q(t) is the analytically continued second integral of $K(\tau)$,

$$Q(t)=\int_0^\infty d\omega\frac{G(\omega)}{\omega^2}\{\coth(\beta\omega/2)[1-\cos(\omega t)]+i\,\mathrm{sen}(\omega t)\}$$

RDM exact master equation The expectation value $\langle\sigma_z\rangle_t = P(t)$ is related to the number of extra Cooper pairs in the island. For the general initial state Eq. (10) it takes the form

$$P(t) = \frac{p_R - p_L}{2}\mathrm{Tr}[\sigma_z(t)\sigma_z] + \frac{1}{2}\mathrm{Tr}[\sigma_z(t)] + a\mathrm{Tr}[\sigma_z(t)\sigma_x] + b\mathrm{Tr}[\sigma_z(t)\sigma_y]. \tag{11}$$

Within the path-integral method, a set of exact non-Markovian master equations for the elements of the RDM can be obtained[14]. In particular

$$\dot{P}(t) = 2\left[aK_B^{(a)}(t,t_0) + bK_B^{(a)}(t,t_0)\right] + \int_{t_0}^{t} dt'\left[K_A^{(a)}(t,t') - K_A^{(s)}(t,t')P(t')\right] \tag{12}$$

The kernels $K^{(s/a)}_{A,B}(t,t')$, defined in Ref. [14], contain the effect of dissipation and of the external bias. Effects of the preparation in an off-diagonal state of the RDM are in the inhomogeneous part. They are crucial at short times, whenever the dynamics exhibits underdamped coherent oscillations. For weak-damping, which is the interesting regime for QC, and for static bias, $\varepsilon(t) = \varepsilon_0$, P(t) is found to be, ($t_0=0$)

$$P(t) = P_\infty + P_1 e^{-\Gamma_r t} + P_2\cos(\nu t)e^{-\Gamma t} + P_3 sin(\nu t)e^{-\Gamma t} \tag{13}$$

The asymptotic (equilibrium) value and the other coefficients P_i are given in Ref. [14]. The oscillation frequency is $\nu = [\Delta^2_{ren} + \varepsilon^2_0]^{1/2}$ where $\Delta_{ren} = \Delta(\Delta/\omega_c)^{\alpha/(1-\alpha)}$. Transient effects of the initial preparation are evident in Fig. 2(b) The incoherent relaxation rate Γ_r and the dephasing rate Γ are $\Gamma_r = \left(\Delta^2_{ren}/2\nu^2\right)J(\nu)\coth(\beta\nu/2)$ $\quad \Gamma = \Gamma_r/2 + 2\pi\alpha(\varepsilon_0/\nu)^2 k_B T$.

In the presence of a high-frequency a.c.-field modulating the bias, $\varepsilon(\tau) = \varepsilon_0 + \varepsilon_1\cos(\Omega\tau)$ ($\Omega \gg \Delta, \nu, \Gamma, \Gamma_r$) calculations can be performed using a field-average of eq.(12)[13, 14]. For $\varepsilon_0 \ll \Omega$, and $J_0(\varepsilon_1/\Omega) \neq 0$, the system behaves like in the case of a static bias, but with a reduced effective tunneling matrix element $\Delta_{eff,0} = \Delta J_0(\varepsilon_1/\Omega)$. Because Γ and Γ_r are proportional to $\Delta^2_{eff,0} \leq \Delta^2$, both the dephasing rate Γ and the relaxation rate Γ_r can be strongly reduced by a suitably chosen driving field. Analogous features are found when a resonant ($\varepsilon_0 = n\Omega$) a.c.-field is applied. In this case, when $J_n(\varepsilon_1/\Omega)$ does not vanish the decay rates are proportional to $\Delta^2_{eff,n} = \Delta^2 J^2_n(\varepsilon_1/\Omega) \leq \Delta^2$.

6. CONCLUSIONS

Recently a.c. measurements were performed[15] using the set-up of Ref. [6]. It was observed that an a.c. field renormalizes the tunneling amplitude but it does not affect decoherence. This indicates that decoherence in the set-up of Ref. [6] is mainly due to other sources. On the other hand the a.c. tuneable renormalization of the tunneling amplitude may be used for dynamical fine tuning of the couplings and of time evolution. This is important for procedures where a precise switch off of the couplings is needed. It is finally worth mentioning that the formalism described above can be extended to the case of coupling with a spin-bath[7] which models background charges.

ACKNOWLEDGMENTS

We acknowledge G. Giaquinta, M. Grifoni, G.M. Palma, J. Siewert who were involved in stages of this work. This work has been supported by CRUI-DAAD under the Vigoni program.

REFERENCES

1. Leggett, A.J., 1986. In *Directions in Condensed Matter Physics*, (G. Grinstein and G. Mazenko Eds.), World Scientific, Singapore, p. 187.
2. Ekert, A., and Jozsa, R., 1996, *Rev. Mod. Phys* **68**: 733; Bennett, C. H., October 1995, *Physics Today*, p.24; Di Vincenzo, D.,1995, *Science* **270**: 25; Steane, A., 1998, *Rept. Prog. Phys.* **61**: 117.
3. Shor, P.W., 1997, *SIAM Journ. Comput.* **26**: 1487; Grover, L.K., 1997, *Phys. Rev. Lett.* **79**: 325.
4. Makhlin, Y., Schön, G., and Shnirman, A., 1999, *Nature* **398**: 305.
5. Averin, D.V., 1998, *Sol.State Comm.* **105**: 659; Mooij, J.E. *et al.*, 1999, *Science* **285**: 1036.
6. Nakamura, Y., Pashkin, Yu. A., Tsai, J.S., 1999, *Nature* **398**: 786.
7. Weiss, U., 1999, *Quantum Dissipative Systems*, World Scientific, Singapore.
8. Schön, G., Zaikin, A., 1990, *Phys. Rep.* **198**: 237; Mooij, J. E., Schön, G., 1992, in *Single Charge Tunneling*, NATO ASI Series, Vol. B 294, (H. Grabert and M.H. Devoret, eds.), Plenum Press, New York; 1991, *Z. Phys. B - Condensed Matter* **85**.
9. Barone, A. and Paternó, G., 1982, *Physics and applications of the Josephson effect*, J. Wiley & Sons; see also the contribution of A. Barone to this volume.
10. Bouchiat, V., *et al.*, 1999, *Physica Scripta* **T76**: 165.
11. Fazio, R., Palma, G.M., and Siewert, J., Dec. 20th, 1999, *Phys. Rev. Lett.*
12. Leggett, A. J., *et al.*, 1987, *Rev. Mod. Phys.* **59**: 1.
13. Grifoni M., and Hänggi, P. 1998, *Phys. Rep.* **304**: 229; Hänggi, P., Talkner, P., and Borkovec, M., 1990, *Rev. Mod. Phys.* **62**: 251.
14. Grifoni, M., Paladino E., and Weiss, U., 1999, *Eur. Phys. J. B* **10**: 719.
15. Y. Nakamura, private communication.

Addressing Single Sites Of A CO_2-Laser Optical Lattice

F. S. CATALIOTTI, R. SCHEUNEMANN, T. W. HÄNSCH, AND M. WEITZ
Max-Planck-Institut für Quantenoptik, Hans-Kopfermann-Str. 1, D-85748 Garching, Germany and Sektion Physik der Universität Munchen, Schellingsstr. 4, D-80799 München, Germany

1. INTRODUCTION

The lattice potential created by retroreflecting a CO_2 laser beam at 10.6 μm offers many intriguing prospects for the manipulation of cold atoms. The possibility of individually controlling single lattice sites, exploited in this report, together with the long coherence times allowed by the negligible photon scattering rate, opens the way for the likely realisation of fault tolerant quantum logic gates[1-3]. Furthermore the Lamb-Dicke regime is accessible allowing for the possible implementation of dark state cooling schemes such as Raman sideband cooling[4-6] which could lead to the achievement of Bose-Einstein condensation (BEC) by optical means only.

The interaction of Bose-Einstein condensates with optical standing waves provides a wide variety of intriguing phoenomena, many of which are explored in the present volume. Optically induced Bragg diffraction have been used to coherently split a condensate, measure its coherence or to observe matter wave dispersion[7]. Quantum tunnelling in optical lattices has allowed the realisation of a novel type of output coupler for atom lasers and presents many intriguing possibilities for the study of Josephson junctions[8,9]. Matter wave amplification[10] and the possible creation of squeezed states for matter waves[8] also rely on the application of optical lattices to BEC.

Bose-Einstein Condensates and Atom Lasers
Edited by Martellucci *et al.*, Kluwer Academic/Plenum Publishers, 2000

Other aspects available to the manipulation of cold atoms with optical lattices include very efficient schemes for quantum error correction[11] and fault-tolerant computing[12] which can be straightforwardly implemented due to the inherent possibility of parallel operation[13]. However, he possibility to selectively address and manipulate single qubits is central to the operation of quantum logic systems. This is difficult to achieve in conventional optical lattices, where the spatial period is of the order of half the wavelength of an atomic absorption line. Recently, imaging a phase grating with near-resonant light has created a potential of larger period, so that individual lattice sites could be resolved[14]. However, to image lattices with counterpropagating trapping beams, a much larger wavelength is required.

Here we report on the optical imaging of rubidium atoms trapped in an extremely far detuned optical lattice formed by an infrared standing wave near 10.6 μm. Moreover, we have been able to manipulate atoms confined in the single lattice sites by illuminating them with a pulse of resonant light. Preparation and readout of individual qubits should hence be possible in this kind of optical lattice. Additionally in our lattice we have reached the Lamb-Dicke regime in all spatial dimensions together with unprecedented filling rates. We plan to implement Raman sideband cooling for the possible realisation of an all-optical BEC.

2. QUASI STATIC DIPOLE TRAP

The trapping potential in optical lattices is based on the interaction of an induced atomic dipole moment with an off-resonant laser field. When the laser frequency is tuned to the red side of an atomic transition, the atoms are attracted towards the maximum field intensity and the trapping potential is given by $U = 1/2\alpha|\mathbf{E}|^2$. Here α is the atomic polarisability and $|\mathbf{E}|$ is the amplitude of the light field.

Since the frequency of the trapping laser in our experiment is more than an order of magnitude below that of the lowest atomic dipole transition in rubidium, the polarisation α can be approximated by its static value, and spontaneous scattering is largely suppressed[15].

For our experimental parameters, the expected time to spontaneously scatter a photon is above 600 s. As in our earlier work[16], we use a retroreflected CO_2-laser beam near 10.6 μm to realise an optical lattice with a large lattice constant. The atoms are polarisation gradient cooled into pancake shaped microtraps spaced by 5.3 μm. The one-dimensional arrangement minimises the heating rate due to reabsorption of fluorescence photons from neighbouring lattice sites.

3. EXPERIMENTAL APPARATUS

Our experimental set-up shown in Fig. 1 is also described elsewhere[17]. Briefly, in our vacuum system a combination of an ion pump and titanium sublimation pump yields a background pressure below 1×10^{-9} mbar. A resistively heated alkali metal dispenser provides a compact and controllable source of thermal rubidium atoms[18]. A single mode CO_2-laser generating an output power up to 50 W near 10.6 μm. Its infrared beam passes through an acousto-optic modulator (AOM), which is used both for optical isolation and to control the beam intensity. Two adjustable ZnSe lenses inside the vacuum chamber and a retroreflecting mirror are used to form a standing wave with a beam waist of typically 35 μm. With an input power of 14 W at this beam waist the calculated trap depth[16] is $U_0/\hbar = 43.5$ MHz, which corresponds to a temperature of 1.4 mK. Atoms of ^{85}Rb are collected and pre-cooled in a magneto-optical trap (MOT) operated with diode laser sources.

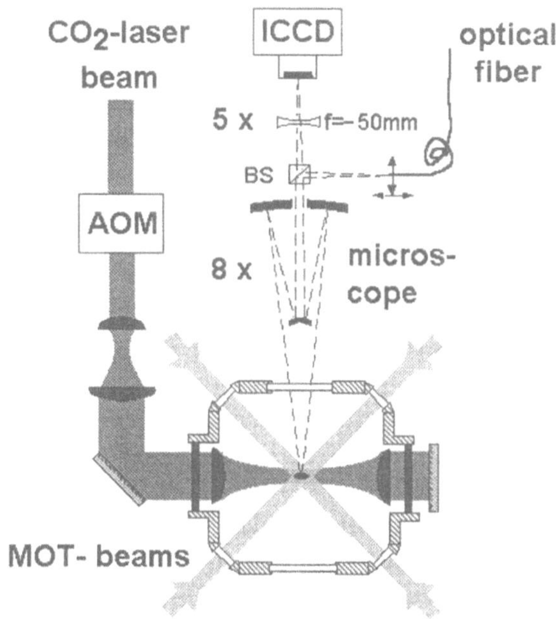

Figure 1. Schematic of the experimental apparatus

A Questar long distance microscope QM100 placed 10 cm away from the trap centre outside the vacuum system images the trapped atoms. For alignment purposes and for taking the time-of-flight (TOF) pictures, we have placed an intensified CCD camera behind this microscope. This results in a magnification of the lattice by a factor of eight. To image the optical lattice, we have inserted an additional biconcave lens of f = -50 mm focal length between the microscope and camera, and this results in an enlargement of the image by an additional factor of five. With this total magnification of 40 the rescaled pixel size is 0.6 μm and the field of view is 346 μm × 230 μm. Light from an additional diode laser resonant with the $5S_{1/2}$, F = 3 to $5P_{3/2}$, F = 4 cycling transition can be sent through the core of an optical fiber and is imaged via a beam-splitter through the microscope onto the sample. The optical set-up here represents that of a confocal microscope and allows the illumination of only a small area of the lattice.

4. LATTICE PROPERTIES

In a typical experimental run, we collect about 5×10^7 ^{85}Rb atoms during a MOT loading period of 3 s. The atoms are then compressed for 20 ms in a temporal dark MOT[19] realised by detuning the cooling laser 13 linewidths to the red side of the cooling transition and by simultaneously reducing the repumping laser intensity by a factor of 10. During this phase the CO_2-laser is switched on and the focus of the beam is superimposed on the MOT region. At the end of the dark MOT phase all resonant beams are extinguished with a combination of AOMs and mechanical shutters and the magnetic field is switched off. The number of atoms which remain trapped in the optical lattice, as well as their spatial extension, are measured by pulsing on the MOT beams and imaging the fluorescence onto both a calibrated photodiode and an intensified CCD camera. Typically we load around 1×10^6 ^{85}Rb atoms into the lattice, distributed over roughly 100 lattice sites, corresponding to a population of about 7×10^4 atoms per site in the central region. We have observed a lifetime of 3.4 s limited mainly by collisions with the thermal rubidium background gas.

The vibrational frequencies have been measured by parametrically exciting the atoms as described previously[16]. Briefly, the CO_2-laser beam intensity is modulated by the AOM, and significant vibrational excitation occurs if the modulation frequency equals twice a trap vibrational frequency. The induced trap loss, resulting in a reduced fluorescence, is recorded with the intensified CCD camera. This allows spatial resolution of the fluorescence. Since the oscillation frequencies depend strongly on the position of the micro-traps along the symmetry axis of the lattice this is of importance here. With the typical parameters of 14W and 35 μm we measure oscillation frequencies in the central trap region of ν_z = 54(5) kHz in the axial and ν_r = 4.2(5) kHz in the radial direction. The Lamb-Dicke limit for resonant excitation, corresponding to an oscillation frequency above ν_{rec} = 3.8 kHz, is therefore fulfilled in all three spatial dimensions.

We have determined the temperature of the atoms by switching off the trapping potential and observing the ballistic expansion of the cloud. Such a TOF method at the quoted trap depth yields a temperature of 70 μK for ^{85}Rb, almost a factor five above the results of our previous[16,20] taken with lower trapping beam intensities. If we decrease the CO_2-laser intensity, the measured temperature reduces to 10 μK and the atomic phase space density $n\lambda_{dB}^3$ reaches its maximum value of about 1/300. On the other hand, when increasing the trapping beam laser power the atomic temperature rises to values approaching the Doppler temperature limit of 140 μK. We attribute this trapping beam intensity dependent temperature to the fact that upper and lower electronic states are ac Stark shifted by different amounts in the presence of the CO_2 laser field. The static polarizability of the upper state is about a factor of 2.5 above that of the ground state. For larger laser power the differential ac Stark shift becomes comparable to the ^{85}Rb upper state hyperfine splitting and the efficiency of sub-Doppler laser cooling is strongly reduced.

5. SINGLE LATTICE SITE MANIPULATION

After optimisation of the trap parameters as described, we proceeded to image the optical lattice. For these experiments, the power of the CO_2-laser beam was reduced to 4 W to achieve lower temperatures. After loading the lattice and holding the atoms for 100ms the CO_2-laser beam was switched off, and the MOT lasers were pulsed on for 20μs while accumulating the atomic fluorescence with the ICCD-camera. We have chosen the detuning of the cooling laser to be resonant with the cycling transition and also added repumping light. Fig. 2A shows the image of 50 accumulated recordings after subtracting the constant background due to spurious scattered light and dark counts. The atoms are localised in periodically spaced pancake-like micro-traps of 5.3 μm period, each of these sites now containing up to 4×10^3 atoms. Assuming a Gaussian density distribution in the trap, we have measured a radial and axial HWHM of $\sigma_r = 6.8(7)$ μm and $\sigma_z = 0.9(1)$ μm. If we assume $k_BT/2 = 2\pi^2 m v_i^2 \sigma_i^2$ for atoms in a harmonic potential, the theoretical radial width is $\sigma^{theo}_r = 5.2$ μm using the measured temperature of T = 52 μK . The calculated axial width is $\sigma^{theo}_z = 0.4$ μm, which indicates that the experimentally measured axial width is mainly determined by the finite resolution of our imaging system. Fig. 2B represents a cross section along the lattice axis by integrating Fig. 2A along the radial direction. From this plot, we can evaluate a contrast defined as $C = (I_{max} - I_{min}) / (I_{max} + I_{min})$ of 32(5) %. From this contrast and the measured axial width, one can estimate the spatial resolution of our imaging system to be 1.9 μm. The measurements show that it is possible to distinguish atoms in neighbouring lattice sides and thus read out the information of individual quantum bits stored in the antinodes of this 1D optical lattice.

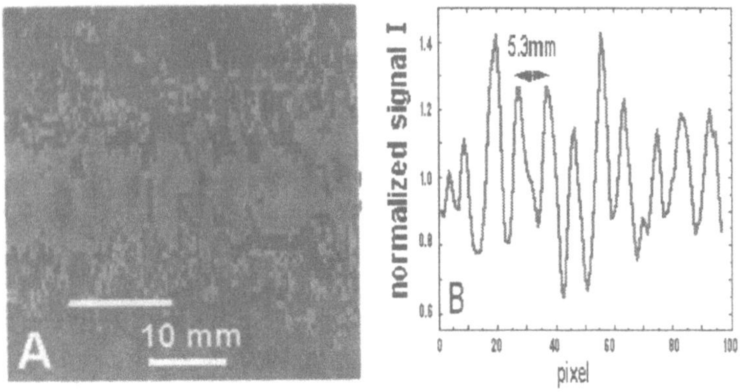

Figure 2. (A) Image of rubidium atoms trapped in a one-dimensional optical lattice with a period of 5.3 μm. The axes of the infrared CO_2-laser beam is oriented horizontally. (B) Cross section of this image obtained by integrating the measured atomic fluorescence perpendicular to the trapping beam axis.

Fig. 3A shows the image of the lattice of ^{85}Rb atoms illuminated by the MOT beams similar to that already described. The CO_2 trapping laser was now left on during the entire cycle. This allowed illumination times as long as 100 μs. In the absence of the trapping field the contrast of the images vanishes within 30 μs due to the thermal expansion of the cloud. The MOT cooling beams were resonant with the $5S_{1/2}$, F=3 to $5P_{3/2}$, F=4 transition at the bottom of the central potential wells. Fig. 3B depicts an image taken by illuminating a single trapping site for a period of 100 μs with around 10 μW of light, resonant with the $5S_{1/2}$, F=3 to $5P_{3/2}$, F=4 cycling transition at the bottom of the trap, through the fiber and by the MOT repumping beams. The exposure shows atoms localised in one distinct potential well of the standing wave, with the two neighbouring lattice sites suppressed by a factor of approximately 2.3. Note that the rest of the lattice is still filled, but is not visible here. This shows that in principle it is possible to address single qubits in an optical lattice. In order to investigate whether the neighbouring wells were being perturbed by the focused laser beam, the following procedure was used. After loading the atoms into the trap, we applied a 10 μs long pulse of light through the fiber with the same frequency, but with 20 times higher intensity. Again the MOT repumping beams were used to provide the necessary repumping light. Fig. 3C depicts the image of the lattice after interaction with such a pulse using the MOT-beams for exposure of the picture. The population of a single lattice site has been almost completely removed, while the short pulse affects atoms in the neighbouring sites much less. By varying the position of the optical fiber along the axial direction of the lattice we could address different lattice site within our optical field of view, which comprises around 50 lattice sites. While at present we used an optical imaging system optimised for the visible spectral region, the optical resolution could be further improved with a system optimised for the atomic fluorescence wavelength of 780 nm for the rubidium D2 line. Alternatively, one could use shorter wavelength transitions for the fluorescence imaging, e.g. the 5S-6P line of the rubidium atom near 420 nm.

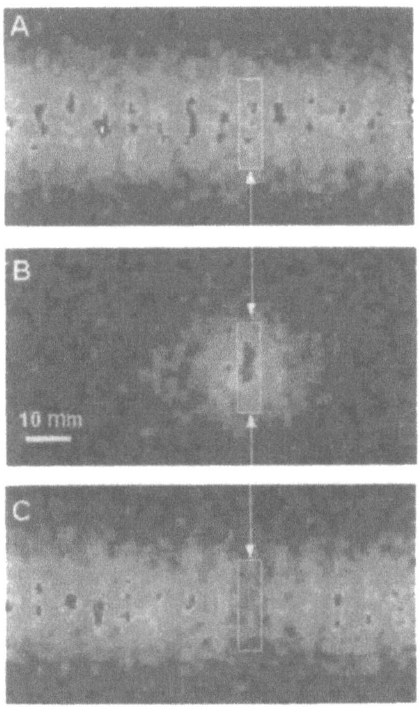

Figure 3. Images of the lattice after the following manipulations: (A) No manipulation of the atoms in the micro-traps. (B) Only atoms from a single lattice site are illuminated during the exposure with a focused laser beam. (C) Image of the lattice after removing one lattice site before the exposure.

6. PERSPECTIVES

In the future we wish to implement Raman sideband cooling in our optical lattice. We plan to follow the strategy outlined in Ref. [6] where the

Raman transitions are stimulated by the lattice laser. This cooling mechanism is one-dimensional but due to the high atomic densities peculiar to our lattice the elastic collision rate should be high enough to allow sympathetic cooling of the other dimensions. If the cooling process would work all the way to the ground state we could end up in producing an array of independent condensates in a completely new regime given the high frequencies obtained in our trapping potential.

We also wish to explore the possibility of performing quantum logic operations with this far-detuned optical lattice. To avoid rapid decoherence, the atoms must also be cooled to the vibrational ground state of the lattice. Coupling of different atoms should be possible via controlled cold collisions[2] or by inducing coherent dipole-dipole interactions[1]. In both the proposed schemes, the efficient interaction of atoms in different lattice sites requires a state-dependent lattice geometry, as can be achieved by additional closer-resonant optical beams with detuning comparable to the atomic fine structure, i.e. tuned between the $P_{1/2}$ and $P_{3/2}$ levels. The CO_2-laser lattice allows a very controlled loading of the near-resonant manipulation lattice, while the possibility to *spatially resolve* the individual qubits can be maintained. After preparing the qubits in the CO2-laser lattice, the retroreflected 10.6 μm beam is extinguished and the closer resonant light switched on. The atoms are then transferred into the near-resonant standing wave, and conditional dynamics is now possible. The described scheme should enable the realisation of entangled two- and more particle quantum states with the possibility of addressing individual atoms. The realisation of highly parallel quantum gates in an optical lattice would represent an important step towards a fault-tolerant quantum computer.

7. CONCLUSIONS

To conclude, we have optically resolved an optical lattice based on the infrared radiation of a CO_2-laser operating near 10.6 μm. This type of optical lattice has exciting prospects for quantum logic experiments. In addition our 1-D geometry minimises reabsorption of spontaneously emitted photons. This could allow for the possibility of reaching Bose-Einstein condensation by optical cooling alone.

ACKNOWLEDGMENTS

This work is supported in parts by the Deutsche Forschungsgemeinschaft. F. S. Cataliotti is on leave from: Dipartimento di Fisica Università di Firenze, L.go E.Fermi 2, I-50125 Firenze, Italy.

REFERENCES

1. Brennen, G., Caves, C., Jessen, P., and Deutsch, I., 1999, *Phys. Rev. Lett.* **83**: 1060.
2. Jaksch, D., Briegel, H.-J., Cirac, J.I., Gardiner, C.W., and Zoller, P. 1999, *Phys. Rev. Lett.* **82**: 1975.
3. Hemmerich, A., 1999, *Phys. Rev. A* **60**: 943.
4. Hammann, S., Haycock, D., Klose, G., Pacs, D., Deutsch, I., and Jessen, P., 1998, *Phys. Rev. Lett.* **80**: 4149.
5. Perrin, H., Kuhn, A., Bouchoule, I., and Salomon, C., 1998, *Europhys. Lett.* **42**: 395.
6. Vuletic, B., Chin, C., Kerman, A., and Chu, S., 1999, *Phys. Rev. Lett.* **81**: 5768.
7. Helmerson, K., in this volume.
8. Kasevich, M., in this volume.
9. Smerzi, A., in this volume.
10. Ketterle, W., in this volume.
11. See, e.g.: Steane, A.M., 1998, *Rep. Prog. Phys.* **61**: 117.
12. Steane, A., 1996, *Proc. Roy. Soc. A* **452**: 2552; Caldebank, A.R., and Shor, P.W., 1996, *Phys. Rev. A* **54**: 1098.
13. Briegel, H.-J., Calarco, T., Kaksch, D., Cirac, J.I., Zoller, P., quant-ph/9904010.
14. Boiron, D., Michaud, A., Fournier, J.M., Simard, L., Sprenger, M., Grynberg, G., and Salomon, C., 1998, *Phys. Rev. A* **57**: R4106.
15. Takekoshi, T., Knize, R.J., 1996, *Opt. Lett.* **21**: 77, Takekoshi, T., Yeh, J.R., and Knize, R.J., 1995, *Opt. Comm.* **114**: 421.
16. Friebel, S., D'Andrea, C., Walz, J., Weitz, M., and Hänsch, T.W., 1998, *Phys. Rev A* **57**: R20.
17. Scheunemann, R., Cataliotti, F.S., Hänsch, T.W., and Weitz, M., submitted to *Phys. Rev. Lett.*
18. Fortagh, J., Grossmann, A., Hänsch, T.W., and Zimmermann, C., 1998, *J. Appl. Phys.* **84**: 6499.
19. Ketterle, W., Davis, K.B., Joffe, M.A., Martin, A., and Pritchard, D.E., 1997, *Phys. Rev. Lett.* **70**: 2253.
20. Friebel, S., Scheunemann, R., Walz, J., Hänsch, T.W., and Weitz, M., 1998, *Appl. Phys. B* **67**: 699.

Scissors Mode And Superfluidity Of A Trapped Bose-Einstein Condensed Gas

O. M. MARAGÒ, S. A. HOPKINS, J. ARLT, E. HODBY,
G. HECHENBLAIKNER, AND C. J. FOOT
Clarendon Laboratory, Department of Physics, University of Oxford, Parks Road, Oxford, OX1 3PU, United Kingdom.

1. INTRODUCTION

The relationship between Bose-Einstein condensation (BEC)[1] and superfluidity has been studied extensively in liquid helium[2], but only recently has it been possible to examine it in condensates of dilute alkali metal vapours by manufacturing a vortical wavefunction in a spinor superposition state[3] and by the evidence of a critical velocity[4]. However neither case examine the distinctive properties of superfluids in rotating potentials i.e. transverse excitations, which has lead to much interesting physics in the case of helium. In the dilute alkali metal condensates various phenomena which imply the occurrence of superfluidity have been observed e.g. collective modes of excitation[5] and demonstrations of the coherence of the wavefunction[6]. In a recent theoretical paper D. Guéry-Odelin and S. Stringari[7] describe how the superfluidity of a trapped BEC may be demonstrated directly and we report the results of such an experiment. Guéry-Odelin and Stringari analyse the so-called scissors mode in which the atomic cloud oscillates with respect to the symmetry axis of the confining potential and they point out that the scissors mode has been used in nuclear physics to demonstrate the superfluidity of neutron and proton clouds in deformed nuclei[8,9].

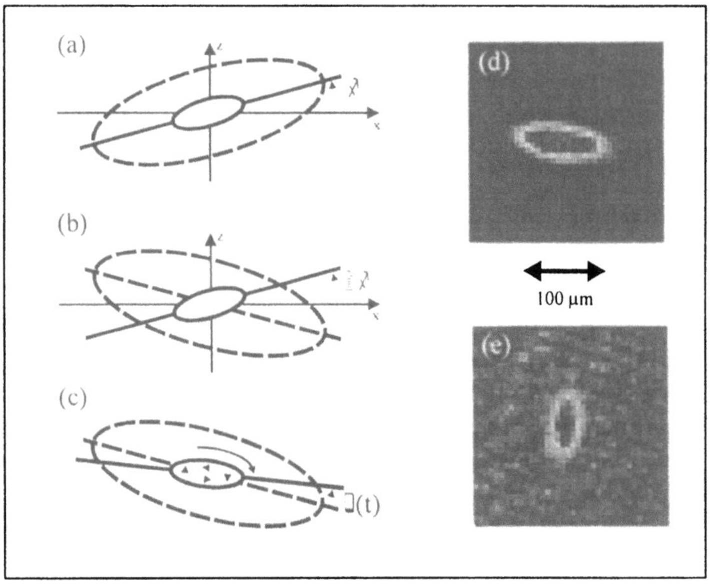

Figure 1. The method of exciting the scissors mode by a sudden rotation of the trapping potential. The solid lines indicate the shape of the atomic cloud and its major axes. The dashed lines indicate the shape of the potential and its major axes. (a) The initial situation after adiabatically ramping on the field in z direction, with cloud and potential aligned. (b) The configuration immediately after rotating the potential, with the cloud displaced from its equilibrium position. (c) The large arrow indicates the direction of the scissors mode oscillation and the smaller arrows show the expected quadrupolar flow pattern in the case of a BEC. The cloud is in the middle of an oscillation period. (The angles have been exaggerated for clarity.). (d) Image of the thermal cloud in the trap. (b) Image of the condensate after 15 ms of time of flight. The condensate expands most rapidly along the direction which is initially most tightly confined, leading to a 90° difference in orientation as compared to the thermal cloud.

The full theoretical analysis of the scissors mode is given in[7] and we only outline the key points here. The starting point is a BEC in an anisotropic harmonic potential with three different frequencies $\omega_x \cong \omega_y < \omega_z$. The scissors mode may be initiated by a sudden rotation of the trapping potential through a small angle as indicated in Fig. 1. In the subsequent motion, the cloud is not deformed provided that the change in the potential is too small to excite shape oscillations. For a thermal gas both rotational and irrotational fluid flow occur in the scissors mode and the normal fluid is predicted to exhibit two frequencies corresponding to those forms of motion. For the BEC there is only irrotational flow because of its single valued wavefunction and therefore it only exhibits one frequency, which is different from either of the frequencies observed for the thermal cloud.

2. EXPERIMENTAL RESULTS

In our experiment the trapping potential is created by a time-averaged orbiting potential (TOP) trap[10] which is a combination of a static quadrupole field, of gradient B_Q' in the radial direction, and a rotating field $\mathbf{B}_T(t)$ in the x-y plane. A term $\mathbf{B}_z(t)$, along the z direction, is added to the usual field. The effect of this additional term is to rotate the symmetry axes of the potential through an angle φ in the xz-plane and to reduce the oscillation frequency in the z direction from its value when $B_z=0$. Thus simply switching on $\mathbf{B}_z(t)$ also changes the cloud shape and so excites quadrupole mode oscillations. To avoid this we first adiabatically modify the usual TOP trap to a tilted trap and then quickly change $B_z(t)$ to $-B_z(t)$. This procedure rotates the symmetry axes of the trap potential by 2φ without affecting the trap oscillation frequencies (Fig. 1).

Our apparatus for producing BEC of ^{87}Rb is described in[11]. The following experimental procedure was used to excite the scissors mode both in the thermal cloud and in the BEC. Laser cooled atoms were loaded into the magnetic trap and after evaporative cooling the trap frequencies were ω_x = 90±0.2 Hz and $\omega_z = \sqrt{8}\,\omega_x$. The trap was then adiabatically tilted in 1 s by linearly ramping $\mathbf{B}_z(t)$ to its final value, corresponding to $\varphi = 3.6°$ and a reduction of the trap frequency ω_z by 2 %. Suddenly reversing the sign of $\mathbf{B}_z(t)$ in less than 100 μs excites the scissors mode in the trapping potential with its symmetry axes now tilted by $-\varphi$, as described above. The initial orientation of the cloud with respect to the new axis is $\theta_0=2\varphi$, so this angle is the expected amplitude of the oscillations (Fig. 1 (b)). The angle of the cloud was extracted from a 2-dimensional Gaussian fit of the absorption profiles such as those shown in Fig. 1 (d) and (e).

For the observation of the thermal cloud the atoms were evaporatively cooled to 1 μK which is about 5 times T_c, the temperature at which quantum degeneracy is observed. At this stage there were 10^5 atoms remaining with a peak density of $n_0 \cong 2\times10^{12}$ cm^{-3}. The scissors mode was then excited and pictures of the atom cloud in the trap were taken after a variable delay. The results of many runs are presented in Fig. 2 (a) showing the way the thermal cloud angle changes with time. The model used to fit this evolution is the sum of two cosines, oscillating at frequencies ω_1 and ω_2. From the data we deduce $\omega_1/2\pi$=338.5±0.8 Hz and $\omega_2/2\pi$=159.1±0.8 Hz. These values are in very good agreement with the values 339±3 Hz and 159±2 Hz predicted by theory[7]; which correspond to $\omega_1=\omega_z+\omega_x$ and $\omega_2=\omega_z-\omega_x$ We measured ω_x and ω_z by observing the center of mass oscillations of a thermal cloud in the untilted TOP trap and calculating the modification of these frequencies caused by the tilt. The amplitudes of the two cosines were found to be the same, showing that the energy is shared equally between the two modes of oscillation.

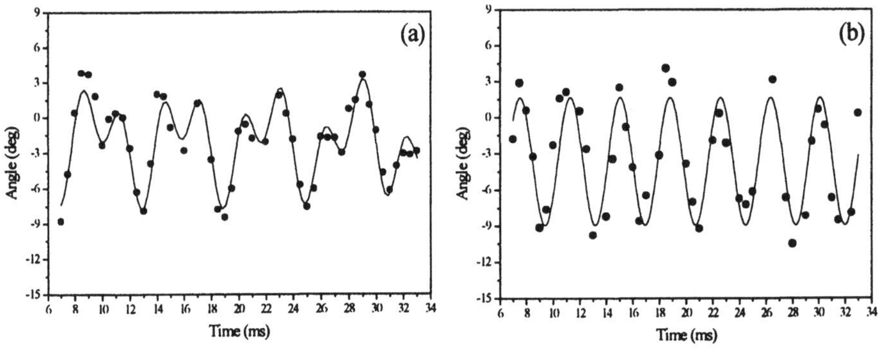

Figure 2. (a) The evolution of the scissors mode oscillation with time for a thermal cloud. For a classical gas the scissors mode is characterised by two frequencies of oscillation. The temperature and density of our thermal cloud are such that there are few collisions, so no damping of the oscillations is visible. (b) The evolution of the scissors mode oscillation for the condensate on the same time scales as the data in (a). For the BEC there is an undamped oscillation at a single frequency ω_c. This frequency is not the same as either of the thermal cloud frequencies.

To observe the scissors mode in a Bose-Einstein condensed gas, we carry out the full evaporative cooling ramp to well below the critical temperature, where no thermal cloud component is observable, leaving more than 10^4 atoms in a pure condensate. After exciting the scissors mode we allow the BEC to evolve in the trap for a variable time and then use the time-of-flight technique to image the condensate 15 ms after releasing it from the trap. The repulsive mean-field interactions cause the cloud to expand rapidly when the confining potential is switched off, so that its spread is much greater than the initial size. The aspect ratio of the expanded cloud is opposite to that of the original condensate in the trap, so that the long axis of the time-of-flight distribution is at 90° to that of the thermal cloud as shown in Fig. 1 (e). However this difference in the orientation does not affect the amplitude of the angle of oscillation. The scissors mode in the condensate is described by an angle oscillating at a single frequency ω_c Fig. 2 (b) shows some of the data obtained by exciting the scissors mode in the condensate. Consistent data, showing no damping, was recorded for times up to 100 ms. From an optimised fit to all the data we find a frequency of $\omega_c/2\pi$=265.6±0.8 Hz which agrees very well with the predicted frequency of 265±2 Hz from $\omega_c = \sqrt{(\omega_x^2 + \omega_z^2)}$. The aspect ratio of the time-of-flight distribution is constant throughout the data run confirming that there are no shape oscillations and that the initial velocity of a condensate does not have a significant effect.

These observations of the scissors mode clearly demonstrate the superfluidity of Bose-Einstein condensed rubidium atoms in the way predicted by Guéry-Odelin and Stringari[7]. Direct comparison of the thermal cloud and BEC under the same trapping conditions shows a clear difference in behaviour between the irrotational quantum fluid and a classical gas.

ACKNOWLEDGMENTS

This work was supported by the EPSRC and the TMR program (No. ERB FMRX-CT96-0002). O.M. Maragò acknowledges the support of a Marie Curie Fellowship, TMR program (No. ERB FMBI-CT98-3077).

REFERENCES

1. Anderson, M.H., *et al.*, 1995, *Science* **269**: 198; Davis, K.B., *et al.*, 1995, *Phys. Rev. Lett.* **75**: 3969. For a review see: Bose-Einstein Condensation. In *Atomic Gases*, Proceedings of the International School of Physics ``Enrico Fermi", 1999, (M. Inguscio, S. Stringari and C.E. Wieman, eds.), IOS Press, Amsterdam.
2. Tilley, D.R., and Tilley, J., 1991, *Superfluidity and Superconductivity*. Adam Hilger Ltd, Bristol and Boston, 3rd ed.
3. Matthews M.R., *et al.*, 1999, *Phys. Rev. Lett.* **83**: 2498.
4. Raman, C., *et al.*, 1999, *Phys. Rev. Lett.* **83**: 2502.
5. Jin, D.S., *et al.*,1996, *Phys. Rev. Lett.* **77**: 420; Mewes, M.-O., 1996, *et al.*, *Phys. Rev. Lett.* **77**: 988.
6. Andrews, M.R., *et al.*,1997, Science **275**: 637; Hall, D.S., Matthews, M.R., Wieman, C.E., and Cornell, E.A., 1998, *Phys. Rev. Lett.* **81**: 1543; Anderson, B.P., and Kasevich, M.A., 1998, *Science* **282**: 1686.
7. Guéry-Odelin, D., and Stringari, S., 1999, *Phys. Rev. Lett.* **83**: 4452.
8. Lo Iudice, N., and Palumbo, F., 1978, *Phys. Rev. Lett.* **41**: 1532; Lipparini, E., and Stringari, S., 1983, *Phys. Lett. B* **130**: 139.
9. Enders, J., *et al.*, 1999, *Phys. Rev. C* **59**: R1851.
10. Petrich, W., *et al.*, 1995, *Phys. Rev. Lett.* **74**: 3352.
11. Arlt, J., *et al.*, 1999, to be published in J. Phys. B, Dec.

Experiments With Potassium Isotopes

C. FORT
European Laboratory for non Linear Spectroscopy – Dipartimento di Fisica, Università di Firenze, Largo E. Fermi 2, I-50125, Firenze, Italy

1. INTRODUCTION

Experiments on cooling and trapping of potassium are strongly motivated by the occurrence of three isotopes: ^{39}K, ^{40}K, ^{41}K with a relative abundance of 93.26%, 0.012% and 6.73%, respectively. ^{39}K and ^{41}K are bosons while ^{40}K is a fermion, therefore potassium is a good candidate to study a degenerate dilute Fermi gas and bosons-fermions mixtures.

A quantum degenerate trapped gas of fermions is relatively unexplored both theoretically and experimentally. The first onset of Fermi degeneracy in a trapped atomic gas of ^{40}K has been very recently reported by B. DeMarco and D. Jin[1]. They were able to cool a sample of 7×10^5 fermionic potassium atoms down to 0.5 T_F (T_F is the Fermi temperature). A sample of trapped quantum degenerate fermions will provide the opportunity to investigate fundamental physical phenomena eventually including pairwise correlation, analogous to Cooper pairing of electrons, and BCS transition to the superfluid state in fermionic system with attractive interactions.

The experimental method to achieve quantum degeneracy in a sample of alkali-metal atoms[2], consists of two different cooling stages. In the first one laser cooling techniques allow to gain many orders of magnitude in the phase space density ρ providing a sample of atoms with $\rho\sim10^{-7}$. The final gap in the phase space density is covered by evaporative cooling in a magnetic trap. To be effective this cooling process requires a high enough[3] ratio between elastic and inelastic collision rates $\gamma_{el}/\gamma_{inel}$. In the case of potassium the laser cooling process is affected by the peculiar features of the

Bose-Einstein Condensates and Atom Lasers
Edited by Martellucci *et al.*, Kluwer Academic/Plenum Publishers, 2000

levels structure, and at the time we started to investigate potassium, little was known on the collisional properties of potassium isotopes.

Something more must be said regarding the possibility to evaporate a sample of fermions. The Pauli exclusion principle inhibits s-wave collisions (the predominant collisional channel which is active at very low temperature) between spin polarized fermion atoms. As a consequence for a sample of spin polarized fermions in a magnetic trap the evaporative cooling will stop at low temperature preventing from reaching quantum degeneracy. Different solutions were proposed to circumvent this problem.

The use of a mixture of fermions in different spin states already succeeded in reaching quantum degeneracy[1]. If two or more spin components are simultaneously trapped, spin-statistics does not prohibit s-wave collisions between the spins at low energy and evaporative cooling is possible. A different solution has been suggested[4] based on the enhancement of p-wave collision rate via an applied dc electric field. In this case the presence of an electric field modifies the collisional properties inducing a non-zero cross section at low energy. Another possible solution would be the use of a mixture of bosons and fermions and cool down the fermion sample through thermalization with evaporating bosons (sympathetic cooling[5]). In spite of the experimental complication and the uncertainties on the collisional properties, the alternative use of a boson-fermion mixture presents rich possibilities of experimental investigations and would also provide an efficient diagnostic of quantum degeneracy of the fermion sample[6]. While in the Bose case a phase transition occurs passing from the classical regime to the degenerate one, a trapped Fermi gas undergoes a gradual crossover between the classical limit and the Fermi sea.

2. POTASSIUM BOSONIC ISOTOPES

As a preliminary step to use the potassium bosons as collisional partners to cool down via sympathetic cooling the fermionic isotope, we studied cooling and trapping of ^{39}K and ^{41}K in a double-MOT set-up (Fig. 1). A double-MOT apparatus allows both efficient loading and long trapping time necessary for the evaporative cooling.

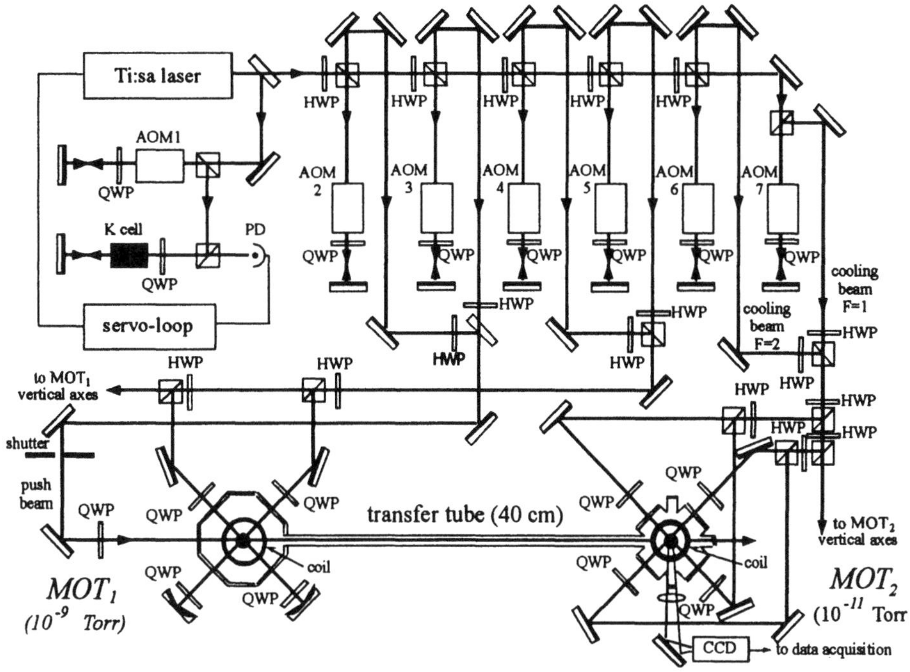

Figure 1. Experimental apparatus for a double-MOT of potassium isotopes. Frequencies of the all the relevant transitions are within ~1.3 GHz, therefore all the laser beams are derived from a single Ti:sapphire laser pumped by a 15 W Ar-ion laser. The laser frequency is offset locked using an acousto-optic modulator (AOM1) in double-pass configuration to a saturated absorption signal obtained in a potassium vapour cell. Using six different AOMs we obtain independent beams for the two MOTs and for the pushing beams necessary for transferring atoms from the first to the second cell. By changing the lock position and the AOM frequencies, we can switch from one isotope to the other. QWP = quarter wave plate, HWP = half wave plate.

The laser cooling process for ^{39}K and ^{41}K is complicated by the structure of the D_2 transition at 766 nm (Fig. 2). The hyperfine spacing of the upper level is comparable to the natural linewidth (6.2 MHz) preventing to isolate a single cooling transition. The laser detuning, typically used to cool and trap other atoms, results in strong optical pumping. The laser field can excite Zeeman sublevels belonging to different states in the multiplet, thus inducing a coherence among them. The transitions are affected by interference mechanisms and different schemes must be used to optimize trapping and cooling processes.

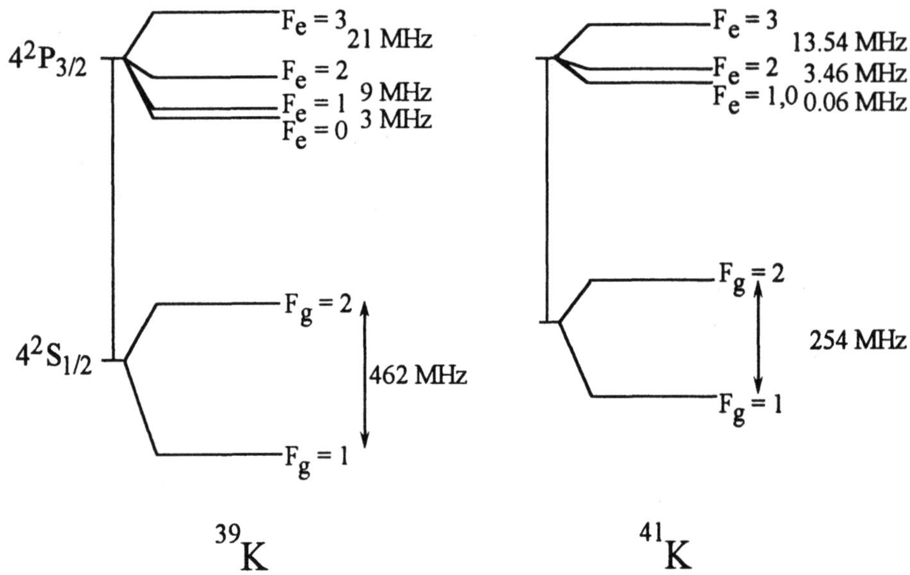

Figure 2. Levels structure of the D_2 transition for the two bosonic isotopes of potassium

The first relevant consequence is that two laser frequencies are needed, separated by the hyperfine splitting of the ground state, both intense and red-detuned with respect to the whole hyperfine structure of the excited state[7]. A detailed description of our studies on the laser cooling process in a MOT both for ^{39}K and ^{41}K can be found in Refs. [8, 9]. In this contest we will discuss only the relevant results concerning the minimum temperature and maximum density observed, since these are the important parameters for the next magnetic trapping and evaporative cooling. Because of the high intensity regime necessary to capture atoms in the MOT, the typical temperature during the loading phase is relatively high: few mK. In order to further cool the cloud we found a regime ("cooling phase") of reduced intensity and detuning applied for few ms after the loading, allowing to decrease by one order of magnitude the temperature. The coldest observed temperature (measured both with the Release and Recapture method and observing the expansion of the cloud with a CCD camera) is ~150 μK corresponding roughly to the Doppler limit. We were never able to observe sub-Doppler temperatures, in agreement with the theoretical analysis predicting a sub-Doppler component of the cooling force only in presence of very stable laser light (both in frequency and in amplitude)[8]. During the "cooling phase" an increase in density was also observed. The peak density ($\sim 10^9$ cm^{-3}) is, however, still lower than those obtained in a standard MOT with effective sub-Doppler cooling. We also tried well established techniques to increase the density in conventional MOTs, like CMOT[10] or darkSPOT[11], without any result.

To summarise, at the end of the laser cooling cycle we are able to collect in the second MOT 10^8-10^9 atoms with a maximum density of $\sim 10^9$ cm^{-3} and a minimum temperature of 150 μK. This results in a phase space density $\rho=10^{-8}$. These numbers have to be compared with those typical at the end of the laser cooling stage in a rubidium BEC experiment where one can collect 10^9 atoms with a density of $\sim 10^{11}$ cm^{-3} and a temperature of few tens of μK ($\rho=10^{-6}$).

In order to assess the effectiveness of evaporative cooling, one has to evaluate the elastic collisions rate $\gamma_{el}=n\,\sigma\,v$ (where n is the density, σ is the elastic collision cross section and v is the relative velocity of two colliding atoms) at the end of the laser cooling cycle. At low temperature, where s-wave collisions are the predominant collisional channel, $\sigma = 8\pi a^2$ (*a* is the scattering length). At the time we were facing this problem, theoretical predictions of *a* based on photoassociative spectra of ^{39}K were contradictory (see Table 1[12,13]). Then we decided to estimate experimentally the collisional rate, loading a cloud of cold ^{39}K or ^{41}K in a quadrupole magnetic trap. We loaded the trap at a low gradient (B'_z=18 G/cm) where the influence of the gravitational potential cannot be neglected, then we increased the gradient at B'_z=270 G/cm and we observed the relaxation of the cloud to equilibrium. The measured time necessary to the cloud to relax to equilibrium is $\sim$10 s for ^{41}K and $\sim$16 s for ^{39}K. Since the lifetime of our trap is $\tau \sim 60$ s, one can estimate $\gamma_{el}/\gamma_{inel} \sim 10$, which is too small to start an efficient evaporation. Our very preliminary results have found confirmation in recent papers (see Table 1[14,15]).

In Ref. [14] a new analysis of photoassociation spectra of ^{39}K gives a relatively small value of the scattering length *a* for ^{41}K and an even smaller one for ^{39}K. Furthermore the expected sign the ^{39}K scattering length is negative. Ref.15 reports the first direct measurement of |*a*| for the fermionic isotope (^{40}K) in a magnetic trap, from which values for the bosonic isotopes are predicted and found in good agreement with Ref. [14].

In the near future new works will appear[16,17] confirming the negative scattering length for ^{39}K (Table 2). The flourishment of papers on collisional properties of potassium isotopes is an evidence of the interest around this atom.

As for the cooling of the two potassium bosonic isotopes, we conclude that having a relatively low density and high temperature at the end of the laser cooling stage and collisional parameters not favourable, an efficient evaporation would need a very long trapping lifetime (at least 1000 s). Therefore it seems a reasonable choice to try to use a "simpler" atom like

^{87}Rb as partner for the sympathetic cooling of the potassium fermionic isotope. We converted our double-MOT apparatus to cool and trap rubidium.

After the optical cooling stage, including a CMOT and a molasses phase, we measured a phase-space density $\rho = 10^{-7}$. The evaporative cooling has been then performed in a 4-coils Ioffe-Pritchard type magnetic trap driven at a current of 240 A. The cigar shape harmonic potential for ^{87}Rb atoms in $|F_g = 2, M_F = 2>$ level has an axial frequency $\nu_z = 12.6$ Hz and a radial frequency $\nu_r = 60$ Hz. After the transfer of atoms in the magnetic trap we start evaporation with an elastic collision rate $\gamma_{el} = 30\ s^{-1}$, that combined with the lifetime of our trap gives $\gamma_{el}/\gamma_{inel} = 1800$. After ~30 s of evaporation a Bose-Einstein condensate of 2 10^5 atoms with a peak density of 7 10^{13} cm^{-3} formed[18]. We observed the BEC transition at $T_C = 200$ nK.

Table 1. Triplet scattering length for potassium isotopes expressed in Bohr radii unit

Isotopes	H.Boesten et al.[12] 1996	R.Côté et al.[13] 1998	J.Bohn et al.[14] May 1999	B.DeMarco et al.[15] 1999
39-39	$-1200<a_T<-60$	81.1±2.4	-17±24	$-80<a_T<-28$
		1.7±4.4	40-40	$136<a_T<176$
41-41	$25<a_T<60$	286±36	65^{+13}_{-8}	$49<a_T<62$
39-40		47.5±2.3	$460^{+330}_{-\infty}$	$a_T>500$ or $a_T<-900$
39-41		5.1±4.1	-205^{+140}_{-40}	$140<a_T<185$
40-41		-162±36		$83<a_T<99$

Table 2. Triplet scattering length for potassium isotopes expressed in Bohr radii unit

	J.Burke et al.[16]	C.Williams et al.[17]
39-39	$-42<a_T<+8$	$-60<a_T<+15$

3. POTASSIUM FERMIONIC ISOTOPES

Regarding ^{40}K, we stress that it is, together with ^6Li, one of the two stable fermionic isotopes among alkali atoms. For ^6Li laser cooling is limited by the large photon-recoil velocity and the unresolved excited-state hyperfine splitting resulting in temperatures of the order of 200 μK[19,20]. Furthermore the ground hyperfine states $F_g = 9/2, 7/2$ of ^{40}K present a rich Zeeman structure with nine low-field seeking spin state magnetically trappable.

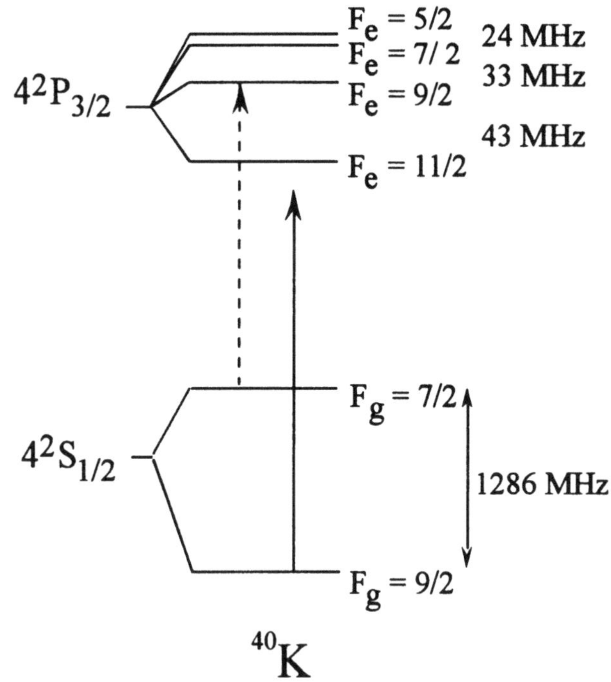

Figure 3. Levels structure of the D_2 transition for the fermionic isotopes of potassium. Note that the hyperfine structure is inverted both in the ground and in the excited state.

We realized the first MOT for ^{40}K in a natural abundance sample[21]. As a consequence of the very low natural abundance of this isotope, we observed a MOT with only $\sim 10^4$ atoms.

The cooling and trapping of ^{40}K proved to be simpler than for the other isotopes. The hyperfine structure is indeed slightly larger and, more important, inverted (Fig. 3). In this case is possible to operate the MOT recovering the usual configuration with the trapping laser red-detuned with respect to the $F_g = 9/2 \rightarrow F_e = 11/2$ cycling transition and the repumping light tuned to the $F_g = 7/2 \rightarrow F_e = 9/2$ transition.

We measured the temperature of the MOT obtaining, due to the poor signal to noise ratio,

only an estimation of T~50 μK, well below the Doppler limit. A better measurement of temperature both in a MOT and in a molasses of ^{40}K has been done in our group very recently[22] loading the MOT from an enriched sample (3%). Using the enriched sample the MOT collects up to $5\ 10^7$ atoms, with a peak density of $5\ 10^{10}$ cm^{-3}, providing high enough signal to measure the temperature with the Time Of Flight (TOF) method. The measured temperature (Fig. 4) confirmed that sub-Doppler mechanisms are effective in a MOT of ^{40}K.

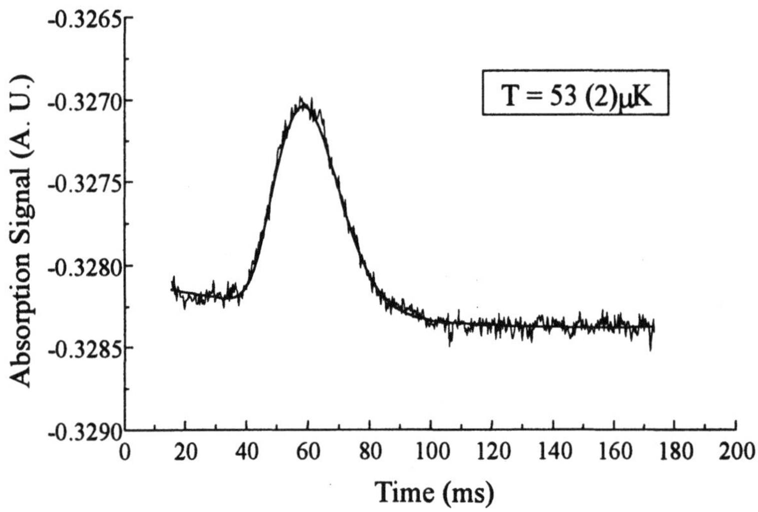

Figure 4. Typical Time Of Flight signal for a sample of ^{40}K atoms cooled in a MOT

The expected behaviour of temperature proportional to the intensity of the laser field and to the inverse of the detuning was observed. A complete study of the temperature was done also in the molasses where a minimum temperature T~15±5 μK was measured, a factor of 10 below the Doppler-cooling limit.

These results for the laser cooling of the fermionic isotope of potassium are very encouraging. The low temperature attainable provides a suitable environment for confining ^{40}K in a optical dipole trap where all the magnetic levels can be trapped. In particular, a tightly confining standing-wave trap could be used to increase the number-density of cold fermionic ^{40}K atoms, and reduce their temperature by means of Raman techniques[23]. Recently, it has been demonstrated[23] that this kind of trap are a very useful tool in investigating the occurrence of Feshbach resonances in presence of magnetic field. The importance of Feshbach resonances, which can change the character of the interaction between atoms, would be fundamental once achieved quantum degeneracy in a gas of fermionic atoms. Changing the interaction between ^{40}K atoms from repulsive to attractive, would provide the possibility to investigate BCS-like transitions.

To summarize, even if potassium is a "difficult" atom to take to quantum degeneracy, it is a rich atom presenting two bosons isotopes one with positive, the other one with negative scattering length, and a fermionic isotope which quantum degeneracy has been already achieved.

ACKNOWLEDGMENT

The work was done at the European Laboratory for non Linear Spectroscopy in Florence and I would like to thank all the people involved: M. Prevedelli, F. S. Cataliotti, G. Modugno, G. Roati, F. Minardi, M. Inguscio, L. Ricci, G. M. Tino, J. R. Ensher, and E. A. Cornell.

REFERENCES

1. DeMarco, B., and Jin, D.S., 1999, Onset of Fermi degeneracy in a trapped atomic gas, *Science* **285**: 1703.
2. Anderson, M.H., Ensher, J.R., Matthews, M.R., Wieman, C.E., and Cornell, E.A., 1995, Observation of Bose-Einstein condensation in a dilute atomic vapor, *Science* **269**: 1989.
   Davis, K.B., Mewes, M.O., Andrews, M.R., van Druten, N.J., Durfee, D.S., Kurn, D.M., and Ketterle, W., 1995, Bose-Einstein condensation in a gas of sodium atoms, *Phys. Rev. Lett.* **75**: 3969.
   Bradley, C.C., Sackett, C.A., and Hulet, R., 1997, Bose-Einstein condensation in lithium: observation of limited condensate number, *Phys. Rev. Lett.* **78**: 985.
3. Ketterle, W., and van Druten, N.J., 1996, Evaporative cooling of trapped atoms, *Adv. At. Mol. Opt. Phys.* **37**. 181.
4. Geist, W., Idrizbegovic, A., Marinescu, M., Kennedy, T.A.B., and You, L., Evaporative cooling of trapped fermionic atoms, cond-mat/9907222.
5. Myatt, C.J., Burt, E.A., Ghrist, R.W., Cornell, E.A., and Wieman, C.E., 1997, Production of two overlapping Bose-Einstein condensates by sympathetic cooling, *Phys. Rev. Lett.* **78**: 586.
   Geist, W., You, L., and Kennedy, T.A.B., 1999, Sympathetic cooling of an atomic Bose-Fermi gas mixture, *Phys. Rev. A* **59**: 1500.
6. Vichi, L., Inguscio, M., Stringari, S., and Tino, G.M., 1998, Quantum degeneracy and interaction effects in spin-polarized Fermi-Bose mixtures, *J. Phys. B* **31**: L899.
7. Williamson III, R.S., and Walker,T.,1995, Magneto-optical trapping and ultracold collisions of potassium atoms, *J. Opt. Soc. Am. B* **12**: 1393.
   Santos, M.S., Nussenzveig, P., Marcassa, L.G., Helmerson, K., Flemming, J., Zilio, S.C., and Bagnato, V.S., 1995, Simultaneous trapping of two different atomic species in a vapor-cell magneto-optical trap, *Phys. Rev. A* **52**: R4340.
   Wang, H., Gould, P.L., and Stwalley, W.C., 1996, Photoassociative spectroscopy of ultracold ^{39}K atoms in a "dark spot" vapor cell magneto-optical trap, *Phys. Rev. A* **53**: R1216.
8. Fort, C., Bambini, A., Cacciapuoti, L., Cataliotti, F.S., Prevedelli, M., Tino, G.M., and Inguscio, M., 1998, Cooling mechanisms in potassium magneto-optical traps, *Eur. Phys. J. D* **3**: 113.
9. Prevedelli, M., Cataliotti, F.S., Cornell, E.A., Ensher, J.R., Fort, C., Ricci, L., Tino, G.M., and Inguscio, M., 1999, Trapping and cooling of potassium isotopes in a double-magneto-optical-trap apparatus, *Phys. Rev. A* **59**: 886.
10. Petrich, W., Anderson, M.H., Ensher, J.R., and Cornell, E.A., 1994, Behavior of atoms in a compressed magneto-optical trap, *J. Opt. Soc. Am. B* **11**: 1332.

11. Ketterle, W., Davis, K.B., Joffe, M.A., Martin, A., and Pritchard, D.E., 1993, High densities of cold atoms in a dark spontaneous-force optical trap, *Phys. Rev. Lett.* **70**: 2253.
12. Boesten, H.M.J.M, Vogel, J.M., Tempelaars, J.G.C., Verhaar, B.J., 1996, Properties of cold collisions of ^{39}K atoms and of ^{41}K atoms in relation to Bose-Einstein condensation, *Phys. Rev. A* **54**: R3726.
13. Côtè, R., Dalgarno, A., Wang, H., Stwalley, W.C., 1998, Potassium scattering lengths and prospect for Bose-Einstein condensation and sympathetic cooling, *Phys. Rev. A* **57**: 4118R.
14. Bohn, J.L., Burke, J.P., Green, C.H., Wang, H., Gould, P.L., and Stwalley, W.C., 1999, Collisional properties of ultracold potassium: consequences for degenerate Bose and Fermi gases, *Phys. Rev. A* **59**: 3660.
15. DeMarco, B., Bohn, J.L., Burke, J.P., Holland, M., and Jin, D.S., 1999, Observation of p-wave threshold law using evaporatively cooled fermionic atoms, *Phys. Rev. Lett.* **82**: 4208.
16. Burke, J.P., Greene, C.H., Bohn, J.L., Wang, H., Gould, P.L., and Stwalley, W.C., December 1999, Determination of ^{39}K scattering lengths using photoassociation spectroscopy of the O^-_g state, *Phys. Rev. A* **60**.
17. Williams, C.J., Tiesinga, E., Julienne, P.S., Wang, H., Stwalley, W.C., and Gould, P.L., December 1999, Determination of the scattering lengths of ^{39}K from 1_u photoassociation line shapes, *Phys. Rev. A* **60**.
18. Fort, C., Prevedelli, M., Minardi, F., Cataliotti, F.S., Ricci, L., Tino, G.M., and Inguscio, M., Collective excitations of a ^{87}Rb Bose condensate in the Thomas-Fermi regime, submitted to *Europhys. Lett.*
19. Ghaffari, B., Gerton, J.M., McAlexander, W.I., Strecker, K.E., Homan, D.M., and Hulet, R.G., Laser-Free Slow Atomic Source, to be published in *Phys. Rev. A.*
20. O'Hara, K.M., Granade, S.R., Gehm, M.E., Savard, T.A., Bali, S., Freed, C., and Thomas, J.E., 1999, Ultrastable CO_2 Laser trapping of Lithium Fermions, *Phys. Rev. Lett.* **82**: 4204.
21. Cataliotti, F.S., Cornell, E.A., Fort, C., Inguscio, M., Marin, F., Prevedelli, M., Ricci, L., and Tino, G.M., 1998, Magneto-optical trapping of fermionic potassium atoms, *Phys. Rev. A* **57**: 1136.
22. Modugno, G., Benko, C., Hannaford, P., Roati, G., and Inguscio, M., Sub-Doppler laser cooling of fermionic ^{40}K atoms, submitted to *Phys. Rev. A.* cond-mat 9908102.
23. Vuletic, V.,Chin, C., Kerman, A.J., and Chu, S., 1998, Degenerate Raman Sideband Cooling of trapped Cesium Atoms at Very High Atomic Densities, *Phys. Rev. Lett.* **81**: 5768.
24. Vuletic, V., Kerman, A.J., Chin, C., and Chu, S., 1999, Observation of Low-Field Feshbach Resonances in Collisions of Cesium Atoms, *Phys. Rev. Lett.* **82**: 1406.

Equilibrium State And Excitations In Trapped Fermi Vapours

A. MINGUZZI AND M. P. TOSI
INFM and Scuola Normale Superiore, Pza dei Cavalieri 7, I-56126 Pisa, Italy

1. INTRODUCTION

Cold dilute Fermi gases can be obtained experimentally with similar techniques as those employed to cool bosons. Among alkali gases only ^{6}Li and ^{40}K are (stable) fermionic isotopes. Experiments aimed at reaching quantum degeneracy of fermionic species are in progress at JILA[1,2], ENS and Rice University; ^{40}K has already been cooled down to a temperature of 0.3 T_F in a magnetic trap[3].

In dilute neutral Fermi gases the kinetic energy has a major role as compared to their bosonic counterparts. For what concerns interactions, due to the effect of statistics, *s*-wave collisions are allowed only between fermions belonging to different hyperfine levels. From experiments and calculations we know that the *s*-wave scattering length is small and repulsive for ^{40}K and strong and attractive for ^{6}Li. Due to this property, ^{6}Li is expected to undergo a transition to a superfluid phase at very low temperatures.

The analysis of the dynamical behavior of the Fermi gas may become a useful tool for the study of the interactions in the system and also a possible way to detect the superfluid transition[4,5].

In this Chapter we will present a theoretical description of the dynamical properties of a trapped mufti-component gas in the normal state at zero temperature and a calculation of its spectral modes in some relevant cases. The spectrum will be finally compared with that of a superfluid Fermi gas.

Bose-Einstein Condensates and Atom Lasers
Edited by Martellucci *et al.*, Kluwer Academic/Plenum Publishers, 2000

2. DYNAMICAL PROPERTIES OF HOMOGENEOUS NORMAL FERMI GASES

The spectrum of excitations of a normal (*i.e.* non superfluid) degenerate Fermi gas is much richer than that of a Bose gas. This is due to the fact that while in a Bose system the spectrum of single-particle and collective excitations coincide (theorem of Gavoret and Nozières[6]), in the fermionic case these excitations have different nature and also different frequencies. *Single-particle* excitations are described for a weakly interacting system in terms of the excitation of a particle out of the Fermi sphere, with the simultaneous creation of a hole; while *collective* excitations involve the coherent motion of many fermions, which can result in a density fluctuation or in a transverse-current fluctuation, and also in a spin wave in the case of a two-component system. The most rigorous definition of these modes for an interacting system is given in terms of the poles of the single-particle Green's function or of the appropriate two-particle Green's function, respectively.

Before embarking in the description of the dynamical behavior of the system one has also to take into account the role of collisions: depending on whether the collisions are very frequent on the time scale of an oscillation (collisional regime) or very rare (collisionless regime) the theory has to be formulated accordingly.

For a homogeneous non-interacting Fermi gas in the collisionless regime, the spectrum of excitations is given by the particle-hole continuum. In the plane (ω, k) these excitations occupy the area between the two parabolas $\omega_1 = k^2/2m + kv_F$ and $\omega_2 = k^2/2m - kv_F$ where v_F is the Fermi velocity, related to the Fermi energy as $\varepsilon_F = mv_F^2/2$. In addition to this broad feature, in the case of repulsive interactions the spectrum of the collisionless gas also contains a phonon with velocity $c_0 \geq v_F$, which is usually referred to as *zero sound.* This mode is instead not stable in the case of attractive interactions, when it decays rapidly into the continuum of particle-hole excitations.

In the collisional regime the long-wavelength density fluctuations of the Fermi gas are the usual compressional phonons (*first sound*), with a speed given by $c_1 = v_F/\sqrt{3}$ in the regime of weak interactions. We remark that at low temperatures the collisional regime can be realized only in the presence of impurities (*i.e.* particles of another species).

From the Landau theory of normal Fermi liquids[7] we know that zero sound involves an anisotropic oscillation of the Fermi surface while first sound, being dominated by collisions, is characterized by an isotropic breathing of the Fermi sphere.

3. ONE-COMPONENT TRAPPED FERMI GAS

Current experiments are usually performed in magnetic traps which provide a spatial confinement for the gas. The trapping potential $V_0(\mathbf{r})$ can be modelled as a harmonic potential along the three space directions: $V_0(\mathbf{r}) = m\omega_f^2 r^2/2$.

Due to the presence of the confinement, we expect the spectrum of collective modes of the cloud to be discrete: in the following we shall identify in the trapped system the equivalent of the zero sound and first sound modes described above for the homogeneous gas.

We consider as a first example the case of a fully spin-polarized Fermi cloud. This can be well approximated as a non-interacting gas due to the Pauli principle. Since we want to describe a system with a large number of fermions ($N \geq 10^3$) we can safely employ the semiclassical local-density approximation for the equilibrium density profile[8,9]:

$$n_0(\mathrm{r}) = \theta(\varepsilon_F - V_0(\mathrm{r}))\left[A^{-1}(\varepsilon_F - V_0(\mathrm{r}))\right]^{3/2} \tag{1}$$

where $A = \hbar^2(6\pi^2)^{2/3}/2m$. This is usually called the "Thomas-Fermi" profile.

3.1 Collisionless Regime

To discuss the dynamics of a confined Fermi cloud in a regime where collisions have little influence, we employ the equation for the Wigner distribution function $f_\mathrm{p}(\mathrm{R},t) = \int dr\langle\Psi^+(\mathrm{R}+\mathrm{r}/2,\mathrm{t})\rangle\exp(i\mathrm{p}\cdot\mathrm{r}/\hbar)$ in the form of a Boltzmann equation with the collision integral set to zero; *i.e.*

$$\partial_t f_\mathrm{p}(\mathrm{R},t) + \frac{\mathrm{p}}{\mathrm{m}}\cdot\nabla_\mathrm{R} f_\mathrm{p}(\mathrm{R},t) - \nabla_\mathrm{R} V_0(\mathrm{R})\cdot\nabla_\mathrm{p} f_\mathrm{p}(\mathrm{R},t) = 0 \tag{2}$$

This equation is analogous to the Vlasov-Landau transport equation for quasi-particles in a homogeneous Fermi fluid in the collisionless regime[7] and is expected to be valid down to very low temperature provided that the cloud contains a large number of fermions.

In searching for solutions of Eq. (2), we focus on the shape-deformation modes which can be monitored by measuring the mean square radius of the cloud as a function of time. Within this class of modes, we adopt a variational Ansatz as already proposed by Bijlsma and Stoof[10] for a Bose cloud.

This amounts to choosing $f_p(\mathrm{R},t)$ in the form of the equilibrium solution Φ_{eq} (p,R) after resealing both the space and the momentum variable through time-dependent factors in each geometrical direction. More precisely, for ($i = x, y, z$) we introduce variational parameters α_i (t) describing the deviation of the mean square sizes of the cloud from their equilibrium values as $\alpha_i(t)=\left[\langle R_i(t)^2\rangle / \langle R_i^2\rangle eq\right]^{1/2}$. The form of $f_p(\mathrm{R},t)$ is taken as

$$f_p(R,t)=\left(c_x c_y c_z\right)\Phi_{eq}\left(\sqrt{c_i}\alpha_i\left(p_i - mR_i\dot{\alpha}_i/\alpha_i\right), \sqrt{c_i}R/\alpha_i\right) , \qquad (3)$$

the time-independent coefficients c_i being introduced in Eq. (3) in order to ensure that the initial momentum distribution is isotropic. The equilibrium form of the Wigner function for the degenerate Fermi gas subject to spherical confinement can in turn be chosen as a generalized Fermi sphere, *i.e.* as $\Phi_{eq}(\mathrm{p,R})=\theta\left(\varepsilon_F - p^2/2m - m\omega_f^2R^2/2\right)$.Of course, this yields the same equilibrium density as that reported earlier in Eq. (1).

We emphasize that in the present treatment we are assuming that, even though collisions have negligible influence on the dynamics of the cloud, they must have been active in the past in order to bring the cloud to the state of thermal equilibrium.

The equations of motion for the $\alpha_i(t)$ in each spatial direction are obtained by taking the moments of Eq. (2) with respect[10] to $R_i p_i$:

$$\ddot{\alpha}_i + \omega_f^2\alpha_i = \omega_f^2/\alpha_i^3 \qquad (4)$$

It is worth noticing that for a non-interacting gas Eq. (4) follows directly from the dynamical scaling Ansatz (3) and from the virial theorem: it does not require a specific assumption on the shape of the variational function.

Eq. (4) can be solved analytically in the non-linear regime[11]. We are interested here in the small-oscillation regime. By linearizing Eq. (4) one obtains an oscillation of the cloud at frequency $\omega = 2\omega_f$, corresponding to the frequency of a single-particle excitation in the cloud.

This result is the analogue of the zero sound in the homogeneous system, where for vanishing values of the coupling strength the deformation of the Fermi surface associated to this mode involves only a few (quasi)particle excitations[7].

3.2 COLLISIONAL REGIME

We describe a non-interacting Fermi gas by a set of equations of motion for its physical variables, *i.e.* density $n(\mathrm{r}, t)$ current $j_i(\mathrm{r}, t)$, kinetic tensor $\Pi_{i,j}(\mathrm{r}, t,)$ and so on, which are defined microscopically as the moments of the Wigner distribution function f_{p} (r, t) with respect to its variable p (see *e.g.* [12]).

The first two of these equations have the form of conservation laws for particle number,

$$\partial_t n(\mathrm{r}, t) = -\nabla \cdot \mathrm{j}(\mathrm{r}, t) \tag{5}$$

and for the momentum,

$$m\partial_t j_i(\mathrm{r}, t) = -\nabla_j \Pi_{ij}(\mathrm{r}, t) - n(\mathrm{r}, t)\nabla_i V_{ext}(\mathrm{r}, t). \tag{6}$$

The next equation is related to energy conservation, but is omitted here since we shall consider only isothermal fluctuations.

In the collisional regime we can employ for the kinetic stress tensor an adiabatic local-density approximation

$$\Pi_{ij}(\mathrm{r}, t) = \delta_{ij} P(\mathrm{r}, t) = \delta_{ij} \frac{2}{5} A[n(\mathrm{r}, t)]^{5/3} \tag{7}$$

which allows the closure of Eqs. (5) and (6) upon expressing the pressure P(r, t) of the inhomogeneous Fermi gas in terms of its density. Eq. (7) is valid in the regime of small oscillations and for a system whose inhomogeneity is sufficiently weak: more explicitly, we are assuming that the length scale for the variation of the density profile in space is large relative to the inverse Fermi wavenumber k_F^{-1}.

By combining Eqs. (5) and (6) and linearizing and Fourier-transforming them with respect to time, we are left with an eigenvalue differential equation with regular singular points. This can be solved in a series form by adopting the Fuchs method[13]. For the case of spherical harmonic confinement the solution yields the following dispersion relation[11]:

$$\omega(n, l) = \omega_f (l + 4n(n + l + 2)/3)^{1/2} \tag{8}$$

where l is the total angular momentum quantum number and n is the number of radial nodes. The dispersion relation (8) correctly displays the sloshing mode solution ($n = 0$, $l = 1$, $m = 0$) at a frequency $\omega = \omega_f$, in agreement with the generalized Kohn theorem[14].

The modes corresponding to the dispersion relation (8) are the equivalent of first sound in the trapped system: indeed it is easy to show that in the homogeneous limit Eqs. (5) and (6) have as solution a sound mode with velocity $c = v_F/\sqrt{3}$.

As a final comment we remark that while these solutions resemble those obtained for a Bose-condensed cloud at $T = 0$ in the (bosonic) Thomas-Fermi approximation[15], there is a fundamental difference between the two cases. In the Fermi gas the density fluctuations must vanish continuously at the Fermi radius in order that the Kohn theorem be satisfied, while in the Bose cloud they vanish discontinuously at the Bose radius. In the Bose gas within the Thomas-Fermi approximation the kinetic energy is neglected and therefore the density fluctuation is allowed to present a discontinuous jump at the border; this instead is not possible in the Fermi gas where the kinetic energy term is dominant.

4. TWO-COMPONENT FERMI GAS

A two-component dilute Fermi gas has already been realized experimentally by trapping ^{40}K in two different hyperfine levels inside a magnetic trap[16].

In the two-component system the interactions are no longer negligible, since s-wave collisions between fermions having different internal states are allowed. We investigate here the role of interactions in the equilibrium state and in the dynamical properties of the system[17].

4.1 Phase Diagram

We begin by discussing the properties of the ground state of the two-component Fermi gas. We shall consider for the moment the normal state.

The equilibrium density profiles are obtained in the local-density approximation by the solutions of the coupled equations for the densities of each component $n_\sigma(\mathrm{r})$:

$$n_\sigma(r) = \theta[\varepsilon_\sigma - V_\sigma(\mathrm{r}) - f n_{\bar\sigma}(\mathrm{r})]\{A^{-1}[\varepsilon_\sigma - V_\sigma(\mathrm{r}) - f n_{\bar\sigma}(\mathrm{r})]\}^{3/2}, \qquad (9)$$

where $\bar\sigma$ denotes the component different from σ. In Eq. (9) $V_\sigma(\mathrm{r})$ are the static confining potentials, $f = 4\pi\hbar^2 a_{\uparrow\downarrow}/m$ the inter-component interaction strength related to the scattering length $a_{\uparrow\downarrow}$ and ε_σ are the chemical potentials, to be determined from the condition $N_\sigma = \int d^3r n_\sigma(\mathrm{r})$. We are working under the hypothesis that the N_σ 's are fixed, *i.e.* these equations do not allow for redistributions of population in the two hyperfine states.

The strength of the interactions in a symmetric system $(N_\uparrow = N_\downarrow)$ can be measured by the parameter $k_F^0 a_{\uparrow\downarrow}$, where k_F^0 is the non-interacting Fermi wavevector. For harmonic trapping we have $k_F^0 a_{\uparrow\downarrow} \cong 1.2 N^{1/6} a_{\uparrow\downarrow}/a_{ho}$, where N is the total number of fermions and $a_{ho} = (\hbar/m\omega_f)^{1/2}$ is the harmonic oscillator length.

The current experiments on ^{40}K are performed in the weak-coupling regime. In this case the density profiles are very close to those of a non-interacting gas. The profiles can be approximately described by the analytic Ansatz

$$n(\mathrm{r}) = \left(8N/\pi^2 R_F^3\right)\left(1 - r^2/R_F^2\right)^{3/2} \theta\left(1 - r^2/R_F^2\right), \tag{10}$$

where R_F is still the true Fermi radius in the interacting mixture, related to the Fermi energy as $\varepsilon_F = m\omega_f^2 R_F^2/2$. The form (10) is adjusted to preserve normalization to N as well as the value of R_F. This Ansatz will allow us to treat analytically the dynamical properties of density fluctuations in the gas.

By increasing the value of the coupling strength in the case of repulsive interactions while preserving the spherical symmetry, a spontaneous symmetry breaking occurs towards a situation where the two components are spatially separated along the radial direction. This phase separation is driven by the competition between kinetic energy, which disfavours localization and repulsive interaction energy, which favours the spatial separation of the two components. At high enough values of the coupling strength the gas diminishes its total energy through phase separation (see Fig. 1). From the numerical solutions of Eq. (9) we have found that the onset of the transition occurs at a value of the coupling strength $N^{1/6} a_{\uparrow\downarrow}/a_{ho} \cong 1$ Although this is far from the present experimental parameters, one may think of exploiting Feshbach resonances to tune the value of $a_{\uparrow\downarrow}$ towards such a regime. The transition described above is fully analogous to a ferromagnetic transition[7] and belongs to the category of "quantum phase transitions" since it is driven by the interactions and occurs at zero temperature[18].

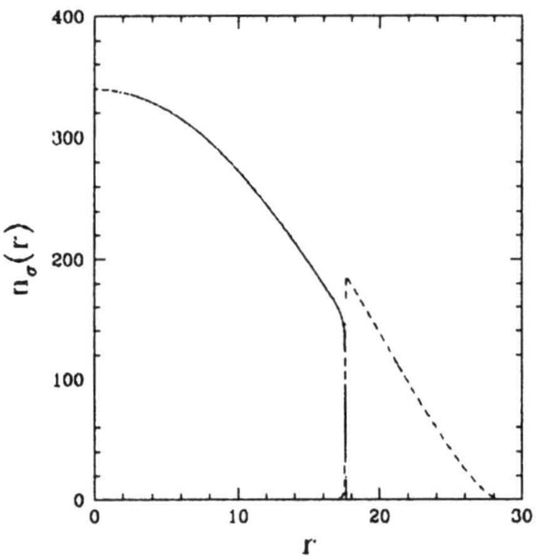

Figure 1. Equilibrium density profiles (in units of a_{ho}^{-3}) of the two components of a symmetric Fermi vapour, in the phase-separated regime for a value of the coupling strength $Na_{\uparrow\downarrow}/a_{ho} = 1.8$.

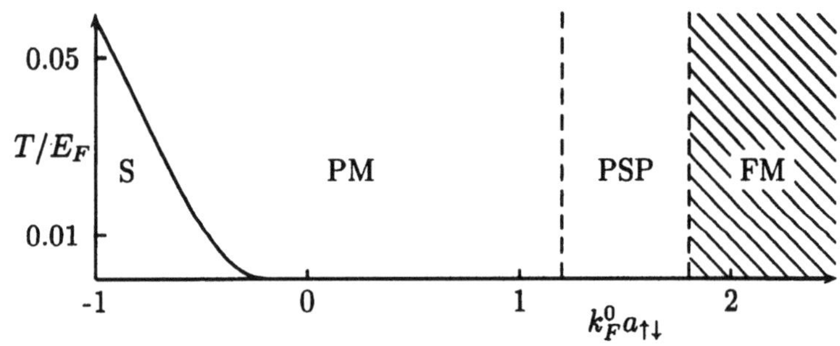

Figure 2. A sketch of the expected phase diagram for asymmetric two-component Fermi gas. A magnetic analogy is used to label the regions of different overlap of the two components: complete (paramagnetic, PM), partial (partially spin-polarized, PSP) and zero (ferromagnetic, FM) overlap. S indicates the superfluid region.

For negative values of the coupling strength the system is instead expected to undergo a pairing transition towards a superfluid state[19]. In the weak-coupling regime the proposed pairing mechanism is an *s*-wave BCS-like pairing between fermions belonging to different hyperfine levels[20]. This yields a critical temperature $T_{BCS} = 0.28\varepsilon_F \exp\left(-\pi/\left(2k_F^0 |a_{\uparrow\downarrow}|\right)\right)$. Other pairing mechanisms have also been investigated[21,22].

In summary, a rich phase diagram is predicted for asymmetric two-component Fermi gas in a magnetic trap (see Fig. 2).

4.2 Collective Modes

The collective modes of a two-component, harmonically trapped Fermi gas in the collisionless regime have been investigated by Vichi and Stringari[23]. The frequencies of the modes of the cloud are slightly shifted with respect to the multiples of the bare trap frequency under the effect of the interactions.

We focus here on the collisional regime for a symmetric vapour (*i. e.* $N_\uparrow = N_\downarrow$ and $\omega_{f\uparrow} = \omega_{f\downarrow}$) at weak interaction strength. In this case we are able to obtain an analytic solution for the modes.

The equations of motion for the density fluctuations of each component are given by a generalization of Eq. (5) and (6) to include the effect of mean field interactions as

$$m\frac{\partial^2 n_\sigma(\mathrm{r})}{\partial t^2} = \nabla^2\left[\frac{2}{5}An_\sigma^{5/3}(\mathrm{r},t)\right] + \nabla\cdot\{n_\sigma(\mathrm{r},t)\nabla[V_\sigma(\mathrm{r}) + fn_{\bar{\sigma}}(\mathrm{r},t)]\} \quad (11)$$

In the case of a symmetric mixture Eq. (11) lead to separate equations of motion for the total density fluctuations $n(\mathrm{r},t) = n_\uparrow(\mathrm{r},t) + n_\downarrow(\mathrm{r},t)$ and for the concentration fluctuations $M(\mathrm{r},t) = n_\uparrow(\mathrm{r},t) - n_\downarrow(\mathrm{r},t)$. These equations can be solved analytically in the small-oscillation regime with the same technique adopted for the one-component gas, if we make use of the analytical Ansatz (10) for the equilibrium profiles. This allows us to take into account the interactions through the ratio of the non-interacting to the interacting Fermi energy $C = (\varepsilon_F^0/\varepsilon_F)^2$.

In the case of spherical harmonic confinement, the dispersion relation for the total density fluctuations is

$$(\omega_{nl}/\omega_f)^2 = l + 2n + n(3 - C)(2n + 2l + 1)/3 \quad (12)$$

while that for the concentration fluctuations is

$$(\omega_{nl}/\omega_f)^2 = (l + 2n)(2C - 1) + n(5C - 3)(2n + 2l + 1)/3 \quad (13)$$

For the total density fluctuations all the surface modes ($n = 0$) are not affected by the interactions, and the $l = 1$ result agrees with the Kohn theorem.

It is interesting to observe that the dispersion relation in Eq. (12) for the total density fluctuations, in the limit $C \cong 1$, has also been found by Baranov and Petrov for the low-lying excitations of a trapped BCS-paired superfluid in the collisionless regime[5]. This can be understood as follows. In a superfluid the continuum of quasi-particle and quasi-hole excitations is lifted by the opening of the gap: there are no such excitations below an energy of twice the gap Δ. However, since the compressibility is not affected by the phase transition[24], first sound can propagate without damping in the window below 2Δ. Baranov and Petrov have generalized this result to the case of a trapped system.

In a superfluid gas this mode is called the Bogolubov-Anderson mode[25,26]: its existence is typical of a condensed neutral system and is predicted by the Goldstone theorem[27].

CONCLUSIONS

In summary, in this Chapter we have given a brief overview on the dynamical properties of a trapped Fermi gas, both in the collisionless and in the collisional regime. While in a one-component system interactions are negligible (and nevertheless the system presents a rich spectrum), in the two-component case they can influence in a relevant way the equilibrium state and the dispersion relation. We have described mainly the case of a normal fluid, but we have pointed out that the long-wavelength behavior of a superfluid can be described in terms of the properties of the normal fluid in the collisional regime.

ACKNOWLEDGMENTS

It is a pleasure to acknowledge discussions with M. Baranov, G. Ferrari, A. Griffin and R. Hulet.

REFERENCES

1. DeMarco, B., and Jin, D. S., 1999, Onset of Fermi Degeneracy in a Trapped Atomic Gas, *Science* **285**: 1703.

2. Mewes, M. O., Ferrari, G., Schreck, F., Sinatra, A., and Salomon, C., 2000, Simultaneous Magneto-Optical Trapping of Two Lithium Isotopes, *Phys. Rev. A* **61**: 011403R.
3. Holland, M., DeMarco, B., and Jin, D., 1999, Evaporative Cooling of a Two-Component Degenerate Fermi Gas, *cond-mat/9911017.*
4. Bruun, G., and Clark, C. W., 1999, Detection of the BCS transition of a trapped Fermi Gas, *cond-mat/9906392.*
5. Baranov, M. A., and Petrov, D. S., 1999, Low energy collective excitations in a superfluid trapped Fermi gas, *cond-mat/9901108.*
6. Gavoret, J., and Nozières, P., 1964, Structure of the Perturbation Expansion for the Bose Liquid at Zero Temperature, *Ann. Phys. (N.Y.)* **28**: 349.
7. Pines, D., and Nozières, P., 1966, The Theory of Quantum Liquids, Benjamin, New York, vol. 1.
8. Molmer, K., 1998, Bose Condensates and Fermi Gases at Zero Temperature *Phys. Rev. Lett.* **80**: 1804.
9. Schneider, J., and Walks, H., 1998, Mesoscopic Fermi gas in a harmonic trap *Phys. Rev. A* **57**: 1253.
10. Bijlsma, M., and Stoof, H. T. C., 1999, Collisionless modes of a trapped Bose gas, *Phys. Rev. A* **60**: 3973.
11. Amoruso, M., Meccoli, I., Minguzzi, A., and Tosi, M., 1999, Collective excitations of a trapped degenerate Fermi gas, *Eur. Phys. J. D* **7**: 441.
12. Singwi, K. S.,and Tosi, M. P., 1976, Correlations in Electron Liquids *Solid State Phys.* **36**: 177.
13. Bender, C. M., and Orszag, S. A., 1978, Advanced Mathematical Methods for Scientists and Engineers, McGraw-Hill, New York.
14. Dobson, J. F., 1994, Harmonic-Potential Theorem: Implications for Approximate Many-Body Theories, *Phys. Rev. Lett.* **73**: 7244.
15. Stringari, S., 1996, Collective Excitations of a Trapped Bose-Condensed Gas, *Phys. Rev. Lett.* **77**: 2360.
16. DeMarco, B., Bohm, J. L., Burke, Jr., J. P., Holland, M., and Jin, D. S., 1999, Measurement of p-Wave Threshold Law Using Evaporatively Cooled Fermionic Atoms, *Phys. Rev. Lett.* **82**: 4208.
17. Amoruso, M., Meccoli, I., Minguzzi, A., and Tosi, M., 2000, Density profiles and collective excitations of a trapped two-component Fermi vapour, *Eur. Phys. J. D* **8**: 361.
18. Sondhi, S. L., Girvin, S. M., Carini, J. P., and Shahar, D., 1997, Continuos quantum phase transitions, *Rev. Mod. Phys.* **69**: 315.
19. Stoof, H. T. C., Houbiers, M., Sackett, C. A., and Hulet, R. G., 1996, Superfluidity of spin-polarized ^{6}Li, *Phys. Rev. Lett.* **76**: 10.
20. Houbiers, M., Ferweda, R., Stoof. H. T. C., McAlexander, W. I., Sackett, C. A., and Hulet, R. G., 1997, Superfluid state of atomic ^{6}Li in a magnetic trap, *Phys. Rev. A* **56**: 4864.
21. Baranov, M. A., Kagan, Y., and Kagan, M. Y., 1996, On the possibility of a superfluid transition in a Fermi gas of neutral particles at ultralow temperatures, *JETP Lett.* **64**: 301.
22. Combescot, R., 1999, Trapped ^{6}Li: A High Tc Superfluid?, *Phys. Rev. Lett.* **83**: 3766.
23. Vichi, L., and Stringari, S., 1999, Collective oscillations of an interacting trapped Fermi gas, cond-mat/9905154.
24. Leggett, A. J., 1965, Theory of a Superfluid Fermi Liquid I. General Formalism and Static Properties, *Phys. Rev.* **140**: 1869.

25. Anderson, P. W., 1958, Random Phase Approximation in the Theory of Superconductivity *Phys. Rev.* **112**: 1900.
26. Bogolubov, N. N., Tolmachev, V. V., and Shirkov, D. V., 1959, A new method in the theory of superconductivity, Consultants Bureau, New York.
27. Lange, R. V., 1966, Nonrelativistic Theorem Analogous to the Goldstone Theorem, *Phys. Rev.* **146**: 301.

Photoassociative Spectroscopy Of Cs_2

C. DRAG, B. LABURTHE TOLRA, D. COMPARAT, A. FIORETTI, A. CRUBELLIER, O. DULIEU, F. MASNOU-SEEUWS, S. GUIBAL, AND P. PILLET
Laboratoire Aimé Cotton, CNRS, bât. 505, Campus d'Orsay, 91405 Orsay, France

1. INTRODUCTION

During the last fifteen years, the laser manipulation of neutral atoms has known impressive developments. The experimental techniques of laser cooling of atoms in the mK-μK range and below, as well as the trapping of neutral atomic samples, based on radiative forces, are now well established. Their extension to molecules is however very difficult because of the lack of two-level optical pumping scheme for recycling population[1]. During the last years, molecules have hardly been concerned by the impressive developments in laser cooling. One can quote the deflection of a molecular beam[2] or the demonstration by Djeu and Whitney[3] of laser cooling by spontaneous anti-Stokes scattering, introduced long ago by Kastler[4] as "luminorefrigeration". The latter method presents however poor efficiency and poor control. An interesting specific scheme for the formation of cold molecules is to form cold molecules by molecular photoassociation (PA) of two cold atoms[5]. PA has been demonstrated for alkali atoms[6-10] from Li to Cs, and more recently for hydrogen atom[11]. In this process, two free cold atoms absorb resonantly one photon and produce an excited molecule in a well-defined ro-vibrational state. The excited photo-associated molecules are translationally cold. De-excitation of the photo-associated molecules appears thus as an obvious way to form cold ground state molecules. The cesium dimer presents four attractive long-range Hund's case (c) states below the $6S_{1/2}+6P_{3/2}$ limit (see Fig. 1): 1_g, 0_u^+, 0_g^-, and 1_u (transition from the ground

Bose-Einstein Condensates and Atom Lasers
Edited by Martellucci *et al.*, Kluwer Academic/Plenum Publishers, 2000

state to the attractive 2_u states is electric dipole forbidden), which can be populated by photoassociation. On the contrary to the states 1_g and 0_u^+, which correspond to pretty deep molecular potential wells very repulsive at short inter-atomic distances, we notice that the 0_g^- and 1_u states present a double-well shape for the molecular potential curves. The ro-vibrational levels inside the outer long-range well can be populated by PA. An interesting property is the resonant character of this process and its consequence for application to high resolution molecular spectroscopy.

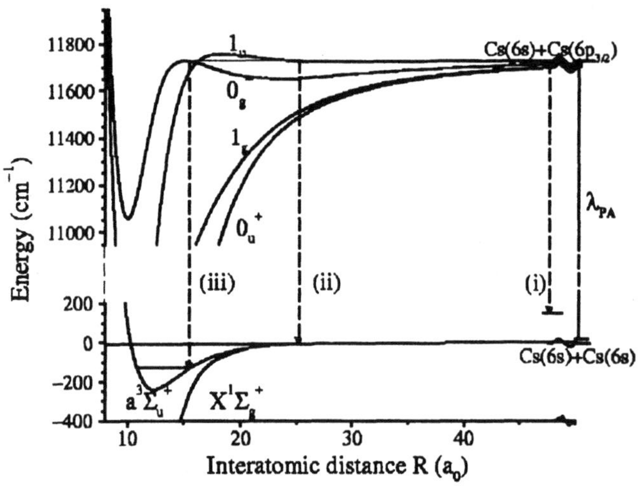

Figure 1. Relevant potential curves of the Cs2 molecule. Line (i) represents spontaneous emission towards continuum states, with dissociation of the molecule; lines (ii) and (iii) represent spontaneous emission towards bound states at the Condon points, with formation of stable ground state cold molecules.

We will report here the experiments concerning on the molecular PA of cold cesium atoms, which offer an ideal system for the formation of cold ground state Cs_2 molecules. The paper is organised as follows. Sect. 2 is devoted to the description of the experiments. The obtained PA spectra are then described and analysed in Sect. 3. The analysis of the intensities of the resonance lines gives access to collision parameters such as scattering lengths necessary to predict the stability of a Bose-Einstein condensate. In Sect. 4, we report the evidence for the formation of ground-state cold molecules. We describe the measurement of the temperature of the cold molecules and we give an estimate for the formation rate. Finally, in Sect. 5 we conclude and we analyse the perspectives in the research field of cold molecules.

2. EXPERIMENTAL SETUP

The principle of the experiment is described in several references[10,12-15]. The cold atom source is provided by the use of a Cs vapour loaded magneto-optical trap (MOT). The trapped, cold Cs atoms are illuminated with a cw laser to produce the photoassociative transitions; PA is achieved by continuously illuminating the cold Cs atoms with the beam of a Ti:Sapphire laser pumped by an Argon ion laser. The maximum available power in the experiment zone is 600 mW, focused on a spot with 400 μm diameter. The power of the PA laser is gradually reduced for detuning close to of the $6s_{1/2}, F=4 \rightarrow 6p_{3/2}, F'=5$ atomic transition in order to avoid perturbations of the MOT operation. The frequency scale is calibrated using a Fabry-Perot interferometer, and the absorption lines of iodine. Two kinds of detection are used for observing the photoassociation process. First one observes with a photodiode the fluorescence yield from the trap, which allows one to analyse the trap losses. Second, in the case of formation of cold ground-state molecules, we used photo-ionisation of the translationally cold Cs_2 molecules in Cs_2^+ ions, which are detected through a time of flight mass spectrometer.

3. PHOTOASSOCIATIVE SPECTRA DETERMINATION OF TRIPLET SCATTERING LENGTH

The fluorescence and the Cs_2^+ ion spectra are recorded as a function of the PA laser frequency. A summary of our data is shown in Fig. 2 with typical spectra obtained by using a MOT atomic sample. The origin of the energy scale is fixed at the $6s_{1/2}, F=4 \rightarrow 6p_{3/2}, F'=5$ atomic transition. For detuning smaller than 0.1 cm^{-1}, the MOT is destroyed by the PA laser. The fluorescence and the Cs_2^+ ion spectra are very different. Clearly first we observe resonance lines up to a PA laser detuning of 80 cm^{-1} in the case of the Cs_2^+ ion spectrum and only of 40 cm^{-1} for the trap-loss one. Second the density of resonance lines in the trap-loss spectrum is much more important. We will come back on the reasons of these differences in Sect. 4 of this article. The Cs_2^+ ion spectrum exhibits the vibrational progressions of the 0_g^- and 1_u states which are studied in detail in previous papers[12,14]. In this section, we will focus on the spectroscopic analysis of the 0_g^- state. The Cs_2^+ ion spectrum exhibits 133 well resolved structures assigned as the vibrational progression of the 0_g^- state, starting at $v = 0$. The rotational structure, shown for $v = 10$ in the inset of the Fig. 2, is resolved up to $J = 8$

for most of the vibrational levels[i] below $v = 74$. The energies of the spectral lines have been fitted with a Rydberg-Klein-Rees (RKR) and near dissociation expansion (NDE) approach[12] giving, for the outer well, an effective potential curve with a 77.94±0.01 cm^{-1} depth and an equilibrium distance R_e=(23.36±0.10)a_0. This approach provides a good knowledge for the vibrational wave-functions and for the inner and outer turning points of the classical vibrational motion up to $v = 74$. We remark the modulation of the line intensities, which is due to the variations of the Franck-Condon factors of the transitions between the initial state and the final ro-vibrational levels of the 0_g^- state. It has been noticed (see for instance[16]) that this modulation can be used for the determination of collisional parameters such as scattering lengths. The existence of minima in the spectrum reflects the nodes of the radial wavefunction[10,17,18] $\psi(R)$. Considering the case of a s-wave wavefunction, the asymptotic behaviour of $\psi(R)$ varies as: $sin(k(R-a(k)))$ where k is the modulus of the wave-vector associated to the relative motion of the two atoms and $ka(k)$ is a phase-shift. The collision parameter, a, so-called scattering length is defined as the limit of $a(k)$ at zero temperature ($k = 0$). The analysis of the intensity modulation in the PA spectra to determine the scattering lengths has been used for several alkalis[16,19,20]. To determine the scattering length of the molecular triplet state of the cesium dimer, we have taken again the Cs_2^+ ion spectrum by considering doubly polarised atoms prepared in Zeeman sublevel $F = 4$, $m_F = +4$. The colliding atoms in such a state are only coupled with the molecular Cs_2 triplet ground state. The study of the variation of the maximum intensities of the resonance lines leads to the possibility to determine the triplet scattering length, a_T.

Nevertheless the value of the C_6 parameter is not known very accurately. This is linked to the uncertainty of the Cs polarizability measured experimentally, which is introduced in the model potential leading to a dispersion of 10% for the calculated[21-23] C_6. The difficulty for an accurate determination of the scattering length is linked to is its large value, which can be widely modified for a small uncertainty in the molecular parameter, C_6. The precise analysis in progress of the ensemble of the minima leads to the determination of both scattering length and molecular parameter, presently evaluated around $a \approx -750\ a_0$ and $C_6 \approx 6600$ a.u..

[i] The large number of observed rotational levels is due to a co-operative effect between the cooling laser and the PA laser[15]. In fact if the cooling is switched off during the PA phase, at a temperature, $T \leq 30\mu K$, only s-wave has to be considered in the experiment with the excitation of only $J = 0$ an $J = 2$ rotational levels, at $T \cong 200\mu K$ s-, p- and d-waves are essentially present and the excitation of rotational state up to $J = 4$ is possible.

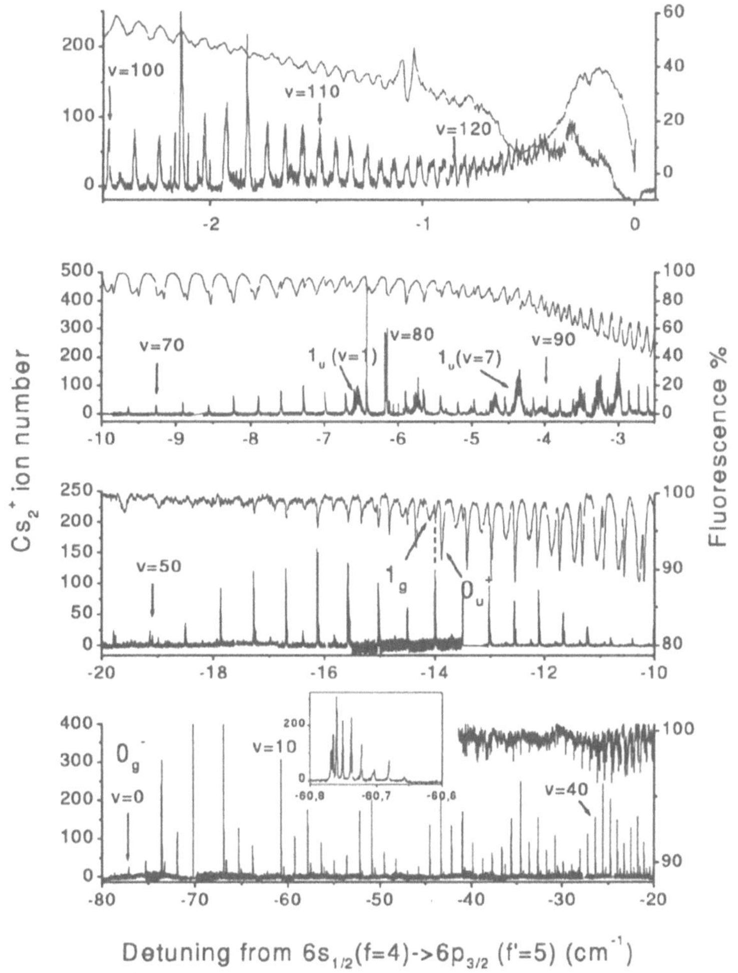

Figure 2. Cs_2^+ ion signal (lower signal) and trap fluorescence yield (upper signal) versus detuning of the PA laser. In the inset, details of the rotational structure of the 0_g^-, $v = 10$ is shown. The dashed lines indicates the correspondence of vibrational levels of the 0_g^- on both spectra; 1_g and 0_u^+ states are only observed on fluorescence spectrum. Notice the different scales for the axis. The spectra are the result of a large number of scans.

4. COLD MOLECULES

The trap loss spectrum is very different from the Cs_2^+ ion spectrum. We observe the expected vibrational progressions in the trap loss spectrum, as it is shown in the energy range 12-16 cm^{-1} of the Fig. 2. Three vibrational progressions assigned to the long-range 1_g, 0_u^+ and 0_g^- states are visible. About 80 lines for each vibrational are well resolved in the range 2-40 cm^{-1} of the Fig. 2. It is less clear for the 1_u state, which does not seem to be present in the fluorescence spectrum. This is probably due to the fact that we have to consider low vibrational levels, leading after spontaneous emission to a pair of atoms with too small relative kinetic energy to escape outside of the atomic trap. Only the vibrational progressions of the 0_g^- and 1_u states are present in the Cs_2^+ ion spectrum The detection is here sensitive up to a detuning range of 80 cm^{-1} for the PA laser. To understand this difference we analyse the Cs_2^+ ion signal by considering the following temporal sequence. We apply now the PA laser beam during a duration of 15 ms and we delay the ionising laser pulse (7 ns width) compared to the switch on of the PA laser. We observe that the Cs_2^+ ion signal decreases with a characteristic time of the order of 10 ms. This time is five orders of magnitude larger than the radiative lifetime of any singly excited molecular state with electric-dipole allowed transition to the ground state. Indeed, it is of the order of the time during which molecules can move significantly out of the trap because of the gravity. This result clearly indicates that ions are not produced by direct photo-ionisation of PA excited molecules, but by photo-ionisation of the ground-state molecules. We have here the proof for the formation of cold ground-state Cs_2 molecules.

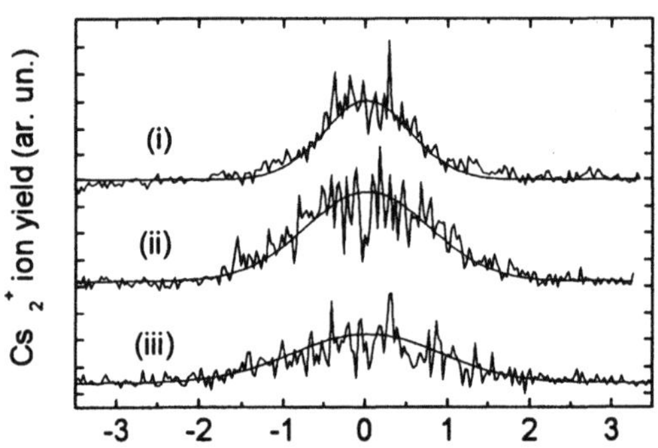

Figure 3. Ballistic expansion of the falling molecular cloud. The ground state cold Cs_2 molecules are photo-ionised in Cs_2^+ ions, using the dye laser focused on a spot of 300 μm diameter. The horizontal scale represents the horizontal position of the photo-ionising beam and the vertical scale the corresponding ion yield. This spatial analysis of the Cs_2^+ ion signal is performed at the MOT position (i) and 1.90 mm (ii) and 3.80 mm (iii) below the MOT position. The cold atomic sample is photo-associated during a time of 3 ms.

These cold ground-state Cs_2 molecules are not trapped by the MOT and can be detected below the trap zone[10,13]. We show on Fig. 3 the ballistic expansion of the falling molecular cloud allowing a measure of the temperature of the molecular cloud. The theoretical fit of the experimental data gives access to the determination of a temperature as low as 20^{+15}_{-5} μK.

The efficiency of the mechanism for the formation of cold molecules comes from the existence of a Condon point at intermediate distance, corresponding to a long-range molecular well. PA happens at long-range distance. If spontaneous emission occurs at a short enough inter-atomic distance (Fig. 1, cases ii or iii), cold ground state molecules can be formed, while spontaneous emission at a long-range distance (case i) leads always to dissociation of the excited molecules. In the case of the 1_g or 0_u^+ states, the vibration of the excited molecules always keeps mostly the two atoms at a too large inter-atomic distance to get the formation of cold molecules after spontaneous decay. In the case 0_g^- or 1_u, the molecule oscillates between long-range and intermediate distances and the formation of stable cold molecules is possible. The formation of translationally cold molecules in the latter cases is due to the particular shape of the external potential wells which offers at the same time an efficient photoassociation rate and a reasonable branching ratio of spontaneous emission towards the ground state. From the measured number of Cs_2^+ ions detected at a detuning corresponding to a given ro-vibrational level, it is possible to estimate the corresponding number of cold molecules produced in the trap. Typically up to 15000 cold molecules per shot are formed in the trap zone. The characteristic time of the stay of cold molecules in the trap zone is 10 ms. One can thus infer a rate of cold molecules formation of about one million per second.

5. CONCLUSION

In conclusion, we have reported PA spectroscopy of Cs_2 for the long-range 0_g^- and 1_u states below the $6s_{1/2}+6p_{3/2}$ dissociation limit. The long-range 0_g^- and 1_u potentials present a Condon point at intermediate distance provided by their double-well shape, which is responsible of the existence of a rather efficient channel in spontaneous emission for the creation of ground state molecules. For the other alkalis, the 0_g^- and 1_u outer wells do not offer such channels to the formation of cold molecules; the situation seems more favourable to the formation of cold molecules for Rubidium, which could present a Condon point at intermediate distance in the case of the 0_g^- state. The use of polarised atoms offers new ways in the PA experiments, in particular for the determination of scattering lengths or Feshbach resonances, which will be very helpful in the future developments of Bose-Einstein condensation of cesium, which is a difficult challenge because of the large negative triplet scattering length and the large cross section of the inelastic collision[27]. PA of cold atoms for formation of stable cold molecules opens a very promising novel field of investigation. Several difficulties should be resolved . The formed molecules are indeed cold translationally[15] (and also rotationally). To get them vibrationally cold is an interesting challenge. Stimulated Raman photoassociation is probably a good way to obtain all the cold molecules in a well-defined ro-vibrational level. Further Raman transition would allow to bring them in the lowest energy level, $v = 0, J = 0$ of the considered molecular ground state. In order to develop applications of cold molecules, it can be necessary to be able to store them. Dipolar or magnetic traps should be settled for this purpose[24,25]. The trapping of translationally cold Cs_2 molecules has already been performed with the use of a CO_2 laser[25]. Starting from colder and denser atomic samples, colder and denser molecular clouds will certainly be obtained in the near future and this could be a way towards Bose-Einstein condensation of molecules. Future developments in the cold molecule field should be similar to those performed in the cold atom research field: molecule optics, molecule interferometry, non-linear optics, high precision spectroscopy, metrology, nano-lithography.... Laser manipulation of molecules has already reached a promising development in molecule optics and molecule interferometry[26].

ACKNOWLEDGMENTS

The authors are grateful to Claude Amiot for stimulating discussions.

REFERENCES

1. Bahns, J.T., Stwalley, W.C., and Gould, P.L., 1996, *J. Chem. Phys.* **104**: 9689.
2. Hermann, A., Leutwyler, S., Wöste, L., and Schumacher, E., 1979, *Chem. Phys. Lett.* **62**: 444.
3. Djeu, N., and Whitney, W.T., 1981, *Phys. Rev. Lett.* **46**: 236.
4. Kastler, A.,1950, *J. Phys. Radium* **11**: 255.
5. Thorsheim, H.R. Weiner, J., and Julienne, P.S., 1987, *Phys. Rev. Lett.* **58**: 2420.
6. Lett, P.D., Helmerson, K., Phillips, W.D., Ratliff, L.P., Rolston, S.L., and Wagshul, M.E., 1993, *Phys. Rev. Lett.* **71**: 2200.
7. Miller, J.D., Cline, R.A, and Heinzen, D.J., 1993, *Phys. Rev. Lett.* **71**: 2204.
8. Abraham, E.R.I., Ritchie, N.W.M., McAlexander, W.I., and Hulet, R.G., 1995, *J. Chem. Phys.* 103, 7773.
9. Wang, H., Gould, P.L., and Stwalley, W.C., 1996, *Phys. Rev. A* **53**: R1216.
10. Fioretti, A., Comparat, D., Crubellier, A., Dulieu, O., F.Masnou-Seeuws, and Pillet, P., 1998, *Phys. Rev. Lett.* **80**: 4402.

11. Mosk, A.P., Reynolds, M.W., Hijmans, T.W., and Walraven, J.T.M., 1999, *Phys. Rev. Lett.* **82**: 307.
12. Fioretti, A., Comparat, D., Drag, C., Amiot, C., Dulieu, O., F. Masnou-Seeuws, and Pillet, P., 1999, *Eur. Phys. J. D* **5**: 389.
13. Comparat, D., Drag, C., Fioretti, A., Dulieu, O., and Pillet, P., 1999, *Journal of Molecular Spectroscopy,* **195**: 229.
14. Comparat, D., Drag, C., Laburthe Tolra, B., Fioretti, A., Pillet, P., Crubellier, A., Dulieu, O., and Masnou-Seeuws, F., 2000, accepted for publication to *Eur. Phys. J. D.*
15. Fioretti, A., Comparat, D., Drag, C., Gallagher, T.F., and Pillet, P., 1999, *Phys. Rev. Lett.* **82**: 1839.
16. Tiesinga, E., Williams, C.J., Julienne, P.S., and Jones, K.M., Lett, P.D., Phillips, W.D., 1996, *J.Res. Natl. Inst. Stand. Technol.* **101**: 505.
17. Pillet, P., Crubellier, A., Bleton, A., Dulieu, O., Nosbaum, P., I. Mourachko, and Masnou-Seeuws, F., 1997, *J. Phys. B* **30**: 2801.
18. Julienne, and P.S., Res. J., 1996, *Natl. Inst. Stand.Technol.* **101**: 487.
19. Abraham, E.R.I., Alexander, W.I., Gerton, J.M., and Hulet, R.G., 1997, *Phys Rev. A* **55**: R3299.
20. Boesten, H.M.J.M., Tsai, C.C., Gardner, J. R., Heinzen, D.J., van Abeelen, F.A., and Verhaar, B.J., 1996, *Phys Rev. Lett.* **77**: 5194.
21. Marinescu, M., Sadeghpour, H.R., and Dalgarno, A., 1994, *Phys. Rev. A* **49**: 982.
22. Patil, S.H.,and Tang, K.T., 1997, *J. Chem. Phys.* **106**: 2298.
23. Maeder, F., and Kutzelnigg, W., 1979, *Chem. Phys.* **42**: 95.
24. Weinstein, J.D., de Carvalho, R., Martin, A., Friedrich, B., and Doyle, J.M., 1998, *Nature* **395**: 148.
25. Takekoshi, Patterson, B.M., and Knize, R.J., 1998, *Phys. Rev. Lett.* **81**: 5105.
26. Bordé, C.J., Courtier, N., du Burck, L., Goncharov, A.N., and Gorlicki, M., 1994, *Phys. Lett. A* **188**: 187.
27. Arndt, M., Ben Dahan, M., Guéry-Odelin, D., Reynolds, M.W., and Dalibard, J., 1997, *Phys. Rev. Lett.* **79**: 625.

Index

Adiabatic elimination; 187; 190; 191; 199; 220
Atom amplification; 14; 20; 24
Atom heterodyne; 238; 239
Atom interferometer; 14; 24; 71; 74; 103; 106; 231; 232; 233; 235; 243
Atom interferometry; 15; 71; 97; 231; 236; 243
Atom laser; 2; 4; 13; 14; 15; 20; 24; 26; 60; 61; 63; 64; 74; 97; 98; 103; 106; 107; 117; 118; 119; 120; 121; 122; 123; 125; 126; 127; 129; 137; 138; 139; 141; 142; 144; 145; 146; 149; 153; 162; 185; 187; 189; 192; 201; 238; 239; 275
Atom optics; 4; 15; 26; 55; 56; 65; 68; 74; 97; 98; 113; 122; 127; 150; 153; 161; 162; 163; 165
BEC; 1; 2; 3; 4; 12; 26; 27; 31; 32; 34; 43; 47; 48; 49; 55; 56; 57; 58; 59; 60; 61; 62; 63; 66; 68; 69; 70; 71; 72; 73; 77; 78; 97; 98; 99; 103, 104; 105; 106; 107; 108; 110; 111; 129; 132; 1333; 153; 178; 189; 190; 191; 192; 194; 196; 197; 198; 201; 202; 204; 228; 229; 238; 239; 250; 257; 259; 275; 276; 285; 286; 287; 288; 289; 295; 296
Born parameter; 83; 84
Bose condensate; 4; 5; 15; 42; 53; 84; 86; 87; 88; 89; 90; 91; 92; 95; 129; 137; 139; 142; 150; 163; 224; 252; 257; 300; 311
Bose-Einstein condensate; 1; 2; 3; 4; 6; 7; 9; 11; 14; 15; 16; 17; 20; 31; 41; 42; 43; 47; 53; 54; 56; 63; 64; 65; 66; 68; 74; 75; 79; 83; 94; 95; 97; 98; 100; 103; 104; 106; 107; 108; 113; 117; 118; 119; 120; 123; 129; 139; 141; 153; 162; 165; 182; 185; 187; 213; 225; 229; 249; 253; 275; 296; 299; 314
Bose-Einstein condensation; 1; 2; 3; 15; 25; 27; 28; 31; 41; 42; 63; 64; 65; 73; 74; 77; 94; 97; 117; 129; 139; 171; 185; 201; 213; 230; 275; 282; 285; 289; 299; 300; 320
Boson Josephson junction; 250; 259
Boson Josephson junction equations; 253; 255; 256
Bragg diffraction; 22; 57; 58; 59; 60; 64; 66; 70; 71; 74; 109; 275
Bragg spectroscopy; 16; 17; 63; 64; 94
Chemical potential; 19; 37; 38; 78; 83; 111; 142; 144; 182; 202; 241; 250; 251; 257; 262; 307
Coherence; 4; 14; 15; 18; 22; 34; 35; 40; 42; 55; 57; 60; 62; 64; 68; 78; 82; 83; 93; 100; 102; 103; 127; 138; 139; 162; 186; 198; 231; 238; 239; 243; 244; 249; 275; 285; 293
Coherence length; 83

Coherent condensate oscillations; 250; 252
Coherent matter waves; 2; 66; 97; 103; 106; 129; 153
Cold collision; 213; 214; 220; 221; 222; 223; 282; 300
Cold molecules; 313; 314; 317; 319; 320
Collapses; 38; 42; 226; 227; 251; 260; 261
Collective excitations; 4; ; 5; 138; 139; 300; 302; 311
Collisional regime; 302; 305; 309; 310
Collisionless regime; 302; 303; 309; 310
Condensate phase; 71
Correlation length; 110
Coupled Gross-Pitaevskii equation; 141; 142; 149; 183; 184; 214
Critical velocity; 4; 12; 13; 258; 285
Diffraction; 14; 22; 57; 58; 59; 60; 62; 64; 66; 68; 70; 71; 74; 109; 133; 275
Dipole trap; 106; 276; 298
Dissipative quantum mechanics; 266
Dissipative system; 274
Dissipative terms; 189; 192; 194
Domains; 11; 32; 41; 188
Doppler broadening; 16; 18
Double condensate; 32; 33; 34; 35; 36; 39; 40; 44; 45
Dynamic structure factor; 16; 17; 78; 79; 82; 83; 84; 85; 88; 89; 90; 91; 92; 93
Eikonal expansion; 83; 84; 85; 90; 91
Electromagnets; 165
Elliptic Jacobian functions; 253
Evaporative cooling; 2; 3; 42; 117; 121; 130; 131; 132; 165; 171; 172; 174; 175; 179; 186; 287; 288; 291; 292; 294; 295; 296; 299; 311
External JE; 251
Fabry-Perot interferometer; 153; 157; 162; 163; 315
Fermion; 163; 291; 292; 300; 301; 302; 303; 306; 307; 308
First sound; 302; 303; 306; 310
Four-wave mixing; 14; 15; 24; 66; 67; 74; 153; 162; 163
Fuchs method; 305
Generalized Kohn theorem; 306
Ginzburg Landau equation; 187
Gravity gradiometer; 232; 236; 237
Gross-Pitaevskii equation; 35; 52; 69; 86; 101; 112; 113; 133; 141; 142; 149; 183; 184; 187; 214; 225; 252
Gyroscope; 24; 232; 234; 235; 237
Healing length; 5; 48; 70; 73; 83; 86
Heisenberg limit; 243; 244; 245
Interference; 14; 17; 19; 22; 24; 31; 32; 34; 35; 36; 37; 38; 39; 40; 41; 42; 44; 63; 74; 85; 90; 91; 92; 94; 100; 101; 102; 108; 109; 110; 113; 121; 122; 127; 138; 139; 145; 148; 159; 173; 221; 222; 231; 232; 234; 235; 236; 239; 240; 242; 243; 244; 293
Internal JE; 251
Iron core; 165; 167; 184
Isotropic trap; 46; 50
Josephson critical current; 251
Josephson effect; 32; 74; 239; 241; 246; 249; 250; 251; 257; 258; 263; 274
Josephson oscillation; 14; 254; 255
Laser cooling; 2; 3; 56; 66; 130; 131; 214; 279; 291; 293; 294; 295; 296; 298; 300; 313
Light scattering; 4; 5; 15; 16; 17; 18; 19; 20; 21; 44; 94; 99
Macroscopic quantum phenomena; 250
Macroscopic quantum self-trapping; 250
Magnetic trap; 4; 6; 7; 33; 42; 44; 48; 56; 61; 64; 94; 99; 100; 102; 103; 104; 105; 106; 110; 111; 112; 117; 120; 123; 124; 125; 129; 130; 131; 132; 133; 134; 136; 138; 165; 169; 171; 172; 180; 213; 214; 215; 216; 217; 218; 219; 220; 225; 229; 287; 291; 292; 295; 296; 301; 303; 306; 309; 311; 320
Magnetic trapping; 2; 7; 36; 118; 120; 121; 122; 123; 124; 132; 294
Matter waves; 2; 14; 15; 16; 20; 21; 22; 24; 25; 55; 60; 65; 66; 67; 74; 97; 102; 103; 106; 112; 113; 117; 129; 138; 153; 154; 162; 275
Mean field interactions; 83; 90; 241; 309
Mean field theory; 106
Metastability; 9; 10; 12; 32
Momentum distribution; 16; 17; 18; 56; 59; 77; 78; 79; 80; 81; 82; 85; 87; 88; 89; 92; 304
Momentum space; 4; 25; 60; 62; 77; 78; 87; 90; 92; 95

Momentum transfer; 16; 17; 18; 57; 58; 59; 61; 62; 79; 83; 84; 88; 89; 90; 91; 93; 219; 239
Multiple condensates; 134; 135
Negative scattering length; 199; 206; 225; 228; 245; 253; 295; 298
Non-linear atom optics; 97
Non-linear optics; 65; 162
Non-rigid pendulum; 250
Normal Fermi liquids; 302
Optical bistability; 154; 161
Optical lattice; 32; 154; 155; 158; 159; 160; 162; 238; 239; 240; 242; 243; 244; 245; 275; 276; 277; 278; 279; 280; 281; 282
Optical limiting; 154; 161
Output coupler; 14; 24; 60; 61; 64; 74; 118; 120; 121; 129; 137; 139; 162; 185; 275
π-states; 250
Particle-hole continuum; 302
Pendulum; 250; 253; 254; 255
Phase coherence; 40; 243; 244; 249
Phase imprinting; 68; 70; 106; 107; 109; 110; 113
Phase separation; 9; 10; 307
Phase-contrast imaging; 44; 45; 99
Phase-contrast microscopy; 43
Photoassociative spectroscopy; 299
Plasma frequencies; 254
Positive scattering length; 245; 253
Potassium; 130; 139; 291; 292; 293; 294; 295; 296; 297; 298; 299; 300
Power spectrum; 197; 198
Quadratic Zeeman effect; 171; 172; 174; 175; 181
Quantum; 249
Quantum computation; 265; 266
Quantum fluid; 4; 11; 32; 35; 289
Quantum logic; 275; 276; 282
Quantum phase engineering; 68; 69; 70
Quantum phase transitions; 245; 307; 311
Rabi oscillations; 33; 44; 45; 53; 136
Raman output coupling; 61; 156
Rate equation; 187; 189
Recoil energy; 58; 62; 82; 83; 240
Relative phase; 31; 32; 34; 35; 37; 38; 39; 40; 41; 42; 45; 47; 54; 68; 93; 159; 239; 242; 243; 244; 250; 252; 253; 257; 261
Revivals; 38; 42; 251; 260
RF knife; 171; 172; 173; 174; 175; 177; 181
Running modes; 255
Scattering length; 5; 8; 24; 25; 33; 35; 36; 42; 78; 80; 84; 110; 143; 182; 184; 187; 188; 194; 199; 205; 206; 225; 226; 228; 229; 239; 245; 253; 254; 257; 262; 295; 296; 298; 300; 301; 307; 314; 315; 316; 319
Scissors mode; 285; 286; 287; 288; 289
Solitons; 47; 49; 53; 65; 66; 68; 69; 70; 71; 72; 73; 74; 75; 106; 110; 111; 112; 153; 162
Spatial distribution; 10; 71; 201; 214; 225; 228
Spin domain; 7; 8; 9; 10; 11; 12
Spinor; 7; 8; 9; 11; 12; 41; 285
Superfluid density oscillations; 250
Superfluid helium; 12; 15; 16; 80; 250
Superfluidity; 2; 12; 26; 53; 106; 107; 285; 289; 311
Superradiance; 20; 21; 23
Sympathetic cooling; 42; 165; 178; 180; 213; 225; 282; 292; 296; 299; 300
Talbot effect; 59; 60; 64
Tunnelling; 11; 12; 129; 239; 240; 241; 242; 243; 244; 245; 275
Two-photon transition; 33
Uncertainty principle; 56; 63; 258
Voltage-current characteristic; 257
Vortices; 7; 26; 41; 43; 47; 48; 49; 52; 53; 68; 69; 74; 85; 94; 95; 106; 107; 108; 109; 110; 113; 188
Wavefunction; 11; 12; 13; 14; 19; 26; 31; 47; 48; 55; 56; 57; 60; 68; 70; 92; 109; 110; 111; 119; 142; 143; 144; 155; 157; 158; 183; 213; 215; 225; 285; 286; 316
Weak link; 250
Wigner distribution function; 303; 305
Zeeman effect; 133; 165; 168; 171; 176
Zero sound; 302; 303; 304

MIX
Papier aus verantwortungsvollen Quellen
Paper from responsible sources
FSC® C105338

If you have any concerns about our products,
you can contact us on
ProductSafety@springernature.com

In case Publisher is established outside the EU,
the EU authorized representative is:
Springer Nature Customer Service Center GmbH
Europaplatz 3, 69115 Heidelberg, Germany

Printed by Libri Plureos GmbH
in Hamburg, Germany